Materials Issues in a Hydrogen Economy

Materials Issues in a Hydrogen Economy

Proceedings of the International Symposium

Richmond, Virginia, USA   12 – 15 November 2007

Editors

Puru Jena
Virginia Commonwealth University

Anil Kandalam
McNeese State University

Qiang Sun
Peking University

World Scientific

NEW JERSEY • LONDON • SINGAPORE • BEIJING • SHANGHAI • HONG KONG • TAIPEI • CHENNAI

Published by

World Scientific Publishing Co. Pte. Ltd.
5 Toh Tuck Link, Singapore 596224
USA office: 27 Warren Street, Suite 401-402, Hackensack, NJ 07601
UK office: 57 Shelton Street, Covent Garden, London WC2H 9HE

British Library Cataloguing-in-Publication Data
A catalogue record for this book is available from the British Library.

MATERIALS ISSUES IN A HYDROGEN ECONOMY
Proceedings of the International Symposium

ISBN-13 978-981-283-801-8
ISBN-10 981-283-801-5

Printed in Singapore by B & JO Enterprise

PREFACE

The growing demand for energy, brought about by the rising standard of living in the developing world and global population growth, has made it imperative that new energy sources beyond fossil fuels be found to sustain and grow the current economy. The limited supply of fossil fuels and its adverse effect on the environment also makes it necessary that these sources should be abundant, renewable, secure, clean, safe, and cost-effective. In this regard hydrogen, the most abundant element in the universe and third most abundant element on the earth, has the potential to meet this growing energy need. In addition, hydrogen offers many advantages over other fuels: it is non-toxic, clean to use, and packs more energy per unit mass than any other fuel. However, hydrogen is not an energy source but an energy carrier. Since it does not occur freely in nature and is gaseous at room temperature and pressure, it needs to be produced and stored. To be economical, the production costs not only have to be lowered, but safe and cost effective means of storing, distributing and using hydrogen must also be found. The public also needs to be educated about hydrogen as an alternate fuel. Critical to the success of a hydrogen economy is our understanding of the interaction of hydrogen with materials and solving numerous materials issues relevant to the production, storage, and distribution of hydrogen and its use in fuel cells.

To address the above complex issues an "International Symposium on Materials Issues in a Hydrogen Economy" was held in Richmond, Virginia from November 12-15, 2007. The symposium dealt with the fundamental science and technology challenges related to the production, storage, distribution and use of hydrogen in fuel cells, and safety concerns. Nearly 150 researchers from Physics, Chemistry, Materials Science, and Engineering communities of 17 countries around the world attended this symposium and shared their ideas and results, delineated outstanding problems, and discussed future research. This book contains the proceedings of this symposium.

The topics will include:

Production and Delivery:	Nuclear, Thermo-chemical, Photo-electrochemical, Photo-biological, and Biomass.
Storage:	Liquid, Compressed Gas, and Solid State (complex light metal hydrides, zeolites, clathrates, metal-

	organic frameworks, carbon and boron-nitride based nanostructures, chemical hydrides).
Fuel Cells:	Polymer electrolyte and Hydro-carbon membranes.
Cross-cutting fields:	Catalysis, Nanostructures, Education, Safety, and Economics

The symposium featured 36 invited speakers and 82 contributed presentations addressing issues in production, storage, distribution, safety, education, and economics. Professor John B. Fenn, Nobel Laureate in Chemistry addressed the opening session. The symposium was sponsored by Virginia Commonwealth University and endorsed by American Physical Society, Materials Research Society, and American Chemical Society.

This symposium would not have been possible without the tireless efforts of the members of the International Advisory Board, the National and Local Organizing Committee and financial support from Virginia Commonwealth University, National Science Foundation, Department of Energy, Dominion Resources, Philip Morris USA, and General Motors. Our special thanks go to Dr. Sa Li for help in preparing the proceedings and to the undergraduate and graduate students and postdoctoral fellows in the Physics and Engineering Departments of Virginia Commonwealth University for volunteering their services during the organization of this symposium, to the conferees for the high quality of their participation, and last but not the least to Mrs. Barbara Martin for her assistance through out the two year period this symposium was in the making.

Richmond, Virginia
January, 2009

P. Jena
A. K. Kandalam
Q. Sun

CONTENTS

Safety and Education

Overview

PROGRESS AND CHALLENGES OF A HYDROGEN ECONOMY

M. S. DRESSELHAUS

Department of Physics, Department of Electrical Engineering and Computer Science, Massachusetts Institute of Technology, 77 Massachusetts Avenue Cambridge, MA 02139, USA

Since the publication of the 2003 report on Basic Energy Needs for the Hydrogen Economy, many important advances in hydrogen research have occurred, a cadre of enthusiastic re-searchers has entered the field with great interest shown by students, and private industry has made significant commitment and investment to this technology worldwide. Concurrently, other energy technologies have made major strides forward. This overview discusses these topics and looks to the future.

1. Introduction

Energy availability for the masses is without doubt a dominant challenge of the 21st century. Driven by increasing world populations, an even faster increase in the per capita energy demand, a decreasing availability of traditional sources of energy through fossil fuels and the increasing concern about the need to curb the increase of CO_2 into the atmosphere, the need for a transformation to a sustainable energy supply from renewable sources has emerged as a dominant challenge of this century. President Bush in his 2003 State of the Union Message identified this as a major challenge of his administration, as have other national leaders worldwide. As a result of the Bush 2003 State of the Union Message, a hydrogen initiative was launched by the US Government Funding Agencies.

As a first step, a workshop was held in the spring of 2003, followed by a committee study which resulted in a report [1] which emphasized, on one hand, the appeal of hydrogen as an energy carrier whose release of energy produces only water as a by product without other pollutants or greenhouse gases, and takes advantage of the high efficiency enabled by hydrogen fuel cells. On the other hand, the report emphasized the challenges for the implementation of the hydrogen economy in terms of the enormous technical challenges to be overcome for its implementation, emphasizing that fundamental breakthroughs would be needed in understanding the physical processes involved in the production, storage and use of hydrogen. Understanding the atomic and molecular processes that occur at the interfaces of materials with hydrogen was identified as crucial to producing the new materials that would be needed for these fundamental breakthroughs to occur. The report goes on to say that the discovery of the new materials, new chemical processes and new synthesis techniques that would be needed could only be achieved by initiating a major basic research program with these objectives. Such a research program was subsequently launched by the Basic Energy Sciences Office of the Department of Energy (DOE) following the recommendations of the report, working in close collaboration with the Energy Efficiency and Renewable Energy Office of the DOE, thereby uniting the basic and applied science thrusts through a highly interdisciplinary effort involving chemistry, physics, biology and engineering, all working together to solve the multitude of challenges and opportunities identified in the report. From these efforts, major research advances have occurred over a short period of time, amplified by the corresponding efforts occurring worldwide. The enthusiastic response of the research community and the great interest of students in joining this effort has been noteworthy, leading to a series of other studies and initiatives in other areas of energy research and development. Concurrently, industry has launched major initiatives so that the playing field is rapidly changing as breakthroughs are occurring in other areas. In the present brief report, emphasis is given to an attempt to identify an evolving role for the hydrogen economy within the larger energy challenge.

2. Strategic Issues

The framework for the hydrogen initiative, based on the 2003 Basic Energy Sciences Report "Basic Research Needs for the Hydrogen Economy" [1], was motivated by the charge to the study committee which focused on a hydrogen economy as an isolated entity and the use of hydrogen for transportation applications, exploiting the superior efficiency of the hydrogen and fuel cell combination relative to gasoline and the internal combustion engine. Based on the DOE hydrogen requirements for the years 2010 and 2015 (Table 1), the technology gaps for hydrogen as an energy carrier were identified (Fig. 1) and research directions for bridging these technology gaps were suggested in the report. In the meantime, the auto industry worldwide has taken a hydrogen based vehicle seriously and has moved rapidly in getting hydrogen fuel cell automobiles on the highways to gain experience with this new technology, using presently available methods for hydrogen production and storage, focusing mainly on hydrogen fuel cell development and the infrastructure needed for carrying out a hydrogen vehicle test program. While methods of hydrogen production from natural gas are presently adequate for automotive needs, the use of a fossil fuel natural gas precursor defeats the long term goal of using a sustainable, renewable energy source to provide the large increase in hydrogen production (20-fold by the estimate in Figure 1) that would be required for transportation use. The development of a renewable route for large scale hydrogen production by methods, such as splitting water in a closed cycle water-hydrogen process or by a biologically inspired process remains a long term challenge where there are presently large opportunities for the research community.

The on-board storage of hydrogen to match US consumer appetites for a 500 km (~300mi) range for their family vehicle has been identified as the greatest challenge to the implementation of a hydrogen economy because even the filling of the present fuel tanks of an automobile with liquid (or solid) hydrogen would fall short of meeting the DOE 2015 targets. The auto industry has taken a different approach toward addressing the consumer appetites and is using increased operating efficiency, hybrid vehicle technology to lower the storage requirements. Using this approach, Toyota has recently demonstrated by a run from Osaka to Tokyo a 550 km (350mi) range for its hydrogen fuel cell vehicles based on presently available compressed hydrogen gas cylinder technology. Although researchers from the auto-industry would like to see the academic community and government supported research laboratories come up with a chemisorbed or physisorbed hydride solution for hydrogen storage, the auto industry does not now see the hydrogen storage problem as a technical show-stopper, though widespread public acceptance of the hydrogen gas cylinder technology has not been seriously tested. On the other hand, the auto industry is looking to the research community for major breakthroughs in renewable hydrogen production, reversible solid state hydrogen storage and higher efficiency hydrogen fuel cells to help make widespread adoption of the hydrogen fuel cell vehicle option a reality by mid-century. The arguments on the central role that new materials will play in these break-throughs, as presented in the 2003 hydrogen report [1] remain valid through the present time. What has changed in the interim is the vital role that industry is now playing and the need for the research community to be in close contact with industrial R&D, and to play a role in the incubation of start-up companies to develop the new technology that will be provided by future suppliers to the auto companies. Thus, one strategic issue for the planning of hydrogen research is the coordination, not only between basic and applied research by the multidisciplinary players, but also to look for opportunities where academic and national laboratory research could have a large impact on future industrial product development.

A second strategic issue concerns scale. Projections of global energy needs imply a doubling in overall energy demand and a tripling of the electricity demands by the year 2050.

Table 1: Requirements for a hydrogen fuel cell automobile

Targeted Factor	2005	2010	2015
Specific energy (MJ kg)	5.4	7.2	10.8
Hydrogen (wt%)	4.5	6.0	9.0
Energy density (MJ/L)	4.3	5.4	9.72
System cost ($/kg/system)	9	6	3
Operating temperature (°C)	-20/50	-20/50	-20/50
Cycle life-time (absorption/desorption cycles]	500	1.000	1.500
Flow rate (g/s)	3	4	5
Delivery pressure (bar)	2.5	2.5	2.5
Transient response (s)	0.5	0.5	0.5
Refueling rate (kp/H2/min)	0.5	1.5	2.0

Source: Milliken (2003).

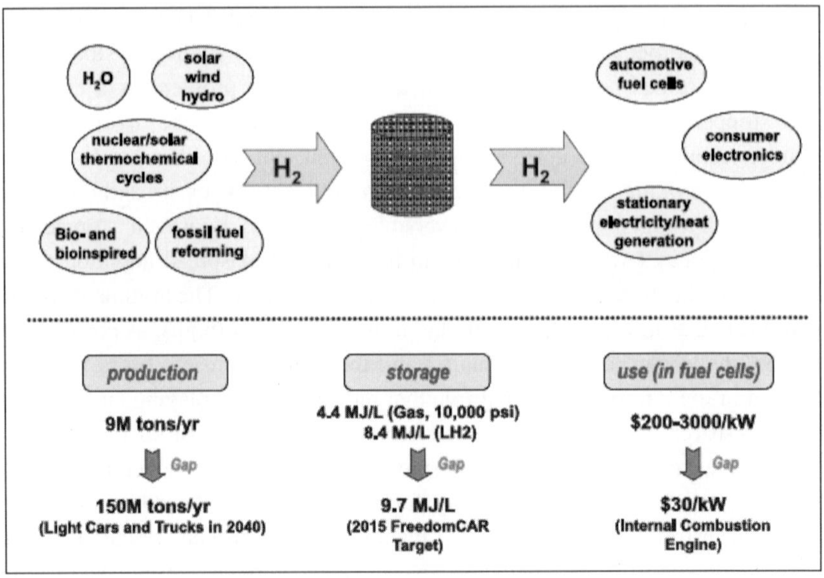

Figure 1: The technology gaps in hydrogen production, storage, and end use in a hydrogen Economy [2].

The only renewable energy source with sufficient capacity to meet these growing energy demands is solar energy. An increase from the present 14TW to 28-30TW by 2050 is expected to come from solar energy used for generating electricity (photovoltaic), providing fuels (biofuels, water splitting, close cycle synfuels), and supplying space and water heating (solar thermal). In this big picture, with solar electric, solar fuel and solar thermal as the energy sources, electricity and hydrogen are cited as complementary energy carriers. When thinking of hydrogen as a chemical carrier of energy, its role in energy storage from the electric grid emerges as an interesting opportunity, as does the generation of close-cycle renewable synfuels using a hydrogen from H_2O and carbon from CO_2 to produce a hydrocarbon fuel using sunlight [3]. The latter research direction, denoted by "transformation and recycling of CO_2 into a new material" was identified in the Declaration issued by the First World Materials Summit held in Lisbon in 2007 [4].

The need for break-throughs with high impact follows from the huge scale of the energy challenge involving a multi-trillion dollar business worldwide. Therefore major emphasis must be given to those research directions which will have the potential for large orders of magnitude impacts. This brings to mind Moore's law which has provided road-maps for the electronics, optoelectronic and magnetic information storage industries for several decades. To have comparable impact on the energy industry, a Moore's law road-map for the Energy Industry is needed. Here new materials will play a vital role, especially nanomaterials, because of the greater ability to modify and control their properties by varying the material's size and composition, their greater surface area to promote catalysis which is based on an exponential $exp(-E/kT)$ dependence, and the independent control of materials parameters which are interdependent in 3D systems.

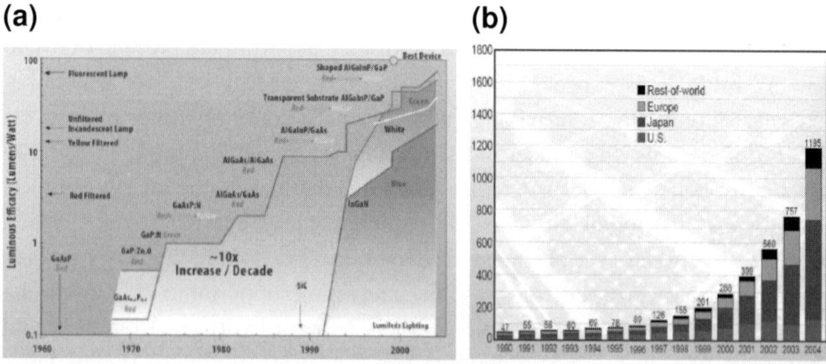

Figure 2: Examples of energy industries showing aspects of Moore's law behavior: (a) solid state lighting efficiency, (b) photovoltaic cell production in MW.

In fact, Moore's law has started to infiltrate the energy industry. One example is solid state lighting where the [lumens/watt] emission from light emitting diodes has followed a Moore's law path in the last 30 years [Figure 2(a)]. This technology now requires half the electrical energy of an equivalent incandescent lamp for a given light output and is expected to have a major impact on the drive toward improving energy efficiency, since residential and industrial lighting currently accounts for 22% of electricity use in the US. Research is actively occurring to improve light quality, to lower cost and to find uses for this transformational technology that are different relative to the technology it replaces. A second example of Moore's law is photovoltaic (PV) cell production [Figure 2(b)] which has had an annual growth rate of ~30% for the last decade, but in which the USA has not been a major player. Recent advances in photovoltaic technology, using three junction devices which capture the solar spectrum very well and using a solar concentrator of 240 suns, have achieved over 40% efficiency in PV conversion [5]. This technology, using 10^{-3} of the "real estate" of conventional solar cells, is well positioned for both scale - up and new applications areas for photovoltaics. Even though the technology is quite complex and requires many semiconducting layers, Spectrolab (a subsidiary of Boeing Corp) has recently released a road-map by which scale-up production of the device for 2010 with over 40% efficiency at a cost of less that \$0.15/kW-hr, with increased performance and lower cost projected for the future. This basic technology could be used for both power generation in power plants or on the rooftops of homes, with a potential for major future impact on electricity production and energy efficiency. Since sunlight is intermittent, there could be interesting opportunities for hydrogen as an energy storage agent to be used in conjunction with this technology.

Another interesting direction where large-scale impacts on energy are occurring is in thermoelectric conversion where increases in the thermoelectric figure of merit and scale-up to samples with higher thermal capacities have been demonstrated. As a result, industrial development in this field is booming with about one million cooling/heating thermoelectric seats sold in 2007 for automotive use. When used in hybrid cars where fuel efficiency is readily monitored, it has been found that the local cooling of passengers by the thermoelectrically equipped seats causes a major decrease in the need for air conditioning for passenger comfort, resulting in a payback of less than 1 year for the thermoelectric car seats, with subsequent cost savings in future fuel consumption [6]. It would be interesting to explore what the effect of thermoelectric car seats would be on the efficiency of hydrogen fuel-cell autos.

The device utilization of the discovery of highly efficient carrier multiplication in semiconductor nanocrystals [7] allowing as many as 6 electron-hole pairs to be produced by a single optical photon incident on a PbSe nanocrystal is now being explored and may eventually result in enhanced photovoltaic device efficiency. If this scientific advance results in improved photovoltaic device efficiency, this may open new opportunities for hydrogen as an energy storage agent.

Finally, high throughput combinatorial screening allows a route for both experiment and theory to scan many variants of multi-component materials by composition, to optimize a material for a given property while at the same time allowing rapid measurement of several other properties of the material in the compositional range where the desired property is optimized. Such capabilities are necessary since a number of properties of a material affect its ultimate device performance, and these properties therefore need to be jointly optimized. For example, a material, which has excellent thermoelectric performance but is toxic, would not make it in the marketplace.

3. Strategies for the Hydrogen Economy

With the principles outlined in §2 in mind, we can identify a number of breakthroughs that have the potential for high impact on the hydrogen economy. As mentioned above, the use of improved catalysts have the potential for high impact because of their $exp(-E/kT)$ dependence. Thus, a promising strategy is the search of new catalysts that lower the energy barriers for chemical reactions, can be made in the optimal small sizes (usually in the 2–5nm range), and can contain cheaper and more plentiful elements. An example where such a specially tailored catalyst has been developed for the hydrogen economy is the Pt_3M catalyst. Density functional theory was first used to establish the concept of using a Pt surface layer of the catalytic particle to rapidly dissociate a hydrogen molecule. The introduction of a first subsurface layer with a PtM composition then provides a mechanism for attaching atomic hydrogen more easily [8]. Such an approach can provide strong binding and also rapid release on hydrogen. Variants of this concept could have an impact on hydrogen production, storage, and use in fuel cells. An implementation of this general concept has recently been made to increasing the catalytic activity of Pt by a factor of 10 in the oxygen reduction reaction by using a surface Pt layer and a subsurface PtNi layer to break the O–O bonds to form O–H bonds. Weak surface bonds prevent the splitting of O–O bonds, while strong surface bonds attract guest species to adhere to the surface, thereby blocking access of other reactants to the catalyst. In the case of the oxygen reduction reaction, the 10-fold increase in catalytic activity for the oxygen reduction reaction which occurs at the anode of hydrogen fuel cells was achieved by using both the (111) crystal orientation of the catalytic particle and its compositional variation [9].

A number of other impressive advances have been made in the laboratory at the research level, and a small number are cited here as examples. One noteworthy example is the identification of a route to increase the tolerance of hydrogen production by a genetically modified Fe–Fe hydrogenase bacterial structure that yields a 100-fold increase in H_2 activity relative to the natural algol enzyme. Simplified and robust analogs of bacterial hydragenase have the potential to lead to the development of a commercial-scale hydrogen production route that may be scalable to large scale production, self-sustaining and cost effective [10].

An interesting approach to lowering the release temperature of hydrogen through increased destabilization is the use of a second compound in a chemical reaction, and for example $LiNH_2 + LiH \text{ -- > } Li_2NH + H_2$ releases hydrogen at ~150°C which is significantly lower than $LiNH_2$ (at 200°C) or LiH (at 500°C). This study is of significant scientific interest. However the storage capacity for the joint reaction is only 6.5%, which could be too low for commercial development [12, 13] On the other hand, the destabilization pair of LBH_4 + MgH_2 with a storage capacity of 11.5% could be more interesting for further commercial development [14].

Some new ideas have recently been introduced into increasing the temperature of operation of PEM membranes and increasing the power density of the fuel cell operation. Some membranes have been developed that conduct protons at temperatures up to 200◦C in the absence of water [15]. A new class of chemically cross linked membranes fabricated at low temperature from liquid recursors significantly enhance proton conductivity by allowing additional acid loading, enhance thermal and mechanical stability by increased cross-linking, while at the same time increasing electrical and chemical exchange with the electrode by enhancing the effective surface area [16].

The advances in hydrogen research are mostly at an early stage with further progress in understanding and in material performance expected in the near term. Applications to industrial products are expected to follow.

4. Concluding Remarks

Because of its special and unique attributes, hydrogen is likely to be one of a mix of future sustainable energy technologies. New materials and nanoscience are necessary to its development as they are to many of the other pertinent energy technologies. The strong interplay between basic and applied sciences, interdisciplinary approaches and the coupling between theory and experiment are all vital. Working closely with industry will be important for identifying research directions with high potential impact. Attention to major advances in other key technologies is equally important for the identification of new priority directions for hydrogen R&D. Because of the highly complementary focus of energy research in different countries, based on their different climatic and cultural constraints, international cooperation and networking should be encouraged and supported. Linking to and coordinating between international groups (such as the World Materials Summit) promoting materials research for energy applications regionally and internationally would be important, so that policy makers worldwide get a clear message about progress in hydrogen research and its potential contribution to the larger picture of providing a sustainable energy supply world-wide.

Acknowledgments

The author acknowledges G. Dresselhaus, V. Berube and M. Hofmann for valuable discussions and assistance with preparation of the manuscript. The MIT author acknowledges support under DE-FG02-05ERR46241.

References

1. G. W. Crabtree, M. S. Dresselhaus, and M. V. Buchanan, Basic Research Needs For the Hydrogen Economy (Office of Basic Energy Sciences, Department of Energy, BES, Washington DC, 2003).
2. G. W. Crabtree, M. S. Dresselhaus, and M. V. Buchanan, Physics Today 57(12), 39–44 (2004). December.
3. Koji Hashimoto, N. Kumagai, K. Izumiya, Z. Kato, Materials and technology for global carbon dioxide recycling for supply of renewable energy and prevention of global warming, 2007.
4. Declaration issued by the First World Materials Summit, Lisbon, Portugal 2007 (see website).
5. R. R. King et al, Appl. Phys. Lett. 90, 183516 (2007).
6. Lou Bell report at the Industrial Physics Forum, Seattle, WA, Oct 2007.
7. R.D. Schaller and V. I. Klimov, Phys. Rev. Lett. 92, 186601, (2004).
8. J. Greeley and M. Mavrikakis, Alloy catalysts designed from first principles, Nature Materials, 3, 810 (2004).
9. V. R. Stamenkovic et al, Science 315, 497 (2007).
10. P. W. King et al Proc. SPIE vol 6340. 63400Y (2006).
11. G. W. Crabtree and M. S. Dresselhaus, MRS Bulletin: Energy Issue page in press (2007).
12. P. Chen, Z. Xiong, J. Luo, J. Lin, K.L. Tan, J. Phys. Chem. B 107, 10967 (2003).
13. J.F. Herbst, L.G. Hector, Jr., Phys. Rev. B 72, 125120 (2005).
14. J.J. Vajo, G.L. Olson, Scripta Mater. 56, 829 (2007).
15. J.A. Asensio, S. Borrs, P. Gmez-Romeroa, Electrochim. Acta 49, 4461 (2004).
16. Z. Zhou, R.N. Dominey, J.P. Rolland, B.W. Maynor, A.A. Pandya, J.M. DeSimone, J. Am. Chem. Soc. 128, 12963 (2006).

Hydrogen Production

A NEW SOLAR METAL SULFATE – AMMONIA BASED THERMOCHEMICAL WATER SPLITTING CYCLE FOR THE PRODUCTION OF HYDROGEN

ALI T-RAISSI,[†] CUNPING HUANG, LIQUN MAO AND NAZIM MURADOV

University of Central Florida, Florida Solar Energy Center
1679 Clearlake Road, Cocoa, FL 32922-5703, USA

All sulfur-family thermochemical water splitting cycles (TCWSCs) rely on concentration and decomposition of sulfuric acid for the oxygen evolution step of the cycle. The sulfuric acid decomposition step presents serious materials and catalyst deactivation challenges. Platinum based catalysts are currently the most active for the H_2SO_4 decomposition, but they deactivate rapidly. To overcome this difficulty metal sulfate based TCWSCs have been developed. However, the metal sulfate based TCWSCs utilize thermal heat input – thus degrading photonic energy. Based upon FSEC's S-NH_3 TCWSC, a new family of hybrid photo/thermo-chemical water splitting cycles is introduced in this paper that employs the quantum portion of the solar spectrum for the production of hydrogen and the thermal energy (*i.e.* IR) portion of solar radiation for generating oxygen. FSEC's metal sulfate – ammonia (MSO_4-NH_3) hybrid photo/thermochemical water splitting cycles are represented by the following reactions:

$SO_2(g) + 2NH_3(g) + H_2O(l) \rightarrow (NH_4)_2SO_3(aq)$	(chemical absorption)	25°C
$(NH_4)_2SO_3(aq) + H_2O \rightarrow (NH_4)_2\, SO_4(aq) + H_2(g)$	(solar photocatalytic)	80°C
$x(NH_4)_2SO_4 + M_2O_x \rightarrow 2xNH_3 + M_2(SO_4)_x + xH_2O$	(solar thermocatalytic)	500°C
$M_2(SO_4)_x(s) \rightarrow xSO_2(g) + 2MO(s) + (x\text{-}1)O_2(g)$	(solar thermocatalytic)	1100°C

Where, M = Zn, Mg, Ca, Ba, Fe, Co, Ni, Mn, Cu.

Chemical equilibrium calculations for the reaction between ZnO and $(NH_4)_2SO_4$ indicate that both $ZnSO_4$ and $ZnO.2ZnSO_4$ can form as the stable reaction products. A series of thermogravimetric/ differential thermal analyses/mass spectrometric (TG/DTA/MS) experiments has been carried out to determine the exact nature of all ZnO + $(NH_4)_2SO_4$ reaction products. Results obtained to date are presented and discussed.

1. Introduction

Production of hydrogen (H_2) is environmentally acceptable if it is produced from a noncarboneaceous feedstock using a renewable energy source. Hydrogen production by splitting water using solar energy is considered the "Holy Grail" of the hydrogen economy. Water splitting can be accomplished either directly

[†] Corresponding author: Tel: (321) 638-1446, Fax: (321) 504-3438, email: ali@fsec.ucf.edu.

(in a single step) or indirectly (via multiple steps). Direct thermal decomposition of water is an energy intensive process that requires temperatures exceeding 2500°C. The main obstacle to this approach is that hydrogen and oxygen (O_2) evolving simultaneously in one reactor can readily recombine to form water – *i.e.* by back reaction. The combination of photovoltaic (PV) cells coupled to water electrolysis often serves as the benchmark solar hydrogen production process by which the performance of other solar based hydrogen generation processes are appraised. PV cell efficiencies vary from 6% for amorphous silicon solar cells to more than 40% for multiple-junction research grade PV cells. Solar cell energy conversion efficiencies for commercially available mono-crystalline silicon PV cells are around 14-16%. The highest efficiency PV cells such as multi-junction cells based on gallium arsenide or indium selenide are still too expensive and not yet economical to use. On the other hand, water electrolysis is a well-developed technology with energy conversion efficiencies in the range of approximately 70-95%, depending on the type of electrolyte used. Therefore the total solar to hydrogen efficiency of a PV-electrolysis system is in the range of 10-15%. We note that the most commonly used PV cells employ only a portion of the solar energy (10 to 12%) while most of the solar thermal heat is unutilized.

Unlike PV-electrolysis, water splitting processes based on the photocatalytic and photoelectrochemical methods provide a single step direct conversion of solar energy into the chemical energy of hydrogen. In the photocatalytic process, a photocatalyst converts the high-energy photons in solar spectrum into electron-hole pairs that promote redox reactions involving water to produce hydrogen and oxygen. In 1998, Khaselev and Turner reported that the hydrogen production efficiency of 12.4% for a monolithic photoelectrochemical-photovoltaic device based on the short-circuit current and the lower heating value of hydrogen [1]. The electrolyte used by Khaselev and Turner was 3 M sulfuric acid aqueous solution and the high cell output could only be maintained for a very short period of time after which the efficiency began to drop precipitously. Licht and co-workers [2] have reported a solar to H_2 energy conversion efficiency of more than 18% using a multi-junction photoelectrode in 1 M $HClO_4$ electrolyte radiated with a lamp equipped with Air Mass 0 (AM0) filter providing a radiant flux of 135 mW/cm^2. However, no information was given by these authors with regard to the lifespan of the photoelectrode used.

Thermodynamically, water splitting requires a minimum of 237.1 kJ per mole of water decomposed at 25°C and 1 atm – corresponding to 1.229 eV.

Considering the prevailing over potentials, water splitting requires upwards of 2.0 V. In a photocatalytic process, this implies the requirement for a wide band gap (greater than 2.0 eV) photocatalyst. There are conflicting requirements for what makes a good photocatalyst for splitting water. In order to utilize large portion of the solar spectrum, a semiconductor with narrow band gap is desirable. However, electron-hole pairs generated by a low band gap semiconductor do not possess sufficient redox potential to allow water splitting. In case of a wide band gap semiconductor such as TiO_2 (band gap energy of 3.0 eV) only a small portion of solar spectrum would be absorbed. Thus, for a semiconductor photocatalyst to be useful for water splitting, it must have several attributes as follows [3]:

its band gap must be wider than about 1.7 eV;
must have a suitable minority band edge and Fermi level that cover both H_2 and O_2 evolution potentials;
be stable in very acidic or very alkaline solutions;
possess high efficiency for conversion of photons to electron-hole pairs; and
electron-hole pairs must be able to rapidly migrate to the semiconductor surface where redox reactions can readily take place thwarting charge recombination.

To date, no such catalyst has been found. Another approach to direct water splitting involves the use of sacrificial reagents – see, for example, the review by Liu and co-workers [4]. The central premise in this approach is that lower potentials would be necessary to evolve hydrogen if a sacrificial reagent is present as opposed to that required for direct unassisted water splitting. Therefore, the efficiency of H_2 or O_2 production from such systems can be significantly higher than direct water splitting. Several redox systems have been extensively investigated including electron donor systems such as: CH_3OH or C_2H_5OH, Na_2EDTA, Na_2SO_3, Na_2S and NaI or KI; as well as the electron scavenger systems, *e.g.* $AgNO_3$ and $Fe(NO_3)_3$. The redox reactions for the photocatalytic production of H_2/O_2 are summarized below:

Photochemical reactions involving electron donors:

$$NaSO_3 + H_2O \rightarrow Na_2SO_4 + H_2,\ \Delta E = 0.93\ V,\ pH = 10\text{-}12 \qquad (1)$$

$$Na2S + H2O \rightarrow Na2S2 + H2,\ \Delta E = 0.524\ V,\ pH = 10\text{-}12 \qquad (2)$$

$$H2SO3 + H2O \rightarrow H2SO4 + H2,\ \Delta E = 0.17\ V,\ pH = 1\text{-}6 \qquad (3)$$

$$CH3OH(aq) \rightarrow CH2O(aq) + H2, \Delta E = 0.13\ V \quad (4)$$

$$NaI + 3H2O \rightarrow NaIO3 + 3H2, \quad \Delta E = 0.43\ V, pH = 10\text{-}13 \quad (5)$$

$$(NH4)2SO3+H2O \rightarrow (NH4)2SO4+H2, \Delta E = 0.52\ V, pH \sim 8.0 \quad (6)$$

Photochemical reactions involving electron acceptors:

$$4Ag^{+}(aq)+2H_2O \rightarrow O_2 + 4H^{+}(aq) + 4Ag(s), \quad \Delta E = 0.42\ V \quad (7)$$

$$4Fe3+(aq)+2H2O \rightarrow O2 + 4H+(aq) + 4Fe^{2+}(aq), \Delta E{=}0.46\ V \quad (8)$$

It should be noted that although redox systems utilizing sacrificial reagents require less energy and can be carried out under milder conditions, a reagent is consumed to produce hydrogen (or oxygen) from water. The H_2SO_3-H_2SO_4 system (Reaction 3) can form a closed cycle if the problem with sulfur formation during H_2SO_3 oxidation is mitigated. The only redox pair listed above that can be made to constitute a thermochemical cycle (TCWSC) with overall reaction being water splitting reaction and co-production of hydrogen and oxygen is Reaction (6). In this paper, we introduce a solar thermochemical water splitting cycle based on Reaction (6) for the hydrogen formation and an oxygen generation reaction barrowed from the sulfur family cycles. In the new cycle, hydrogen and oxygen production steps employ different but complementing sectors of the solar spectrum in order to maximize the overall TCWSC efficiency.

2. Efficiency of thermochemical water splitting cycles

2.1. *Hydrogen production via thermochemical water splitting cycles (TCWSCs)*

Thermochemical water splitting cycles employ two or more chemical reaction steps that taken together form a closed loop with an overall reaction being the splitting of water and co-production of hydrogen and oxygen. Energy is added into one or more steps of the TCWSC. Typically, energy required for splitting water is provided to more than one reaction making up the TCWSC so that each step requires only a portion of the total energy needed to split water ($\Delta H^{o}_{w,liquid}$= 285.9 kJ/mol and $\Delta H^{o}_{w,\,gas}$ = 241.83 kJ/mol at 25°C, 1 atm). Figure 1 depicts a three-step TCWSC in which the total energy (ΔH_w) required is segmented as follows:

$$\Delta H_w = \Delta H_1 + \Delta H_2 + \Delta H_3 \quad (9)$$

Each step requires less energy than that needed for direct water splitting as follows:

$$\Delta H_1 < \Delta H_w;\ \Delta H_2 < \Delta H_{w;}\ \Delta H_3 < \Delta H_w. \quad (10)$$

Clearly, at least two steps are needed to form a TCWSC, namely hydrogen and oxygen evolution steps. In the so called "pseudo TCWSCs", the energy required to perform one of the steps (typically, the oxygen evolving step) exceeds that needed to directly split water (*i.e.* $\Delta H_i > \Delta H_w$). Since more energy than ΔH_w is stored in the products formed from oxygen generation step of the pseudo TCWSCs, hydrogen production step of these cycles can be considerably less energy intensive or even exothermic. Pseudo TCWSCs contain a highly endothermic process for absorbing and storing the solar thermal heat at very high temperatures (above 2000°C) that require mean solar flux concentration ratios, C_R, of 5000 or higher [5]. Unlike direct thermolysis of water that requires high temperature separation of O_2 from H_2, pseudo TCWSCs typically involve separation of O_2 from a solid product (often an oxide) and as such eliminate the need for oxygen and hydrogen separation. Rapid quenching can also mitigate recombination of the products formed. Figure 2 depicts energetics of pseudo TCWSCs having one step that consumes more energy than ΔH_w. Pseudo TCWSCs fall into three categories: nonmetal oxide, metal/metal oxide and metal oxide/metal oxide cycles [6-8].

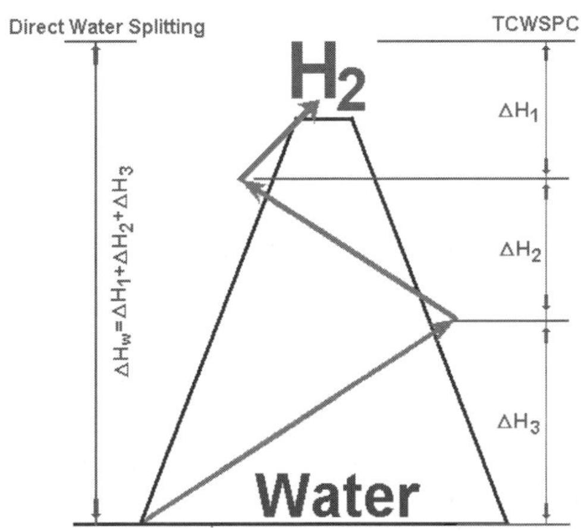

Figure 1. Energetics of TCWSCs.

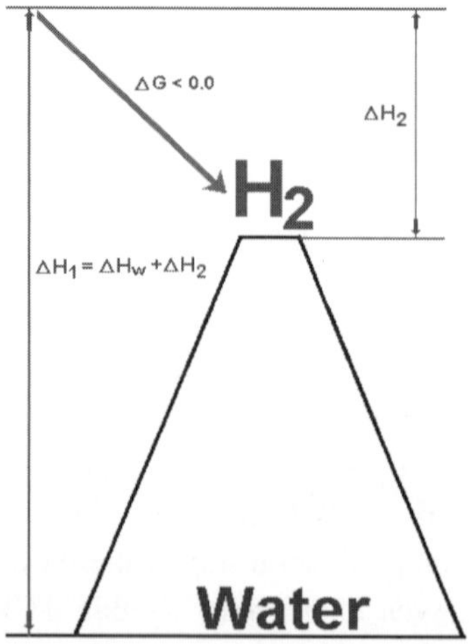

Figure 2. Energetics of pseudo TCWSCs.

Nonmetal oxide cycles:

$$CO2(g) = CO(g) + \frac{1}{2}O2, \quad \Delta H = 283.0 \text{ kJ/mol}, \quad 1700°C \tag{11}$$

$$CO(g)+H2O(g)=H2 + CO2(g), \quad \Delta H=-41.2 \text{ kJ/mol}, \quad 700°C \tag{12}$$

$$SiO2 \rightarrow SiO(g) + \frac{1}{2}O2 \qquad 2977°C \tag{13}$$

$$SiO(g) + H2O \rightarrow SiO2 + H2 \qquad 2656°C \tag{14}$$

Reaction (11) requires higher energy than $\Delta H^o{}_w = 241.83$ kJ/mol.

Metal/metal oxide TCWSCs:

$$M_xO_y = x\,M + \frac{1}{2}y\,O_2; \text{ (endothermic)}, \Delta H^o > \Delta H^o{}_W \tag{15}$$

$$x\,M + y\,H_2O = M_xO_y + y\,H_2; \text{ (exothermic)}, \Delta G < 0 \tag{16}$$

Where, M represents a metal. Basically, any metal that can reduce water and generate hydrogen can be used in a metal/metal oxide based TCWSC. Examples include: Zn/ZnO, Li/Li_2O, Na/Na_2O, K/K_2O, Mg/MgO, Ca/CaO, Mo/MoO_2, W/WO_3, SiO_2/SiO, Sn/SnO_2, FeO/Fe_3O_4, In_2O_3/In_2O, etc. Some metal and metal oxide based pseudo TCWSCs are given below [8]:

$MoO_2(s) \rightarrow Mo + O_2$ 3713°C (17)

$Mo + 2H_2O \rightarrow MoO_2(s) + 2H_2$ 1543°C (18)

$WO_3(s) \rightarrow W + 3/2O_2$ 3910°C (19)

$W + 3H_2O \rightarrow WO_3(s) + 3H_2$ 884°C (20)

$SnO_2 \rightarrow Sn + O_2$ 2650°C (21)

$Sn + 2H_2O \rightarrow SnO_2 + 2H_2$ 600°C (22)

$ZnO \rightarrow Zn + \frac{1}{2}O_2$ 2000°C (23)

$Zn + H_2O \rightarrow ZnO + H_2$ 1100°C (24)

Some low temperature metal/metal oxide cycles do not belong to pseudo TCWSCs [8]:

Hg(g) + H2O → HgO(s) + H2 360°C (25)

HgO(s) → Hg(g) + ½O2 600°C (26)

Cd(s) → H2O → CdO(s) + H2 (electrolytic, 25°C) (27)

CdO(s) → Cd(g) + ½O2 1400°C (28)

These two cycles use heavy metals Hg and Cd and generally viewed as environmentally undesirable cycles.

Metal oxide/metal oxide TCWSCs:

$In_2O_3 \rightarrow In_2O + O_2$ 2200°C (29)

$In_2O + 2H_2O \rightarrow In_2O_3 + 2H_2$ 800°C (30)

$Fe_3O_4(s) \rightarrow 3FeO(s) + \frac{1}{2}O_2$ 2200°C (31)

$3FeO(s) + H_2O \rightarrow Fe_3O_4(s) + H_2$ 400°C (32)

$Ni_{0.5}Mn_{0.5}Fe_2O_4 \rightarrow Ni_{0.5}Mn_{0.5}Fe_2O_{4-x} + \frac{1}{2}\,x\,O_2$ 1100°C (33)

$Ni_{0.5}Mn_{0.5}Fe_2O_{4-x} + x\,H_2O \rightarrow Ni_{0.5}Mn_{0.5}Fe_2O_4 + xH_2$ 600°C (34)

$MnFe_2O_4 + 3CaO + (1\text{-}x)H_2O \rightarrow Ca_3(Fe, Mn)_3O_{8-x} + (1\text{-}x)H_2$ 1000°C (35)

$Ca_3(Fe, Mn)_3O_{8-x} \rightarrow MnFe_2O_4 + 3CaO + \frac{1}{2}(1\text{-}x)O_2$ 600°C (36)

2.2. *TCWSC efficiency*

The overall thermal efficiency ($\eta_{overall}$) (or 1st law efficiency) of a TCWSC is defined as the ratio of hydrogen chemical energy to total energy consumed by the cycle.

$$\eta_{overall} = \frac{n \cdot \Delta H^o{}_f}{\Delta H_{total}} \tag{37}$$

Where n denotes the total mole of H_2 generated by the cycle, $\Delta H^o{}_f$ is enthalpy of water formation and ΔH_{total} is the total energy input to the cycle to produce *n* moles of hydrogen. If the enthalpy formation of water in liquid state is used (at 298 K, ΔH_f = -68.32 kcal/mol = 285.9 kJ/mol), the efficiency calculated is referred as the high heating value (HHV) efficiency, η(HHV). Some argue that the latent heat of condensation cannot be effectively recovered and prefer using the low heating value (LHV) efficiency η(LHV) in which $\Delta H^0{}_f$ is the enthalpy of formation of water vapor at 298 K ($\Delta H^0{}_f$ = -57.41 kcal/mol = 240.2 kJ/mol). The ratio η(HHV)/ η(LHV) = 68.32/54.74 = 1.19. The figure of merit or Carnot efficiency (also, 2nd law efficiency) is defined as:

$$\eta(w) = \frac{n \cdot \Delta G^0{}_f}{\Delta H_{total}} = \frac{237.2 \cdot n}{\Delta H_{total}} \tag{38}$$

Where, $\Delta G^o{}_f$ is to the Gibbs free energy of water formation (237.2 kJ/mol). Since early 1970s, when the concept of TCWSCs was first introduced, numerous methods have been proposed for calculating TCWSC efficiencies [9-11]. Since TCWSCs often contain several reaction steps as well as processes for the material transport and separation, precise determination of the efficiencies has been complicated. Huang and Raissi [12] have shown that efficiency of a TCWSC must be calculated based on a detailed process flowsheet that takes account of material and energy balance as well as precise values of the chemical and physical properties of reactants and products.

Figure 3 depicts a simple flow diagram for a TCWSC. Water is fed into the cycle and hydrogen and oxygen are generated as the only output of the cycle. In addition to the hydrogen and oxygen production steps, there are steps involving separation and recycling thus forming a closed cycle for splitting water into H_2 and O_2. Total energy needed to perform water splitting includes those required to generate H_2 and O_2 (*i.e.* ΔH_1 and ΔH_2), separate reactants from products (ΔH_S) and recycle reactants (ΔE). Then,

$$\Delta H_{Total} = \Delta H_1 + \Delta H_2 + \Delta H_S + \Delta E \tag{39}$$

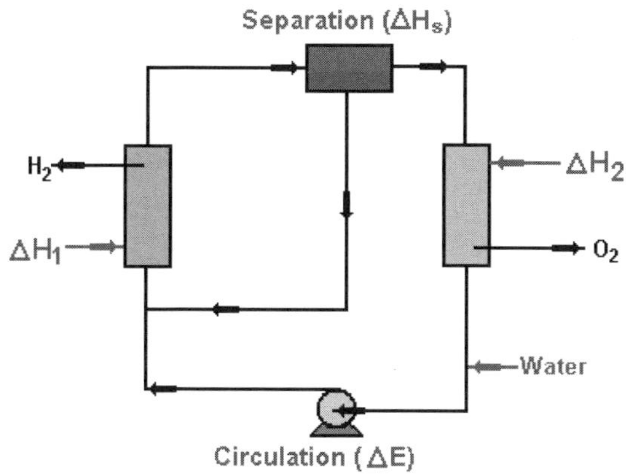

Figure 3. A simple flow diagram depicting TCWSC.

For simplicity, ΔH_S and ΔE may be added to ΔH_1 and ΔH_2 and denoted as $\Delta H(H_2)_{In}$ and $\Delta H(O_2)_{In}$, respectively. Thus, Equation (39) can be written as:

$$\Delta H_{Total} = \Delta H(H_2)_{In} + \Delta H(O_2)_{In} \tag{40}$$

Assuming that the efficiencies for H_2 and O_2 production are $\eta(H_2)$ and $\eta(O_2)$, respectively, we have:

$$\eta(H_2) = \Delta H(H_2)_R / \Delta H(H2)In \tag{41}$$

$$\eta(O2) = \Delta H(O2)R / \Delta H(O_2)_{In} \tag{42}$$

Where, $\Delta H(H_2)_R$ and $\Delta H(O_2)_R$ denote the amount of energy needed to conduct H_2 and O_2 generation reactions, respectively. Thus, the total energy required for H_2 and O_2 production is

$$\Delta H_{Reaction} = \Delta H(H_2)_R + \Delta H(O_2)_R \tag{43}$$

Assuming that *R* is the ratio of the energy input into the O_2 and H_2 production steps, $R = \Delta H(O_2)_{In}/\Delta H(H_2)_{In}$, then overall cycle efficiency, $\eta_{Overall}$, is then

$$\eta_{Overall} = \frac{\Delta H_{Reaction}}{\Delta H_{Total}} = \frac{\Delta H(H_2)_R + \Delta H(O_2)_R}{\Delta H(H_2)_{In} + \Delta H(O_2)_{In}}$$
$$= \frac{\Delta H(H_2)_R}{\Delta H(H_2)_{In} + \Delta H(O_2)_{In}} + \frac{\Delta H(O_2)_R}{\Delta H(H_2)_{In} + \Delta H(O_2)_{In}} \quad (44)$$
$$= \frac{\eta(H_2)}{1+R} + \frac{\eta(O_2)}{1+1/R} = \frac{\eta(H_2) + R\cdot\eta(O_2)}{1+R}$$

In case $\eta(H_2) = \eta(O_2)$, then

$$\eta_{Overall} = \eta(H_2) = \eta(O_2) \neq f(R) \equiv \text{independent of "}R\text{"} \quad (45)$$

In other words, if $\eta(H_2) = \eta(O_2)$, it would not matter how the solar resource is partitioned to supply $\Delta H(H_2)_{In}$ and $\Delta H(O_2)_{In}$ – the cycle efficiency would be the same and equal to $\eta(H_2) = \eta(O_2)$. But, if $\eta(H_2) \neq \eta(O_2)$, Equation (44) implies that, the overall efficiency of the TCWSC ($\eta_{Overall}$) lies between $\eta(H_2)$ (at R= 0) and $\eta(O_2)$ (at $R = \infty$). Therefore, for the maximum overall cycle efficiency, it is necessary that most of the energy input into the cycle occurs at the more efficient step of the cycle. For example, if $\eta(O_2) > \eta(H_2)$, it is necessary that $\Delta H(O_2)_{In} \gg \Delta H(H_2)_{In}$. Likewise, if $\eta(H_2) > \eta(O_2)$, it is desirable that $\Delta H(H_2)_{In} \gg \Delta H(O_2)_{In}$.

2.3. *Sulfuric acid/metal sulfate based TCWSCs*

The sulfur family cycles are widely studied multi-step TCWSCs. The oxygen producing step in the sulfur family cycles is the decomposition of sulfuric acid or a metal sulfate. The energy input for the decomposition of H_2SO_4, calculated using GTT-Technologies' FactSage™ 5.5 thermochemical analysis software, is only about 80.9% of the total energy required for water splitting (*i.e.* 286 kJ/mol) as follows:

$$H_2SO_4 = H_2O + SO_3 \quad \Delta H^o_{298K} = 87.1 \text{ kJ/mol} \quad (46)$$

$$SO_3 = SO_2 + \tfrac{1}{2}O_2 \quad \Delta H^o_{298K} = 144.2 \text{ kJ/mol} \quad (47)$$

For which the overall reaction is:

$$H_2SO_4 = H_2O + SO_2 + \tfrac{1}{2}O_2 \quad \Delta H^o_{298K} = 231.3 \text{ kJ/mol} \quad (48)$$

The remaining 19.1% of the energy required to split water has to be supplied for the H_2 production step. Figure 4 depicts the terrestrial solar direct normal spectral irradiance distribution computed using SMARTS version 2.9.2 model with input file from ASTM Standard Table G 173-03$^{\varepsilon 1}$.

It can be seen that about 80.9% of the total solar irradiance comprising mostly of thermal energy with wavelengths above 520 nm can be utilized for the decomposition of sulfuric acid in the oxygen generation step of the sulfur-family cycles. The remaining 19.1% of the solar irradiance which consists of photonic energy at wavelengths less than about 520 nm will be used for the hydrogen production step of the cycle. In other words, for optimum overall cycle efficiency, it is necessary that the oxygen production step utilizes 80.9% of the solar irradiance as mostly thermal radiation above a wavelength of approximately 520 nm and the hydrogen generation step consumes the remaining 19.1% of solar resource, at wavelengths shorter than 520 nm – within a photolytic and/or photocatalytic reactor. A suitable photocatalyst for carrying out the hydrogen generation step is cadmium sulfide (CdS) for which the optical absorption edge (λ_{edge}) of the bulk material is at 512 nm making it an ideal photocatalyst for conducting the H_2 generation step of a solar-TCWSC.

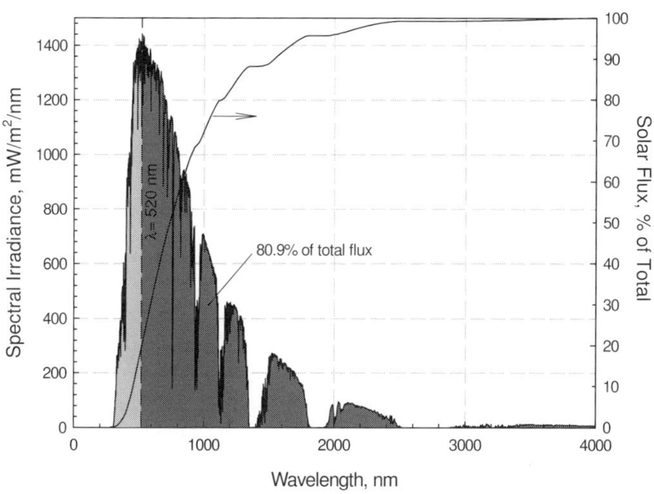

Figure 4. Plot of AM 1.5 direct normal spectral solar irradiance computed using SMATRS version 2.9.2 with input file from ASTM Standard Table G 173-03$^{\varepsilon 1}$.

Decomposition of sulfuric acid presents an efficient means of generating oxygen via a solar thermochemical water splitting cycle pending the required reaction temperatures can be realizable.

Typically, large-scale solar concentrators utilize parabolic reflectors in the form of trough, tower, or dish systems. These solar concentrators are characterized in terms of their mean flux concentration ratio C_R over an area S_a at the receiving focal plane as follows:

$$C_R = q_s/I \tag{49}$$

Where q_s (W/m^2) refers to the solar flux intercepted by unit area of the receiver at the focal plane and I (W/m^2) is the incident normal beam insolation. C_R is often expressed in units of ''suns'' when normalized to I = 1000 W/m^2 [13]. The solar flux concentration ratio typically obtained is at the level of 100, 1000, and 10,000 suns for trough, tower, and dish systems, respectively. The most suitable concentrators for applications involving solar thermochemical water splitting cycles are tower and dish systems.

According to Steinfeld [13], there is a temperature, $T_{optimum}$, for which the TCWSC efficiency is maximum. Assuming a uniform power-flux distribution, $T_{optimum}$ can be determined from the following implicit equation:

$$T^5_{optimum} - (0.75\ T_L)\ T^4_{optimum} - (T_L I C_R/4\sigma) = 0 \tag{50}$$

Where, T_L is the temperature of the thermal reservoir for heat rejection, usually ambient temperature and σ refers to the Stefan–Boltzmann constant ($5.6705 \times 10^{-8}\ Wm^{-2}K^{-4}$). In the case that the TCWSC utilizes only a portion of the solar irradiance (say, above λ) for performing the oxygen production step (see Figure 4), we have

$$T^5_{\lambda,\ optimum} - (0.75\ T_L)\ T^4_{\lambda,\ optimum} - (T_L I_\lambda C_R/4\sigma) = 0 \tag{51}$$

$T_{\lambda,\ optimum}$ refers to the temperature for which the efficiency of oxygen generation step of the solar TCWSC is highest. I_λ (W/m^2) refers to the incident normal beam insolation integrated over wavelength in the range of λ to 4000 nm (see Figure 4).

Figure 5 depicts, the optimum temperatures and maximum achievable efficiencies for the oxygen production step of a sulfur family solar TCWSC as a function of the mean flux concentration ratio at λ= 520 nm. Figure 5 also shows the variation of $T_{optimum}$ vs. λ at constant mean flux concentration ratios in the range of 50 to 10000. The solar insolation values used are taken from Figure (4) for the direct normal spectral irradiance data for the Air Mass 1.5.

Results of Figure 5 indicate that $T_{\lambda,\ optimum}$ for the oxygen generation step of the sulfur family TCWSC utilizing solar irradiance at wavelengths above λ= 520 nm varies between 768°C and 1347°C for uniform power-flux distribution with concentrations in the range of 1000-10,000. For example, at C_R= 1500, $T_{520nm,\ optimum}$ = 852°C – giving a maximum theoretical efficiency (or the 1st law efficiency, η_I) of about 73.33%. In other words, the portion of solar energy that could in principle be captured in the form of chemical energy (decomposition of sulfuric acid and generation of oxygen) is 73.33%. In practice, due to various losses, the maximum efficiency will be lower. Therefore, a solar concentrator (of the tower or dish type) with capability to deliver a mean flux concentration ratio C_R of at least 1500 is needed to carry out the sulfuric acid decomposition reaction for oxygen generation requiring temperatures at or above 852°C (1125 K).

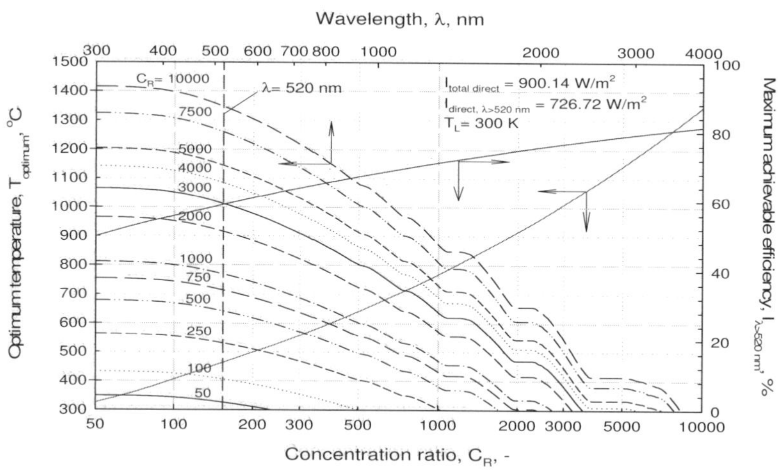

Figure 5. $T_{optimum}$ and maximum achievable efficiencies for oxygen production step of the sulfur family solar TCWSCs as a function of the mean flux concentration ratio, C_R, at λ= 520 nm; and variation of $T_{optimum}$ vs. λ at constant C_R. Direct normal solar irradiance data from Figure (4), AM 1.5.

The exergy efficiency (or the 2nd law efficiency) for the sulfuric acid decomposition step is given by:

$$\eta_{II} = -\acute{n}\Delta G|_{H_2O+SO_2++0.5O_2\rightarrow H_2SO_4}/C_R I_{\lambda>520nm} \qquad (52)$$

Where ΔG refers to the standard Gibbs free energy change for the sulfuric acid decomposition reaction at 298 K and 1 atm (-149.8 kJ/mol). The exergy efficiency is important in determining the merits of any solar thermochemical process. The higher the η_{II}, the lower the required size of the solar installation required for producing a given quantity of product, and lower the plant costs. In the equation above, ń is the molar flow rate of H_2SO_4 consumed which is determined from the definition of the 1st law efficiency as follows:

$$\eta_I = -ń\Delta H|_{H2O+SO2++0.5O2 \rightarrow H2SO4}/C_R I_{\lambda>520nm} \tag{53}$$

In Equation (53), ΔH is the standard enthalpy change for H_2SO_4 decomposition reaction at 298 K and 1 atm (-231.12 kJ/mol). For C_R= 1500, $T_{520nm,\ optimum}$ was calculated to be 1125 K (852°C) – giving a 1st law efficiency of about η_I = 73.33%. Substituting η_I, ΔH, C_R (=1500) and $I_{\lambda>520nm}$ (=726.72 W/m^2) into the equation (53) gives ń = 3.456 $mol/s/m^2$. Knowing ń, from equation (52) we have: η_{II} = 47.52%.

Due to high 1st and 2nd law efficiencies of sulfuric acid based cycles, to date, more than 20 sulfuric acid and/or metal sulfate decomposition based TCWSCs have been reported. Despite difficulties that challenge efficient electrolytic oxidation of sulfur dioxide (SO_2), the Westinghouse hybrid cycle still remains as one of the most studied TCWSCs. The Westinghouse cycle is as follows [14]:

$$SO_2(g) + 2H_2O(l) = H_2 + H_2SO_4(aq) \qquad 77^{o}C \text{ (electrolysis)} \tag{54}$$

$$H_2SO_4(g) = SO_2(g) + H_2O + \tfrac{1}{2}O_2 \qquad 850^{o}C \text{ (thermolysis)} \tag{55}$$

The Westinghouse cycle has many advantages widely reported and discussed in the literature. However, it is known that the Westinghouse cycle is hampered by the low water solubility of SO_2 and challenges presented by the acidity of the SO_2 electrolytic oxidation process [15]. To date, many efforts have been made to improve the efficiency of the electrolytic process for oxidation of SO_2. Prior work has involved the use of a depolarized electrolyzer as well as addition of a third process step - examples include S-I, S-Br and S-Fe cycles given below:

Ispra Mark 13 sulfur-bromine cycle [16]:

$$Br_2(l) + SO_2 + 2H_2O(l) \rightarrow 2HBr(aq) + H_2SO_4(aq) \qquad 77^{o}C \tag{56}$$

$$H_2SO_4(g) \rightarrow SO_2(g) + H_2O(g) + \tfrac{1}{2}\, O_2 \quad 850^{\circ}C \quad (57)$$

$$2HBr(aq) \rightarrow Br_2(aq) + H_2 \text{ (electrolysis)} \quad 77^{\circ}C \quad (58)$$

General Atomics' sulfur-iodine cycle [17]:

$$I_2 + SO_2(g) + 2H_2O(l) \rightarrow 2HI(a) + H_2SO_4(aq) \quad 100^{\circ}C \quad (59)$$

$$H_2SO_4(g) \rightarrow SO_2(g) + H_2O(g) + \tfrac{1}{2}\, O_2 \quad 850^{\circ}C \quad (60)$$

$$2HI \rightarrow I_2(g) + H_2 \quad 450^{\circ}C \quad (61)$$

Sulfur-iron cycle [18]:

$$Fe_2(SO_4)_3(aq)+SO_2 + 2H_2O \rightarrow 2FeSO_4(aq)+2H_2SO_4 \quad 25^{\circ}C \quad (62)$$

$$H_2SO_4(l) \rightarrow SO_2(g) + H_2O(g) + \tfrac{1}{2}\, O_2 \quad 850^{\circ}C \quad (63)$$

$$2FeSO_4(aq) + H_2SO_4(aq) \rightarrow Fe_2(SO_4)_3(aq) + H_2 \quad 25^{\circ}C \quad (64)$$

To make the separation of HI for H_2O easier, Sato and co-workers have proposed a nickel-iodine-sulfur version of S-I cycle [19]:

$$SO_2(aq) + I_2(aq) + 2H_2O(l) \rightarrow 2H_2SO_4(aq) + 2HI(aq) \quad (65)$$

$$2HI(aq)+2H_2SO_4(aq)+2Ni(c) \rightarrow NiI_2(c)+NiSO_4(aq)+2H_2 \quad (66)$$

$$NiI_2(c) \rightarrow NI(c) + I_2(g) \quad (67)$$

$$NiSO_4(c) \rightarrow NiO(c) + SO_3(g) \quad (68)$$

$$SO_3(g) \rightarrow SO_2(g) + \tfrac{1}{2}O_2(g) \quad (69)$$

$$NiO(c) + H_2(g) \rightarrow Ni(c) + H_2O(g) \quad (70)$$

Others include:

$$CO + H_2O \rightarrow CO_2 + H_2 \quad 550^{\circ}C \quad (71)$$

$$CO_2 + SO_2 + H_2O \rightarrow H_2SO_4 + CO \quad 500^{\circ}C \quad (72)$$

$$H_2SO_4(g) \rightarrow H_2O(g) + SO_2(g) + \tfrac{1}{2}O_2 \quad 900^{\circ}C \quad (73)$$

$$SO_2 + H_2O + I_2 \rightarrow SO_3 + 2HI \quad 200^{\circ}C \quad (74)$$

$$SO_3 \rightarrow SO_2 + \tfrac{1}{2}O_2 \quad 900^{\circ}C \quad (75)$$

$$2HI \rightarrow H_2 + I_2 \quad 450^{\circ}C \quad (76)$$

$$2FeSO_4 + I_2 + 2H_2O \rightarrow 2Fe(OH)SO_4 + 2HI \qquad 20^{o}C \qquad (77)$$

$$2HI \rightarrow H_2 + I_2 \qquad 450^{o}C \qquad (78)$$

$$2Fe(OH)SO_4 \rightarrow 2FeSO_4 + H_2O + ½O_2 \qquad 100^{o}C \qquad (79)$$

$$3FeCl_2(s) + 4H_2O \rightarrow Fe_3O_4(s) + 6HCl(g) + H_2 \qquad 650^{o}C \qquad (80)$$

$$Fe_3O_4(s)+Fe_2O_3(s)+6HCl+2SO_2 \rightarrow 3FeCl_2+2FeSO_4+3H_2O \qquad 100^{o}C \qquad (81)$$

$$2FeSO_4 \rightarrow Fe_2O_3(s) + 2SO_2(g) + ½O_2 \qquad 850^{o}C \qquad (82)$$

Although these cycles address some of the challenges associated with water splitting, especially with regard to water solubility of SO_2, they have issues of their own. For example, efficient separation of sulfuric acid from reaction products such as HI, HBr or $FeSO_4$ is challenging. Additionally, the pH of the solutions remains problematic. In fact, this problem becomes more acute due to the generation of other acids such as HI and HBr. For solar driven water splitting, Abanades *et al* [8] screened 280 TCWSCs. They selected 30 TCWSCs as promising which warrant further investigation. Among them, there were nine metal sulfate based TCWSCs – almost ⅓ of all selected cycles since decomposition of H_2SO_4 or MSO_4 presents an effective method for the heat absorbing step of the TCWSCs. The General Atomics' S-I cycle was not among the selected candidates considered suitable for solar interface by Abanades due to the difficulties in separating HI from water. Several examples of metal sulfate cycles are given below:

$$MnSO_4 \rightarrow MnO + SO_2 + ½O_2 \qquad 1100^{o}C \qquad (83)$$

$$MnO + H_2O + SO_2 \rightarrow MnSO_4 + H_2 \qquad 250^{o}C \qquad (84)$$

$$FeSO_4 \rightarrow FeO + SO_2 + ½O_2 \qquad 1100^{o}C \qquad (85)$$

$$FeO + H_2O + SO_2 \rightarrow FeSO_4 + H_2 \qquad 250^{o}C \qquad (86)$$

$$CoSO_4 \rightarrow CoO + SO_2 + ½O_2 \qquad 1100^{o}C \qquad (87)$$

$$CoO + H_2O + SO_2 \rightarrow CoSO_4 + H_2 \qquad 250^{o}C \qquad (88)$$

$$3FeO(s) + H_2O \rightarrow Fe_3O_4(s) + H_2 \qquad 200^{o}C \qquad (89)$$

$$Fe_3O_4(s) + FeSO_4 \rightarrow 3Fe_2O_3(s) + 3SO_2(g) + ½O_2 \qquad 800^{o}C \qquad (90)$$

$$3Fe_2O_3(s) + 3SO_2(g) \rightarrow 3FeSO_4 + 3FeO(s) \qquad 1800^{o}C \qquad (91)$$

$$3FeO(s) + H_2O \rightarrow Fe_3O_4(s) + H_2 \qquad 200^\circ C \qquad (92)$$

$$Fe_3O_4(s) + 3SO_3 \rightarrow 3FeSO_4(g) + \frac{1}{2}O_2 \qquad 800^\circ C \qquad (93)$$

$$FeSO_4 \rightarrow 3FeO(s) + SO_3 \qquad 2300^\circ C \qquad (94)$$

$$Fe_2O_3(s) + 2SO_2(g) + H_2O \rightarrow 2FeSO_4(s) + H_2 \qquad 200^\circ C \qquad (95)$$

$$2FeSO_4(s) \rightarrow 3Fe_2O_3(s) + SO_2(g) + SO_3(g) \qquad 700^\circ C \qquad (96)$$

$$SO_3(g) \rightarrow SO_2(g) + \frac{1}{2}O_2 \qquad 2300^\circ C \qquad (97)$$

$$6Cu(s) + 3H_2O \rightarrow 3Cu_2O(s) + 3H_2 \qquad 500^\circ C \qquad (98)$$

$$Cu_2O(s) + 2SO_2(g) + 1.5O_2 \rightarrow 2CuSO_4 \qquad 300^\circ C \qquad (99)$$

$$2Cu_2O(s) + 2CuSO_4 \rightarrow 6Cu(s) + 2SO_2 + 3O_2 \qquad 1750^\circ C \qquad (100)$$

$$Cu_2O(s) + H_2O(g) \rightarrow Cu(s) + Cu(OH)_2 \qquad 1500^\circ C \qquad (101)$$

$$Cu(OH)_2 + SO_2(g) \rightarrow CuSO_4 + H_2 \qquad 100^\circ C \qquad (102)$$

$$CuSO_4 + Cu(s) \rightarrow Cu_2O(s) + SO_2 + \frac{1}{2}O_2 \qquad 1500^\circ C \qquad (103)$$

$$SO_2 + H_2O + BaMoO_4 \rightarrow BaSO_3 + MoO_3 + H_2O \qquad 300^\circ C \qquad (104)$$

$$BaSO_3 + H_2O \rightarrow BaSO_4 + H_2 \qquad (105)$$

$$BaSO_4(s) + MoO_3(s) \rightarrow BaMoO_4(s) + SO_2(g) + \frac{1}{2}O_2 \qquad 1300^\circ C \qquad (106)$$

The second approach is to introduce a metal oxide as a catalyst to convert low concentration sulfuric acid to metal sulfate which is then decomposed to produce oxygen, sulfur dioxide and metal oxide. Sulfur dioxide and water are sent to acid electrolysis unit for generation of hydrogen and sulfuric acid – closing the cycle. Introducing ZnO into the Westinghouse TCWSC, a new, modified $ZnSO_4$ decomposition based Westinghouse cycle can be written as:

$$SO_2(g) + 2H_2O(l) = H_2 + H_2SO_4(aq) \qquad 77^\circ C \text{ (electrolysis)} \qquad (107)$$

$$H_2SO_4(aq, 50wt\%) + ZnO(s) = ZnSO_4 \cdot H_2O(s) \qquad 80\text{~}350^\circ C \qquad (108)$$

$$ZnSO_4 \cdot H_2O(s) = ZnSO_4(s) + H_2O(g) \qquad 450^\circ C \qquad (109)$$

$$ZnSO_4(s) = SO_2(g) + \frac{1}{2}O_2 + ZnO(s) \qquad 850^\circ C \qquad (110)$$

Similarly, by adding metal oxide catalysts to the Ispra Mark 13 sulfur-bromine cycle, General Atomics' sulfur-iodine cycle and sulfur-iron cycle (Reactions (56) to (64)), a number of new, modified metal sulfate based

TCWSCs can be devised. However, we note that, when energy input for these cycles is solar energy, they can utilize only the thermal energy degrading the photonic portion of solar spectrum to lower grade heat.

At the Florida Solar Energy Center, we have developed a new TCWSC that utilizes the photonic portion of the solar spectrum for the production of hydrogen and the thermal portion of sun light for the generation of oxygen.

3. Solar thermochemical S-NH3 water splitting cycle

3.1. *S-NH_3 hybrid cycle*

As shown in equation (44), TCWSC efficiency is a function of both hydrogen and oxygen production step efficiencies. The efficiency of any solar driven water splitting cycle depends upon the ability of the cycle to utilize as broad of a range of the solar spectrum as possible – i.e. being able to exploit both the photonic (UV/visible) and thermal components of the solar radiation. Degradation (or thermalization) of the high energy portion of solar radiation to thermal heat, as is the case with purely thermochemical water splitting cycles, can lead to lower overall cycle efficiencies. Thermodynamically, the total energy (ΔH) required to produce H_2 by water splitting is $\Delta H = \Delta G + T\Delta S$. At a given temperature T, a process will be more efficient if it can utilize the photonic energy of solar radiation as Gibbs free energy, ΔG, and the remaining mostly thermal component as $T\Delta S$.

Present technologies for the solar production of H_2 by means of direct water splitting either exclusively use the photonic portion of solar spectrum or totally thermalize the entire solar spectrum. For example, direct thermal decomposition of water degrades solar photonic energy to lower grade thermal heat, resulting in an exergy loss. Direct thermolysis of water requires temperatures above 2500°C and in addition, recombination of H_2 and O_2 is a serious issue. Photoelectrochemical (PEC) water splitting has merits over the photovoltaic (PV) plus water electrolysis as it combines photochemical and electrochemical steps into a single process allowing direct H_2 production. However, in the PEC process, only a small portion of the solar spectrum is utilized and the thermal component of sunlight is wasted. Consequently, the PEC energy conversion efficiencies are still at very low levels.

FSEC's hybrid photo/thermo-chemical water splitting cycle employs the quantum portion of the solar spectrum for the production of H_2 and the thermal portion (*i.e.*, IR) portion of solar radiation for O_2 evolution [5,20]. Utilization of the full solar spectrum allows the cycle to reach potentially a higher overall efficiency than is possible with the purely thermochemical water splitting cycles of the past. FSEC's sulfur-ammonia (S-NH_3) hybrid photo/thermochemical cycle is represented by the following four reactions:

$$SO_2(g) + 2NH_3(g) + H_2O(l) \rightarrow (NH_4)_2SO_3(aq) \quad 25^{\circ}C \quad \text{(chemical absorption)} \tag{111}$$

$$(NH_4)_2SO_3(aq) + H_2O \rightarrow (NH_4)_2SO_4(aq) + H_2(g) \quad 77^{\circ}C \quad \text{(photocatalytic step)} \tag{112}$$

$$(NH_4)_2SO_4 \rightarrow 2NH_3(g) + H_2SO_4(g) \quad 252^{\circ}C \quad \text{(thermocatalytic step)} \tag{113}$$

$$H_2SO_4(l) \rightarrow SO_2(g) + H_2O(g) + \tfrac{1}{2}O_2(g) \quad 852^{\circ}C \quad \text{(thermocatalytic step)} \tag{114}$$

Solar thermal energy is used to drive Reactions (113) and (114) for the production of O_2 via decomposition of ammonium sulfate $(NH_4)_2SO_4$ and sulfuric acid H_2SO_4. Reaction (112) is a photocatalytic process in which SO_3^{2-} ions are oxidized to SO_4^{2-} in the presence of UV-visible light, a photocatalyst and water, generating hydrogen. Figure 6 depicts a schematic diagram of the S-NH_3 cycle showing how the thermal, *i.e.* near infrared (NIR) and infrared (IR), and UV-visible portions of solar radiation are resolved using a spectral splitting mirror. The thermal part of the sunlight is then concentrated into a high temperature thermocatalytic reactor/receiver and used for oxygen production, while the photonic (UV and visible light) portion passes through a coating layer driving the photocatalytic hydrogen generation reaction.

Reaction (112) requires approximately 0.52 V potential (vs. NHE) in a 1 M aqueous $(NH_4)_2SO_3$ solution with a pH of 7.8. This potential is about ⅓ of that needed for water splitting (approximately 1.5V). Furthermore, as noted before, Reaction (114) requires 80.9% of the total solar irradiance, comprised of mostly thermal energy with wavelengths above 520 nm. The remaining 19.1% of the solar irradiance, which is photonic energy at wavelengths less than about 520 nm, is then used to carry out the hydrogen production step of the cycle. In the previous section it was shown that this partitioning of solar irradiance was necessary for achieving the highest overall cycle efficiency. In other words, the oxygen production step consumes 80.9% of the solar thermal energy at wavelengths of approximately 520 nm or longer and the H_2 generation step utilizes the remaining 19.1% of solar light having wavelengths shorter than 520 nm. The hydrogen generation step occurs within a photocatalytic reactor. A

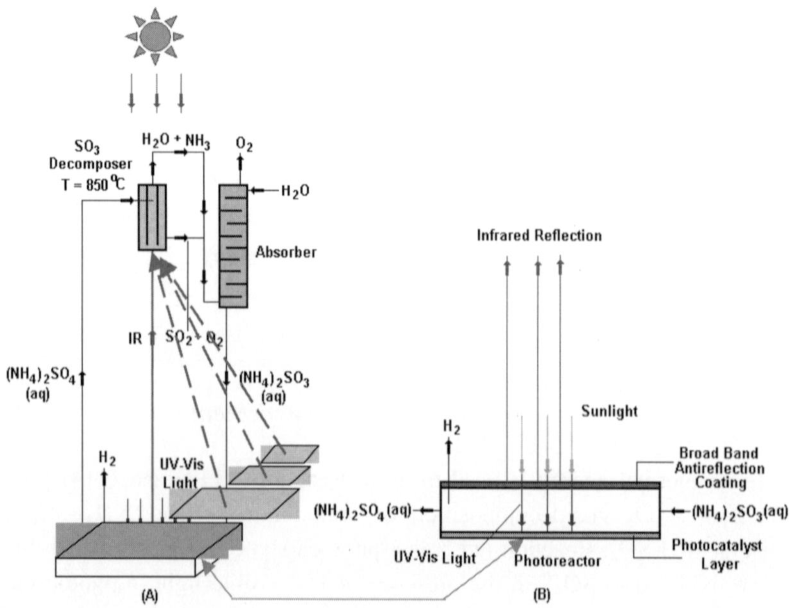

Figure 6. Schematic diagram of S-NH_3 photo-thermochemical water splitting cycle.

suitable photocatalyst for carrying out the hydrogen generation step is cadmium sulfide (CdS) with the optical absorption edge (λ_{edge}) of 512 nm for bulk CdS.

3.2. *Rate of H_2 production for the S-NH_3 TCWSC*

Reaction (111) in the S-NH_3 cycle is a chemical adsorption step involving reaction between an acid gas (SO_2) and an alkaline gas (NH_3) to form aqueous $(NH_4)_2SO_3$. Reaction (112) is a photocatalytic hydrogen production step in which photonic energy is converted to the chemical energy of hydrogen. Reaction (112) can occur by either a visible light photocatalytic process or a UV light photolytic route [21, 22]. Experimental data obtained to date show it is possible to carry out Reaction (112) with an energy conversion efficiency of about 12% using CdS as the photocatalyst. Figure 7 depicts the rate of H_2 production from an aqueous $(NH_4)_2SO_3$ solution using a 1000 W solar simulator fitted with an AM 1.5 global filter. Data of Figure 7 show that the rate of hydrogen production can be increased substantially by using polymer-stabilized platinum doped CdS.

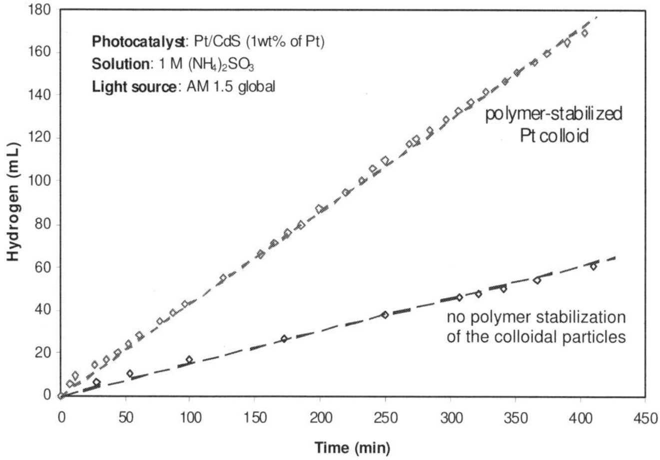

Figure 7. Rate of hydrogen production from aqueous $(NH_4)_2SO_3$ solution (beam area: 33 cm^2, light intensity: ~1.5 kW/m^2, solution pH = 7.5, solution volume = 200 mL, and 0.25 g Pt/CdS photocatalyst).

It should be possible to combine Reactions (113) and (114) into a single step. In fact, Reaction (113) is an intermediate step in which NH_3 is recovered and reacted with SO_2 to form $(NH_4)_2SO_3$ – to be used in the next reaction step.

We employed a Perkin-Elmer Diamond™ TG/DTA system coupled to the Pfeiffer ThermoStar™ benchtop quadrupole mass spectrometer with closed ion source for mass range of 1-300 amu for charting the decomposition of $(NH_4)SO_4$ [23]. Results obtained show that the decomposition of $(NH_4)SO_4$ occurs in two separate and sequential steps – at 250°C and 340°C, depending on the heating rate and material of the sample holder used. No sulfur or nitrogen containing gases were detected during the thermolysis process. This suggests that the S-NH_3 cycle as shown below:

$$2(NH_4)_2SO_4(s) = (NH_4)_2S_2O_7(s) + 2NH_3(g) + H_2O(g) \quad (115)$$

$$(NH_4)_2S_2O_7(s) = 2NH_3(g) + H_2S_2O_7(g) \quad (116)$$

$$H_2S_2O_7(g) = H_2SO_4(g) + SO_3(g) \quad (117)$$

can indeed be made to become a closed TCWSC, with the net reaction being that of water splitting:

$$2(NH_4)_2SO_4(s) = H_2SO_4(g) + SO_3(g) + 4NH_3(g) + H_2O(g) \quad (118)$$

Since most of the energy input into the cycle is used in Reaction (117), the overall cycle efficiency is strongly influenced by the efficiency of the H_2SO_4 decomposition step. Thermocatalytic decomposition of H_2SO_4 has been investigated extensively, and very high process efficiencies have been reported. Splitting the solar irradiance so that it can be input to two separate processes eases the requirement for high process efficiency for the photochemical reaction step of the S-NH_3 cycle. We have studied the thermodynamics, kinetics and flowsheeting of all steps involving sulfuric acid concentration and decomposition, and re-circulation of the un-reacted sulfur trioxide [24, 25].

As noted in the previous section, for C_R= 1500, at 1125 K (852°C), a maximum theoretical efficiency (or the 1st law efficiency, η_I) of about 73.33% is achievable for the H_2SO_4 decomposition step. In other words, the portion of the solar energy that could be captured and used to conduct acid decomposition and O_2 generation is about 73.33%. We also note that at temperatures higher than 1000°C, H_2SO_4 decomposition is no longer kinetically limiting step (*i.e.* there is no need for a catalyst to spur the process to completion). Rather, thermodynamics controls the extent of the conversion.

4. New MSO_4-NH_3 based solar TCWSCs

4.1. *Modified cycles*

As discussed above, FSEC's S-NH_3 cycle also utilizes decomposition of sulfuric acid as the endothermic step for the absorption of solar thermal heat and production of oxygen. However, high temperature concentration and decomposition of sulfur acid presents daunting materials of construction issues. Like the metal sulfate based TCWSCs, it is possible to modify the S-NH_3 cycle and do without the decomposition of H_2SO_4. There are two ways to accomplish this. The first approach is to decompose ammonium sulfate produced in the hydrogen production step of the S-NH_3 cycle (Reaction (111)) to a metal sulfate in the presence of a metal oxide catalyst. The second approach is to convert ammonium sulfate to metal pyrosulfate (*e.g.* ZnS_2O_7).

If a two valance metal oxide MO (*e.g.* ZnO) is introduced into the S-NH_3 TCWSC, a new family of MSO_4-NH_3 based cycles is devised as follows:

$$SO_2(g) + 2NH_3(g) + H_2O(l) \rightarrow (NH_4)_2SO_3(aq) \quad \text{(chemical absorption, 20°C)} \qquad (119)$$

$$(NH_4)_2SO_3(aq) + H_2O(l) \rightarrow (NH_4)_2\,SO_4(aq) + H_2 \quad \text{(solar photocatalytic, 80°C)} \qquad (120)$$

$$(NH_4)_2SO_4(s) + MO(s) \rightarrow 2NH_3(g) + MSO_4(s) + H_2O(g) \quad \text{(solar thermocatalytic, 400°C)} \tag{121}$$

$$MSO_4(s) \rightarrow SO_2(g) + MO(s) + O_2 \quad \text{(solar thermocatalytic, 1100°C)} \tag{122}$$

Where, M = Zn, Mg, Ca, Ba, Fe, Co, Ni, Mn, Cu and Pb. Oxides Fe_2O_3 and Cu_2O can also be included for by slightly modifying Reactions (121) and (122).

Decomposition of metal sulfates, especially $ZnSO_4$, has been reported by a number of researchers [26-28]. For example, T-Raissi and coworkers [26, 27] have given a detailed review of the literature pertaining to the decomposition of $ZnSO_4$. These researchers have also conducted a series of $ZnSO_4$ decomposition experiments at very rapid heating rates in a concentrating solar simulator. Their findings revealed that $ZnSO_4$ can be completely decomposed into SO_2, O_2 and ZnO. Depending on the magnitude of sample heating rates, a small amount of SO_3 may also be formed. At rapid heating rates (1~2°C/s) prevailing within concentrating solar furnaces, formation of SO_3 can be minimized, eliminating a need the separation of SO_2 from SO_3.

Unlike the metal sulfate based TCWSCs that employ reaction between sulfuric acid and a metal oxide (*e.g.* Reaction (108)), the new MSO_4-NH_3 cycles rely on the Reaction (121). Reaction (121) was first reported in 1955 by Dugger and coworkers who developed a process for the recovery of ammonia from ammonium sulfate in a two-stage reaction as follows [29]:

$$(NH_4)_2SO_4 + ZnO \rightarrow 2NH_3 + ZnSO_4 + H_2O \quad 400\text{~}500°C \tag{123}$$

$$ZnSO_4(s) \rightarrow SO_2(g) + ZnO(s) + O_2 \quad 800\text{~}1000°C \tag{124}$$

Experimental data show that all the nitrogen is recovered as NH_3, uncontaminated by sulfur oxides, in the low temperature stage. The major sulfur species formed at high temperatures was sulfur dioxide. In another study, Wentworth [30] has reported an ammonia yield of 99.3% by the following reactions involving ammonium hydrogen sulfate and zinc oxide:

$$2NH4HSO4(l) + 3ZnO(s) \rightarrow 2NH3(g) + ZnO\cdot 2ZnSO4(s) + 2H2O(g) \quad 365 \sim 418°C \tag{125}$$

$$ZnO\cdot 2ZnSO4(s) \rightarrow 2SO2(g) + 3ZnO(s) + O2 \quad 800\text{~}1000°C \tag{126}$$

It is therefore clear that Reaction (121) can be the basis of the new MSO_4-NH_3 TCWSCs.

Just as metal oxides can be used as catalysts for converting ammonium sulfate to metal sulfate, metal sulfates can also be employed for converting ammonium sulfate to ammonium pyrosulfate ($(NH_4)_2S_2O_7$)). Sulfur dioxide and oxygen are products of ammonium pyrosulfate decomposition. Thus, a second class of modified S-NH_3 TCWSCs based on the $M_2S_2O_7$-NH_3 can be devised as follows:

$$SO_2(g) + 2NH_3(g) + H_2O(l) \rightarrow (NH_4)_2SO_3(aq) \quad \text{(chemical absorption, 20°C)} \qquad (127)$$

$$(NH_4)_2SO_3(aq) + H_2O \rightarrow (NH_4)_2\,SO_4(aq) + H_2 \quad \text{(solar photocatalytic, 80°C)} \qquad (128)$$

$$(NH_4)_2SO_4(s) + M_2SO_4(s) \rightarrow 2NH_3(g) + M_2S_2O_7(s) + H_2O(g) \quad \text{(solar thermocatalytic, 400°C)} \qquad (129)$$

$$M_2S_2O_7(s) \rightarrow SO_2(g) + O_2 + M_2SO_4(s) \quad \text{(solar thermocatalytic, 1100°C)} \qquad (130)$$

Where, M = K, Rb, Cs. Reactions (129) and (130) have been described by Wentworth previously [30].

4.2. *Decomposition of metal oxide & ammonium sulfate mixtures*

Reagent grade $(NH_4)_2SO_4$ (Fisher Scientific) and zinc oxide (USP, EM Science) were used without further purification. Deionized water was generated in a two-step purification unit with conductivity of 18.3 MΩ-cm. Tap water was deionized by first passing the liquid through a PRO/RO filtration unit (Labconco) and then through a compact ultra pure water deionization system (Barnstead). As noted before, for the thermal analysis, we used a Perkin Elmer thermogravimetric/ differential thermal analyses (TG/DTA) coupled to a mass spectrometer (Pfeiffer ThermoStar™) with ultra pure grade helium gas (Linde Gas) as a carrier gas. The helium flow rate was set at 150 mL/min, monitored with a rotameter. In some experiments an aluminum sample holder was employed for the measurement of ammonium sulfate decomposition catalyzed by ZnO. The dimensions of the sample holders were: 5.210 mm OD, 4.965 mm high and wall thickness of 0.535 mm. The mixtures of $(NH_4)_2SO_4$ and ZnO were prepared by adding ZnO to aqueous ammonium sulfate solutions and stirring for two hours followed by heating to 50°C under vacuum until completely dehydrated. Samples from dried mixtures of $(NH_4)_2SO_4$ and ZnO were prepared for the TG/DTA/MS analysis at various heating rates and in the temperature range of 50°C to 600°C.

Figure 8 depicts the TG/DTA/MS results for ZnO + $(NH_4)_4SO_4$ mixture with molar ratio of ZnO:$(NH_4)_4SO_4$ = 1:1 at a heating rate of 5°C/min. The MS results show that no O_2, N_2, H_2, HNO_2, NO_2, SO_3 or H_2SO_4 was detectable within temperature range of 50°C to 600°C. The main reaction products determined from the MS measurements were NH_3, H_2O, and small amounts of SO_2 and NO. The results indicate that deammoniation and dehydration of the ZnO + $(NH_4)_2SO_4$ mixture is complex and occurs in several successive stages. TG/DTA curves show that NH_3 is released in five or six separate steps starting

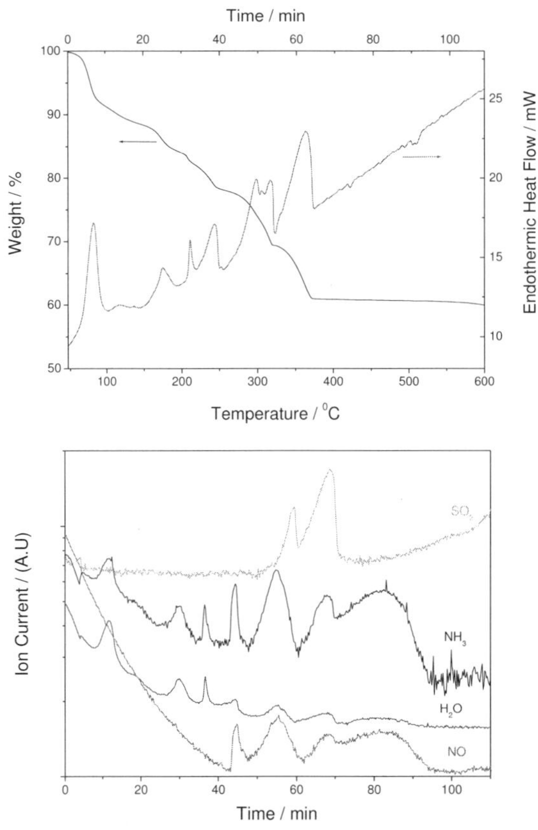

Figure 8. TG/DTA/MS analyses of ZnO + $(NH_4)_2SO_4$ decomposition, mixture molar ratio x= ZnO:$(NH_4)_2SO_4$ = 1:1, heating rate = 5°C/min.

from temperatures as low as 50°C and as high as 500°C. The heating rate has a significant effect on the ammonia release temperature.

Figure 9 depicts the TG/DTA/MS results for ZnO + $(NH_4)_4SO_4$ mixture with molar ratio of ZnO:$(NH_4)_4SO_4$ = 1:1 at a heating rate of 20°C/min. Results of Figure 9 show that when heating rate is increased to 20°C/min, NH_3 evolves at a higher temperature and the extent of NO and SO_2 formed decreases.

Figure 10 depicts the TG/DTA/MS results for ZnO + $(NH_4)_4SO_4$ mixture with molar ratio of ZnO:$(NH_4)_4SO_4$ = 1:1.5 at a heating rate of 20°C/min. Results of Figure 10 show that at high heating rates, the extent of SO_2 and NO formed reduced with the SO_2 peak shifting to higher temperatures.

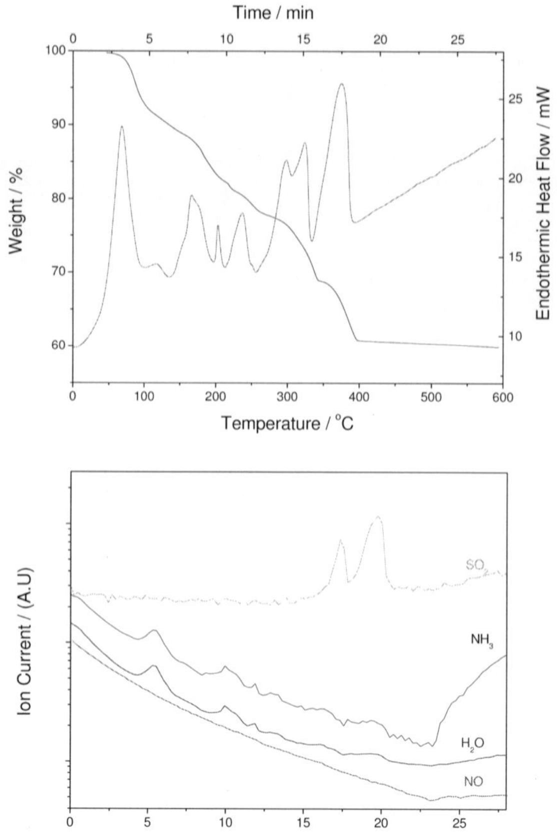

Figure 9. TG/DTA/MS analyses of ZnO + $(NH_4)_2SO_4$ decomposition, mixture molar ratio x= ZnO:$(NH_4)_2SO_4$ = 1:1, heating rate = 20°C/min.

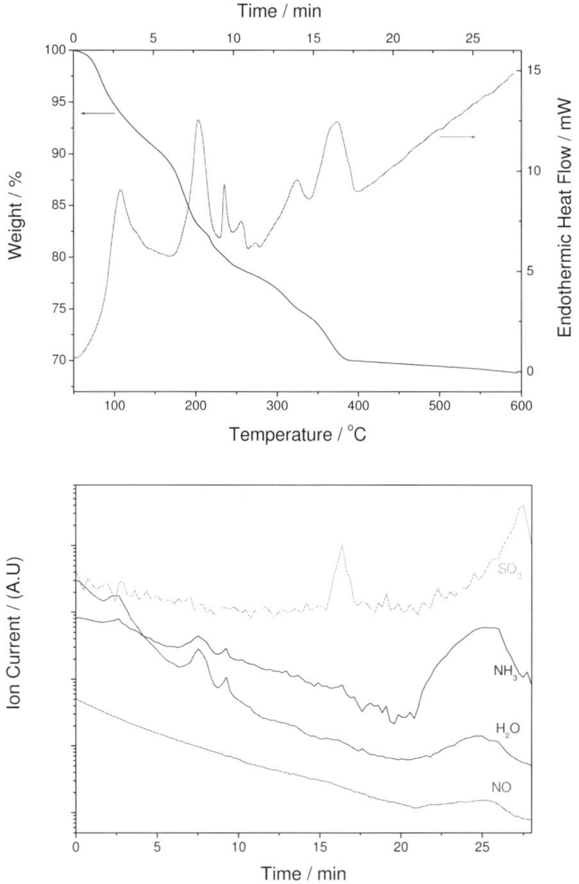

Figure 10. TG/DTA/MS analyses of ZnO + $(NH_4)_2SO_4$ decomposition, mixture molar ratio x= ZnO:$(NH_4)_2SO_4$ = 1.5: 1, heating rate = 20°C/min.

Release of ammonia from a mixture of ZnO and $(NH_4)_2SO_4$ is accompanied by a series of intermediate reactions as discussed by Dugger et al. [29]:

$$(NH_4)_2SO_4(s) = NH_4HSO_4(s) + NH_3(g) \tag{131}$$

$$ZnSO_4(s) + xNH_3(g) = ZnSO_4(s){\cdot}xNH_3 \ (x\text{=1 to 6}) \tag{132}$$

$$ZnO(s) + 2ZnSO_4(s) = ZnO{\cdot}2ZnSO_4(s) \ (\text{excess ZnO}) \tag{133}$$

$$NH_4HSO_4(s) + ZnO(s) = ZnSO_4(s) + NH_3(g) + H_2O(g) \tag{134}$$

$$ZnSO_4{\cdot}xNH_3(s) = ZnSO_4(s) + xNH_3(g) \tag{135}$$

We note that by increasing the ratio of ZnO to $(NH_4)_2SO_4$, more ammonia is released at lower temperatures (see Figures 9 and 10). Also, as indicated by the MS data, less SO_2 is released at lower temperatures (300 to 400°C). By changing the heating method, for example, holding temperature at 200°C for 45 minutes, both SO_2 and NO peaks are significantly reduced. This is shown in the data of Figure 11.

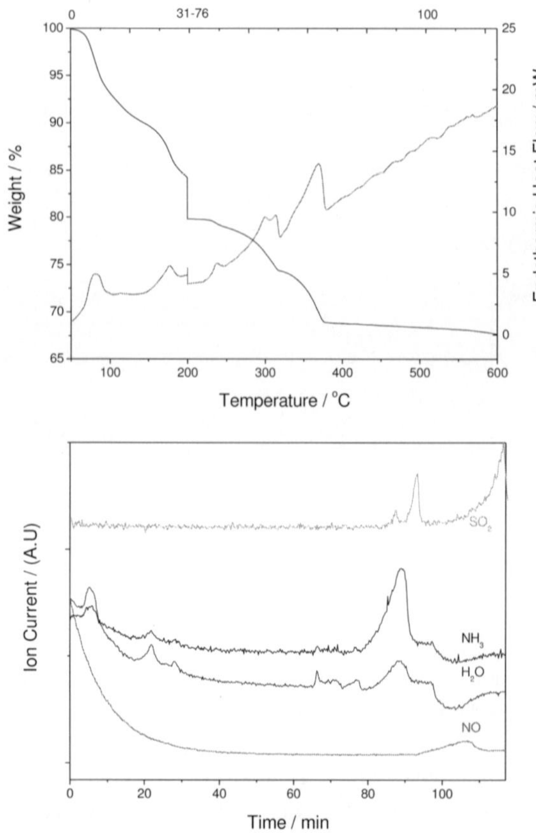

Figure 11. TG/DTA/MS analyses of ZnO + $(NH_4)_2SO_4$ decomposition, mixture molar ratio x= ZnO:$(NH_4)_2SO_4$ = 1.5:1. Samples heated from room temperature to 200°C at heating rate of 50°C/min and held at 200°C for 45 minutes followed by ramping sample temperature to 600°C at 10°C/min.

5. Conclusions

Any thermochemical water splitting cycle consists of at least two main steps: hydrogen and oxygen production steps. A two-step water splitting cycle can not be efficient if energy requirements for these two steps are significantly different. In this paper, we have developed a method for evaluating the overall efficiency of the FSEC developed S-NH_3 and associated solar thermochemical water splitting cycles. Based on the experimental data presented, we have shown that the S-NH_3 TCWSC attains a high 1st law efficiency by splitting the solar spectrum into two sections and using the shorter wavelength photonic portion for CdS photocatalytic H_2 production and the longer wavelength and IR portions of the sunlight, at a mean concentration ratio of 1500 or above, for the thermocatalytic O_2 production from H_2SO_4 decomposition.

Due to the intrinsic difficulties of sulfuric acid decomposition, we have introduced two new classes of solar driven TCWSCs by modifying the original S-NH_3 cycle. They include: (I) 12 metal sulfate-ammonia (MSO_4-NH_3) based TCWSCs and (II) 3 metal pyrosulfate-ammonia ($M_2S_2O_7$-NH_3) based TCWSCs. Our preliminary experimental results of the ammonia released from the ZnO + $(NH_4)SO_4$ mixtures show the feasibility of these new cycles. More experiments are currently underway to determine the reaction mechanisms and the nature of the reaction intermediates and products formed. These experimental and thermodynamic analyses are expected to lead to development of a highly efficient, solar driven water splitting cycle.

Acknowledgment

This research has been funded by the National Aeronautics and Space Administration (NASA) - Glenn Research Center (GRC) under contract NAG3-2751. The authors are grateful to Mr. Timothy Smith, NASA-GRC Program Manager and Dr. David L. Block (FSEC) for their support of this work.

References

1. Khaselev O. and Turner J. A., "A Monolithic Photovoltaic-Photoelectrochemical Device for Hydrogen Production via Water Splitting," *Science*, 280(17), 425-7, **1998**.
2. Licht, S., Wang, B., Mukerji, S., Soga, T., Umeno, M. and Tributsch, H., "Over 18% Solar Energy Conversion for Generation of Hydrogen Fuel; Theory and Experiment for Efficient Solar Water Splitting," *Int. J. of Hydrogen Energy*, 26, 653-9, **2001**.

3. Deutsch, T. G., Koval, C. A. and Turner J. A., "III-V Nitride Epilayers for Photoelectrochemical Water Splitting: GaPN and GaAsPN," *J. Phys. Chem. B*, 110, 25297-307, **2006**.
4. Liu, H., Yuan, J., and Shangguan, W., "Photochemical Reduction and Oxidation of Water Including Sacrificial Reagents and Pt/TiO2 Catalyst," *Energy & Fuels*, 20(6), 2289-92, **2006**.
5. T-Raissi, A., Muradov, N., Huang, C. and Adebiyi, O., "Hydrogen from Solar via Light-Assisted High-Temperature Water Splitting Cycles," *J. Solar Energy Engineering,* **129**, 184-9, **2007**.
6. Bilgen, E., Ducarroir, M., Foex, M., Sibieude, F., and Trombe, F., "Use of Solar Energy for Direct and Two-Step Water Decomposition Cycles," *Int. J. Hydrogen Energy,* 2(3), 251-7, **1977**.
7. Steinfeld, A., "Solar Hydrogen Production via Two-Step Water Splitting Thermochemical Cycle Based on Zn/ZnO Redox Reaction," *Int. J. Hydrogen Energy,* **27**, 611-9, **2002**.
8. Abanades, S., Charvin, P., Flamant, G., and Neveu, P., "Screening of Water-Splitting Thermochemical Cycles Potentially Attractive for Hydrogen Production by Concentrated Solar Energy", *Energy*, 31, 2805-22, **2006**.
9. Bamberger, C.E., Richardson, D.M., "Hydrogen Production from Water by Thermochemical Cycles," *Cryogenics* 16(4), 197-208, **1976**.
10. Bamberger, C. E., "Hydrogen Production from Water by Thermochemical Cycles; a 1977 update," *Cryogenics* 18(3), 170-83, **1978**.
11. Funk, J. E., Conger, W. L., Carty, R. H., "Evaluation of Multi-step Thermochemical Processes for the Production of Hydrogen from Water," *The Hydrogen Economy Miami Energy (THEME) Conference,* Miami Beach, 457, March 18-20, **1974**.
12. Huang, C., and T-Raissi, A., "A Perspective on Thermodynamics and Thermal Efficiency Calculations for Hydrogen Production via Thermochemical Water Splitting Cycles," to appear.
13. Steinfeld, A., "Solar Thermochemical Production of Hydrogen – A Review," *Solar Energy*, 78, 603–15, **2005**.
14. Brecher, L.E., Spewock, S., et al., "Westinghouse Sulfur Cycle for the Thermochemical Decomposition of Water," *Proceedings of the 1st World Hydrogen Energy Conf.,* 1 9A, 1-16, **1976**.
15. Lu, P.W.T. "Technological Aspects of Sulfur Dioxide Depolarized Electrolysis for Hydrogen Production," *Int. J. Hydrogen Energy,* **8**(10), 773-81, **1983**.
16. Beghi, G.E., "A decade of Research on Thermochemical Water Hydrogen at the Joint Research Center, Ispra," *Int. J. Hydrogen Energy,* **11**(12), 761-71, **1986**.

17. Besenbruch, G.E., "General Atomic Sulfur-Iodine Thermochemical Water Splitting Process," *Am. Chem. Soc., Div, Pet, Chem. Preprint*, **271,** 48, **1982**.
18. Takehara, Z., Nogami, M., and Shimizu, Y., *Int. J. Hydrogen Energy,* **14**(4), 233-9. **1989**.
19. Sato, S., Shimizu, S., Nakajima, N., and Ikezoe, Y., "A Nickel-Iodine-Sulfur Process for Hydrogen Production," *Int. J. Hydrogen Energy*, 8(1), 15-22, **1983**.
20. Huang, C., and T-Raissi, A., "A New Solar-Thermochemcial Water Splitting Cycle for Hydrogen Production," *Proc. 15th World Hydrogen Energy Conf.,* Yokohama, Japan, **2004**.
21. Huang, C., Adebiyi, O., Muradov, N., and T-Raissi, A., "UV Light Photolysis of Ammonium Sulfite Aqueous Solution for the Production of Hydrogen," P*roc.16th World Hydrogen Energy Conf.*, Lyon, France, June 13-16, **2006**.
22. Huang, C., Adebiyi, O., Muradov, N., and T-Raissi, A., "Production of Hydrogen via a Sulfur-Ammonia Solar Thermochemical Water Splitting Cycle," *Proc. 16th World Hydrogen Energy Conf.*, Lyon, France, June 13-16, **2006**.
23. Huang, C., Mohajeri, N., Muradov, N., and T-Raissi, A., "Solar Thermocatalytic Decomposition of Ammonium Sulfate," *Proc. of Solar 2007,* Cleveland, Ohio, July 7 to 12, **2007**.
24. Huang, C., and T-Raissi, A., "Analysis of Sulfur-Iodine Thermochemical Cycle for Solar Hydrogen Production. Part I - Decomposition of Sulfuric Acid," *Solar Energy,* **78**(5), 632-46, **2005**.
25. Huang, C., and T-Raissi, A., "Evaluation of Kinetic Models for the Solar Decomposition of Sulfur Trioxide," to appear.
26. Narayan, R., T-Raissi A. and Antal, Jr., M. J., Ind. Eng. Chem. Res. 27, 1050-8 (**1988**).
27. T-Raissi, A., Narayan, R., Mok, W. S. I. and Antal, Jr., M. J., Ind. Eng. Chem. Res. 28, 355-62 (**1989**)
28. Krikorian, O. H. and Shell, P. K., *Int. J. Hydrogen Energy.* 7(6) 463-9 (**1982**).
29. Dugger, G. L., Adams, J. B. and Bart, R., "Ammonium sulfate decomposition", United States Atomic Energy Commission, RMO-2036 (**1955**).
30. Wentworth, W. E., "Thermochemical cycles for energy storage: Thermal decomposition of ZnSO4 systems", Final Topical Report, January 1, 1982-December 31, 1984. Report (**1992**), (NREL/TP-253-4279; Order No. DE91002144).

DEVELOPMENT OF PHOTOCATALYSTS FOR SOLAR HYDROGEN PRODUCTION

AKIHIKO KUDO[a,b]

[a]Department of Applied Chemistry, Tokyo University of Science, Tokyo,1-3 Kagurazaka, Shinjuku-ku, Tokyo 162-8601, Japan, [b]Core Research for Evolutional Science and Technology, Japan Science and Technology Agency (CREST, JST), 4-1-8 Honcho, Kawaguchi-shi, Saitama 332-001, Japan

Photocatalytic water splitting is a challenging reaction because it is an ultimate solution to energy and environmental issues. Recently, many new powdered photocatalysts for water splitting have been developed. For example, a NiO (0.2 wt %)/$NaTaO_3$:La (2%) photocatalyst with a 4.1-eV band gap showed high activity for water splitting into H_2 and O_2 with an apparent quantum yield of 56% at 270 nm. Overall water splitting under visible light irradiation has been achieved by construction of a Z-scheme photocatalysis system employing visible-light-driven photocatalysts, Ru/$SrTiO_3$:Rh and $BiVO_4$ for H_2 and O_2 evolution, and an Fe^{3+}/ Fe^{2+} redox couple as an electron relay. Moreover, highly efficient sulfide photocatalysts for solar hydrogen production in the presence of electron donors were developed by making solid solutions of ZnS with $AgInS_2$ and $CuInS_2$ of narrow band gap semiconductors. Thus, the database of powdered photocatalysts for water splitting has become plentiful.

1. Introduction

Hydrogen is an ideal clean energy as well as a raw material in many chemical industries. Hydrogen is mainly produced by steam reforming of hydrocarbons such as methane in industry. Hydrogen must be produced from water using renewable energy sources such as solar light, if one considers the energy and environmental issues. Photocatalytic water splitting is one of the candidates for solar hydrogen production. When one thinks usage of solar energy one encounters the difficulty due to its low density of the energy. A large area should be used in order to harvest a reasonable amount of solar energy. The photocatalytic water splitting will be advantageous for the large-scale application for the solar hydrogen production because of the simplicity.

Photocatalysis is classified into two categories. One is of application to the improvement of living environment such as anti stain, self-cleaning, and superhydrophilicity [1]. TiO_2 photocatalysts are widely used for these applications and many industrial products have been made. Another direction of the photocatalysis is a light energy conversion as represented by water splitting. Water splitting using the light energy has been studied for a long time using powder and electrode systems since the Honda-Fujishima effect was reported [2,3]. Although it is under a basic research it is still a challenging topic. This research field is remarkably progressing recently. It is no doubt that photocatalytic water splitting will contribute to green sustainable chemistry. The final target of this research field is to achieve an artificial photosynthesis and solar hydrogen production from water.

The present paper focuses on the development of new photocatalyst materials in order to make a photocatalyst library. New materials of photocatalysts that the author and coworkers have found are reviewed.

2. New Photocatalyst Materials for Water Splitting

Table 1 shows photocatalyst materials that the present author and coworkers have found. The photocatalytic reactions are classified into overall water splitting into H_2 and O_2 without any sacrificial reagents, and H_2 or O_2 evolution in the presence of sacrificial reagents. Many tantalates and niobates with wide band gaps show the activities for overall water splitting under UV irradiation [4-16]. In the presence of sacrificial reagents, many metal cation-doped oxides [17-22], valence band-controlled oxides [23-30], and metal sulfides [31-40] show activities for H_2 or O_2 evolution under visible light irradiation. It is important to make such a photocatalyst library in order to look at the overview of materials and get information for design of new materials.

Codoping is sometimes a suitable method to get visible light response. For example, TiO_2 codoped Sb with Cr is active for O_2 evolution whereas only Cr-doped TiO_2 is inactive as usual [17]. The codoping contributes to the compensation of the charge and distortion resulting in the suppression of recombination centers.

A remarkable photocatalyst in Table 1 is Rh-doped $SrTiO_3$ [19]. It is one of the rare oxide photocatalysts that can produce H_2 under visible light irradiation. The visible light response is due to the transition from an electron donor level consisting of Rh 4d orbitals to the conduction band of $STiO_3$ as shown in Fig. 1. This photocatalyst shows the highest activity when the doping amount is ca. 1%. The amount of 1% is not small as a doping level. The Rh dopant forms a discrete level or a mini band.

Another unique photocatalyst is valence band-controlled $BiVO_4$ [24-27]. It can be prepared under ambient condition in aqueous media. It is an environmental-friendly process. It can produce O_2 under visible light irradiation.

The Bi *6s* orbitals in $BiVO_4$ contribute to the formation of the top of valence band (near HOMO). The conduction band is composed of V *3d* orbitals as usual.

Table 1. Photocatalyst library.

UV-responsive photocatalysts	VIS-responsive photocatalysts		
Overall water splitting	H_2 evolution (Sacrificial)	O_2 evolution (Sacrificial)	Overall water splitting
$ZnNb_2O_6$ [4] $Sr_2Nb_2O_7$ [5] $Cs_2Nb_4O_{11}$ [6] $Ba_5Nb_4O_{15}$ [7] $ATaO_3$ (A=Li, Na, K) [8,9] $NaTaO_3$:A (A=Ln, Ca, Sr, Ba) [10,11] ATa_2O_6 (A=Mg, Ca, Sr, Ba) [8] $Sr_2Ta_2O_7$ [5,12] $K_3Ta_3Si_2O_{13}$ [13] $K_3Ta_3B_2O_{12}$ [14] $K_2LnTa_5O_{15}$ [15] $AgTaO_3$ [16]	$SrTiO_3$:Cr,Sb [17] $SrTiO_3$:Cr,Ta [18] $SrTiO_3$:Rh [19] $SnNb_2O_6$ [23] ZnS:Cu [31] ZnS:Ni [32] ZnS: Pb,Cl [33] $NaInS_2$ [34] $AgGaS_2$ [35] $CuInS_2$ - $AgInS_2$-ZnS [36-40] In_2O_3 $(ZnO)_3$ [41]	TiO_2:Cr,Sb [17] TiO_2:Ni,Nb [20] $PbMoO_4$:Cr [22,24] $BiVO_4$ [24-27] Bi_2MoO_6 [28] Bi_2WO_6 [29] $AgNbO_3$ [16] Ag_3VO_4 [30] In_2O_3 $(ZnO)_3$ [41]	$SrTiO_3$:Rh-$BiVO_4$ [42] $SrTiO_3$:Rh-Bi_2MoO_6 [42] $SrTiO_3$:Rh-WO_3 [42]

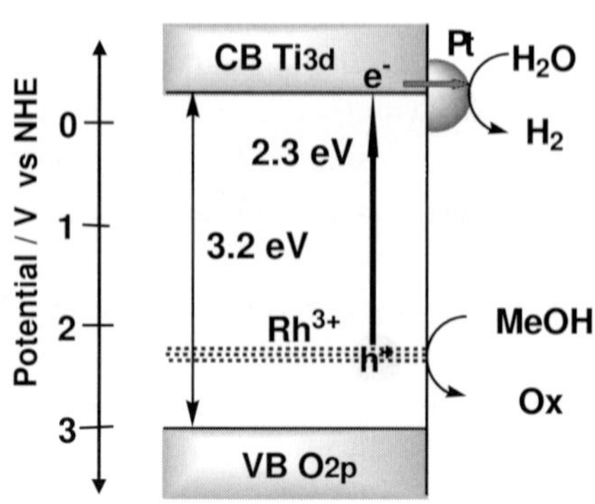

Figure 1. Mechanism of H_2 evolution over Pt/$SrTiO_3$:Rh under VIS light irradiation.

3. Highly Efficient Water Splitting into H_2 and O_2 on Tantalate Photocatalyst

$NiO/NaTaO_3$ is the most active for water splitting among tantalate photocatalysts [43]. The photocatalytic activity of $NiO/NaTaO_3$ increases remarkably with doping of lanthanoid and alkaline earth metal cations [10,11]. The doping of La makes the life time of photogenerated electrons long; that was confirmed by time-resolved infrared absorption spectroscopy [44]. An optimized NiO (0.2 wt %)/$NaTaO_3$:La (2%) photocatalyst shows high activity, with an apparent quantum yield of 56% for water splitting [10]. Under irradiation of the light from a 200-W Xe-Hg lamp, H_2 and O_2 evolve in the form of bubbles, without any sacrificial reagents, as shown in Fig. 2. This photocatalyst responds to only UV light. It should be stressed that this photocatalyst has demonstrated the highly efficient water splitting even using a powdered system.

Figure 2. Water splitting over $NiO/NaTaO_3$:La photocatalyst under UV irradiation
Light Source: 200W Xe-Hg Lamp.

4. Solar Hydrogen Production Using Water and Abundant Sulfur Compounds on Metal Sulfide Photocatalysts

We have examined the photophysical and photocatalytic properties of solid solutions of sulfides based on ZnS according to the band engineering [36-40]. Solid solutions consisting of combinations of $CuInS_2$, $AgInS_2$, and ZnS show the high photocatalytic activities for H_2 evolution from aqueous sulfide and sulfite solutions under visible light irradiation. The diffuse reflectance spectra for $AgInS_2$-$CuInS_2$-ZnS shift monotonically with the composition of the solid solution. This indicates that the levels of the conduction band consisting of Zn 4s-4p and In 5s-5p, and of the valence band consisting of Cu 3d, Ag 4d, and S 3p, shift with the varying composition. $Ru/Cu_{0.25}Ag_{0.25}In_{0.5}ZnS_2$ especially shows excellent activity for the H_2 evolution solution (8L/m2•h) with a solar simulator (AM-1.5). Hydrogen is observed in the form of bubbles as shown in Fig. 3. These sulfide solid solution photocatalysts can utilize visible light of wavelengths up to about 700 nm. The activity is higher than that of the well-known Pt/CdS photocatalyst, which can utilize visible light up to 520 nm. Moreover, toxic elements such as cadmium are not included in the photocatalysts. These photocatalysts will be able to be used for the recovery of hydrogen from water and abundant sulfur compounds in nature, and petroleum and mining industries.

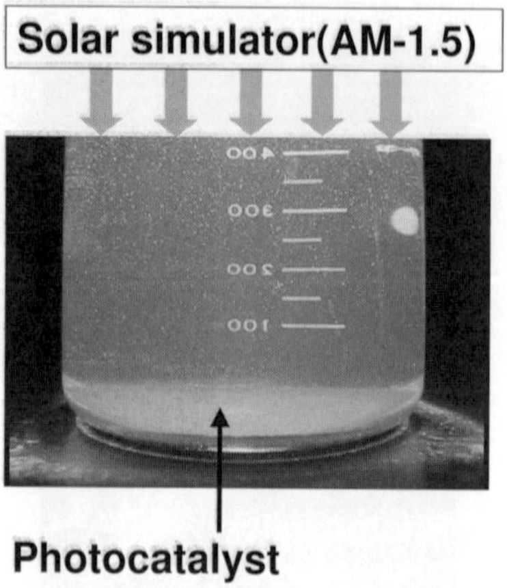

Figure 3. Solar H_2 production on $Ru/Cu_{0.25}Ag_{0.25}In_{0.5}ZnS_2$ photocatalyst from an aqueous K_2SO_3+Na_2S solution using a solar simulator (AM-1.5).

5. Solar Hydrogen Production from Water Using Visible Light Driven Photocatalysts

It is mentioned that $SrTiO_3$:Rh and $BiVO_4$ are remarkable photocatalysts working under visible light irradiation. The system in which $SrTiO_3$:Rh is combined with $BiVO_4$ in the presence of an Fe^{3+}/Fe^{2+} redox couple shows activity for overall water splitting under visible light irradiation according to the scheme as shown in Fig. 4 [42]. This Z-scheme system responds to 520-nm light, corresponding with the absorption edges of $SrTiO_3$:Rh and $BiVO_4$, and is active even with a solar light. In other words, although the efficiency is low, solar hydrogen production from water has been accomplished using a powdered photocatalyst system with visible light response.

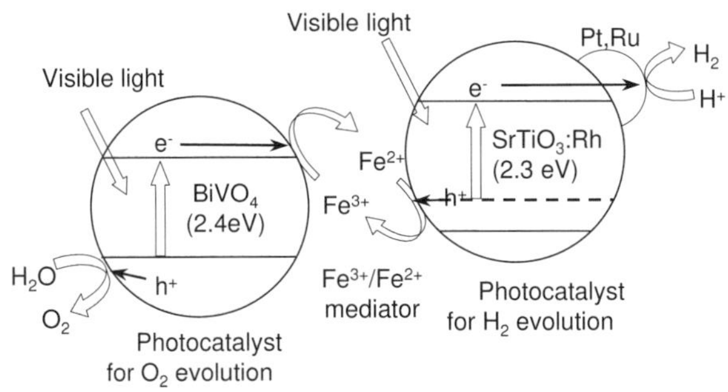

Figure 4. Z-scheme photocatalyst system for solar hydrogen production.

6. Conclusions

The target for efficiency for water splitting into H_2 and O_2 can be said to be 30% in terms of quantum yield at 600 nm in this research field. This means that we have to develop highly active photocatalysts with a 2-eV band gap. At the present stage, although the NiO/$NaTaO_3$:La photocatalyst shows a high quantum yield, it responds to only UV light. The wavelength is far from the target. The sulfide solid solution photocatalysts $AgInS_2$-$CuInS_2$-ZnS show relatively high active for solar hydrogen production in the presence of electron donors, but not for overall water splitting. The new powdered photocatalyst systems, Ru/$SrTiO_3$:Rh-$BiVO_4$, responds to 520 nm for overall water splitting and solar hydrogen production from water. The respondent wavelength is somewhat close to the target. However, the quantum yield is still low compared with the target. We still have to continue the research and make further breakthroughs for solar hydrogen production from water. It will be also important to construct the operating system for photocatalytic hydrogen production, as well as the development of photocatalyst materials. The achievement will lead to an ultimate green sustainable chemistry.

Acknowledgments

This work was supported by the Core Research for Evolutional Science and Technology (CREST) program of the Japan Science and Technology (JST) Agency, and a Grant-in-Aid (No.14050090) for Priority Area Research (No.417) from the Ministry of Education, Culture, Science, and Technology. The author thanks Dr. Kato, Dr. Tsuji, Ms. Omori, and Ms. Konta for their experiments, and Prof. Kobayashi (Kyoto Institute of Technology) and Dr. Shimodaira for DFT calculations.

References

1. A. Fujishima, T.N. Rao, D.A. Tryk, *J. Photochem. Photobiol. C.* **1**, 1 (2000).
2. A. Fujishima, K. Honda. *Bull. Chem. Soc. Jpn.* **44**, 1148 (1971).
3. A. Fujishima, K. Honda. *Nature*. **238**, 37 (1972).
4. A. Kudo, S. Nakagawa, H. Kato. *Chem. Lett.* 1197 (1999).
5. A. Kudo, H. Kato, S. Nakagawa. *J. Phys. Chem. B.* **104**, 571 (2000).
6. Y. Miseki, H. Kato, A. Kudo. *Chem. Lett.* **34**, 54 (2005).
7. Y. Miseki, H. Kato, A. Kudo. *Chem. Lett.* **35**, 1052 (2006).
8. H. Kato, A. Kudo. *Chem. Phys. Lett.* **295**, 487 (1998).
9. H. Kato, A. Kudo. *J. Phys. Chem. B.* **105**, 4285 (2001).
10. H. Kato, K. Asakura, A. Kudo. *J. Am. Chem. Soc.* **125**, 3082 (2003).
11. A. Iwase, H. Kato, A. Kudo. *Chem. Lett.* **34**, 945 (2005).
12. K. Yoshioka, V. Petrykin, M. Kakihana, H. Kato, A. Kudo. *J. Catal.* **232**, 102 (2005).
13. A. Kudo, H. Kato. *Chem. Lett.* 867 (1997).
14. T. Kurihara, H. Okutomi, Y. Miseki, H. Kato, A. Kudo. *Chem. Lett.* **35**, 274(2006).
15. A. Kudo, H. Okutomi, H. Kato. *Chem. Lett.* 1212-1213 (2000).
16. H. Kato, H. Kobayashi, A. Kudo. *J. Phys. Chem. B.* **106**, 12441 (2002).
17. H. Kato, A. Kudo. *J. Phys. Chem. B.* **106**, 5029 (2002).
18. T. Ishii, H. Kato, A. Kudo. *J. Photochem. Photobiol. A.* **163**, 181 (2004).
19. R. Konta, T. Ishii, H. Kato, A. Kudo. *J. Phys. Chem. B.* **108**, 8992 (2004).
20. R. Niishiro, H. Kato, A. Kudo. *Phys. Chem. Chem. Phys.* **7**, 2241 (2005).
21. R. Niishiro, R. Konta, H. Kato, W. J. Chun, K. Asakura, A. Kudo. *J. Phys. Chem. C.* in press.
22. Y. Shimodaira, H. Kato, H. Kobayashi, A. Kudo. *Bull. Chem. Soc. Jpn.* **80**, 885 (2007).
23. Y. Hosogi, Y. Shimodaira, H. Kato, H. Kobayashi, Akihiko Kudo. *Chem. Mater.* in press.

24. A. Kudo, K. Ueda, H. Kato, Mikami. *Catal. Lett.* **53**, 229 (1998).
25. A. Kudo, K. Omori, H. Kato. *J. Am. Chem. Soc. Jpn.* **121**, 11459 (1999).
26. S. Tokunaga, H. Kato, A. Kudo. *Chem. Mater.* **13**, 4624 (2001).
27. J. Yu, A. Kudo. *Adv. Func. Mater.* **16**, 2163 (2006).
28. Y. Shimodaira, H. Kato, H. Kobayashi, A. Kudo. *J. Phys. Chem. B.* **110**, 17790 (2006).
29. A. Kudo, S. Hijii. *Chem. Lett.* 1103 (1999).
30. R. Konta, H. Kato, H. Kobayashi, A. Kudo. *Phys. Chem. Chem. Phys.* **5**, 3061 (2003).
31. A. Kudo, M. Sekizawa. *Catal. Lett.* **58**, 241 (1999).
32. A. Kudo, M. Sekizawa. *Chem. Commun.* 1371 (2000).
33. I. Tsuji, A. Kudo. *J. Photochem. Photobiol. A*, **156**, 249 (2003).
34. A. Kudo, A. Nagane, I. Tsuji, H. Kato. *Chem. Lett.* 882 (2002).
35. A. Kudo. *Int. J. Hydrogen Energy.* **31**, 197 (2006).
36. A. Kudo, I. Tsuji, H. Kato. *Chem. Commun.* 1958 (2002).
37. I. Tsuji, H. Kato, H. Kobayashi, A. Kudo. *J. Am. Chem. Soc.* **126**, 13406 (2004).
38. I. Tsuji, H. Kato, H. Kobayashi, A. Kudo. *J. Phys. Chem. B.* **109**, 7323 (2005).
39. I. Tsuji, H. Kato, A. Kudo. *Angew. Chem., Int. Ed.* **44**, 3565 (2005).
40. I. Tsuji, H. Kato, A. Kudo. *Chem. Mater.* **18**, 1969 (2006).
41. A. Kudo, I. Mikami. *Chem. Lett.* 1027 (1998).
42. H. Kato, M. Hori, Y. Konta, Y. Shimodaira, A. Kudo. *Chem. Lett.* **33**, 1348 (2004).
43. H. Kato and A. Kudo. *Catal. Today.* **78**, 561-569 (2003).
44. A. Yamakata, T. Ishibashi, H. Kato, A. Kudo, and H. Onishi. *J. Phys. Chem. B.* **107**, 14383 (2003).

DIRECT PRODUCTION OF PRESSURIZED HYDROGEN FROM WASTE ALUMINUM WITHOUT GAS COMPRESSOR

TAKEHITO HIRAKI

Center for Advanced Research of Energy Conversion Materials, Hokkaido University, Kita 13 Nishi 8, Kita-ku, Sapporo 060-8628, Japan

N. OKINAKA[1], H. UESUGI[2] AND T. AKIYAMA[1]

[1] *Center for Advanced Research of Energy Conversion Materials, Hokkaido University, Kita 13 Nishi 8, Kita-ku, Sapporo 060-8628, Japan*

[2] *Waseda University, Wasedatsurumaki-cho 513, Shinjuku-ku, Tokyo 162-0041, Japan*

An innovative environment-friendly hydrolysis process for generating high-pressure hydrogen with recycling waste Al has been proposed and experimentally validated. The effect of the concentration of NaOH solution on H_2 generation rate was mainly examined. In the experiments, distilled water and Al powder were placed in the pressure-resistance reactor made of Hastelloy, and was compressed to a desired constant water pressure using a liquid pump. The NaOH solution was supplied by liquid pump with different concentrations (from 1.0 to 5.0 mol/dm^3) at a constant flow rate into the reactor by replacing the distilled water and the rate of H2 generated was measured simultaneously. The liquid temperature in the reactor increased due to the exothermic reaction given by $Al + OH^- + 3H_2O = 1.5H_2 + Al(OH)_4^- + 415.6$ kJ. Therefore, a high-pressure H_2 was generated at room temperature by mixing waste Al and NaOH solution. As the H_2 compressor used in this process consumes less energy than the conventional one, the generation of H_2 having a pressure of almost 30 MPa was experimentally validated together with $Al(OH)_3$—a useful by-product. The energy losses in the proposed system (150.9 MJ) is 55% less than that in the conventional system (337.7 MJ) in which the gas compressor and production of $Al(OH)_3$ consume significantly more energy.

1. Introduction

Thermodynamically, metallic aluminum has a high chemical energy of 788.61 kJ/g [1]. We propose a new system for the treatment of waste aluminum. In previous paper, we demonstrated that aluminum powder can react with water at normal pressure to generate hydrogen, and the temperature dependence of the generation rate conforms to the Arrhenius equation with an activation energy of 69 kJ/mol [2]. The coproduction of hydrogen and aluminum hydroxide from waste aluminum is fairly attractive from the viewpoint of reproduction of latent material energy.

In contrast, a thermodynamic consideration reveals that the following major reaction in the system can proceed extremely rapidly to form products as on the right hand side due to a large equilibrium constant. This suggests the possibility that the aforementioned reaction is strongly exothermic, thereby directly generating high-pressure hydrogen. Ka value was calculated by commercial software code of HSC chemistry 5.11.

$$Al + OH^- + 3H_2O = 1.5H_2 + Al(OH)_4^- + 415.6kJ \quad (1)$$

$$Ka = 3.78\times10^{76} \text{ (at 298 K)}$$

Therefore, hydrogen production according to eq. (1) using waste aluminum shows sufficient potential for eliminating the disadvantages of conventional energy-consuming hydrogen compression process. However, the direct production of high-pressure hydrogen from waste aluminum has not been studied so far in spite of its feasibility from the engineering perspective.

The purpose of this study is, therefore, to produce high-pressure hydrogen by the experimental hydrolysis of aluminum, in which the effect of alkali concentration on the rate of high-pressure hydrogen generation was chiefly examined by using the pressure resistance reactor (autoclave). Further, we studied the energetic life cycle assessment (e-LCA) for the proposed system of waste aluminum treatment by comparison with two conventional systems. In one conventional system, hydrogen is produced by the steam reforming of natural gas and is compressed using a gas compressor, while in the other, aluminum hydroxide is produced in the 1st stage of the Bayer process. The study primarily examines the effect of the change in the hydrogen pressure from 0.1 MPa to 30 MPa on the total energy loss (EXL).

2. Experiment and Method of e-LCA

2.1. Experiment

Figure 1 shows the schematic diagram of the experimental apparatus used. The liquid pump supplies compressed and distilled water or a sodium hydroxide solution into the autoclave. Distilled water is compressed in the reactor, which is made of Hastelloy C-22, to a maximum pressure of 35 MPa, and it is heated to a maximum temperature of 573 K. Prior to the experiments, aluminum powder, having a grain size of 180-425 μm and 99.9% purity, was weighed to 0.5 mol and charged into a metallic filter cage in the cylindrical reactor. Distilled water was then filled up to the controlled regulator level of pressure 10, 20, and 30 MPa, followed by the heating up of the reactor and connecting pipe to the desired temperature. The experiments were initiated by replacing the compressed water with sodium hydroxide solution at concentrations of 1.0 and 5.0 mol/dm^3 (M). In the separator, high-pressure gas pushed the liquid to the lower region. That is, the inflowing liquid was pumped out from the bottom of separator into the storage tank, whereas the gas remained in the upper region of the separator. The weight of the liquid recovered in the storage tank was monitored using an online balance to evaluate the change in gas generation with time. Further, a gas flow meter was used to double check the amount of hydrogen generated. The gas recovered was later introduced into a gas chromatograph for confirming the hydrogen purity. During the experiments, the local pressure and temperatures were measured using a pressure gauge and thermocouples, as shown in Figure 1. The reactor product was also analyzed by X-ray diffractometer.

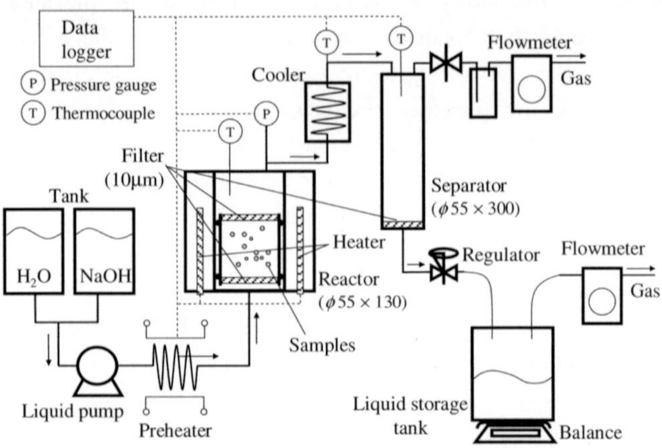

Figure 1. Schematic diagram of the experimental apparatus used for producing high-pressure hydrogen.

2.2. Method of e-LCA

Energy is used to evaluate the qualitative change from the available energy to the unusable one in the form of work. Energy is defined by eq. (2).

$$\varepsilon = H - H_0 - T_0(S - S_0) \tag{2}$$

Energy consists of chemical and physical energies denoted by ε_c and ε_p, respectively, as shown in eq. (3).

$$\varepsilon = e_c + e_p = \varepsilon_c + \varepsilon_T + \varepsilon_p +.\varepsilon_M \tag{3}$$

where ε_c is the standard chemical energy equal to e_c. In contrast, ε_T, ε_p and ε_M are related to temperature, pressure and mixing energies, respectively. They are expressed as follows:

$$\varepsilon_c = \sum n_i \varepsilon_i^0 \tag{4}$$

$$\varepsilon_T = \left(\sum n_i C_{p,i}\right)\left\{T - T_0 - T_0 \ln\left(\frac{T}{T_0}\right)\right\} \tag{5}$$

$$\varepsilon_p = \left(\sum n_i\right) R T_0 \ln\left\{\sum\left(\frac{p_i}{p_0}\right)\right\} \tag{6}$$

$$\varepsilon_M = R T_0 \sum\left[n_i \ln\left\{\frac{n_i}{\left(\sum n_i\right)}\right\}\right] \tag{7}$$

The energy can be calculated for all substances in various states. Many different forms of substances and types of energies are considered in the system. Therefore, the concept of energy is very useful for evaluating the energy efficiency of a system.

The energy loss ε loss in a process can be calculated by the following equation:

$$\varepsilon_{loss} = \varepsilon_{in} - \varepsilon_{out} = \varepsilon_{diff.} + \varepsilon_{diss.} \tag{8}$$

where ε_{in}, ε_{out}, $\varepsilon_{diff.}$ and $\varepsilon_{diss.}$ denote the energy inflow of a system, energy outflow of the system, diffusion energy that is lost outside the system and dissipation energy resulting from an irreversible reaction.

The analysis was based on the following assumptions:

1) The concentration of metallic aluminum in waste aluminum was 15 mass%.
2) Sodium hydroxide for the treatment of waste aluminum was repeatedly used in the following equations:

$$Al + NaOH + 3H_2O = 1.5H_2 + NaAl(OH)_4 \tag{1}$$

$$NaAl(OH)_4 = Al(OH)_3 + NaOH \tag{12}$$

3) The latent environmental burden of waste aluminum was zero.
4) Deionized water was used for producing hydrogen.

5) The construction of both buildings and machines used in the proposed system was not considered because they were not fixed.
6) The transportation of waste aluminum and the residue between the plant and the landfill was not evaluated.
7) Two conventional processes were employed. In one process, aluminum hydroxide was produced by the so-called Bayer process, while in the other, hydrogen was produced by the steam reforming of natural gas.
8) The power generation efficiency was 40% since the remaining 60% was discharged as waste heat.

3. Results and Discussions

3.1. Direct production of high-pressure hydrogen

Figure 2 (a) shows the changes in temperature with time, which were measured using a thermocouple placed in the upper part of the reactor. The data calculated at 10, 20, and 30 MPa were obtained from the piston flow and adiabatic conditions. The entire experimental data revealed a sudden increase in the temperature due to an exothermic reaction caused by the replacement of water by the alkali solution after the induration period. The use of a low-concentration sodium hydroxide solution (1.0M) caused a marginal increase in the temperature where the maximum temperature was only 360 K. Under the assumptions of a piston flow of the sodium hydroxide solution up to the reactor and a uniform temperature in the reactor, the temperature history was roughly estimated by using eq. (1). The temperature increased rapidly due to the exothermic heat when 5.0 M sodium hydroxide solution was used in the experiments, reaching a maximum value of 420 K within a couple of minutes; thus, the calculated data gradually decreased. This was probably caused by the aluminum consumption and heat loss from the reactor. Figure 2 (b) shows the reaction curves for five runs, which were obtained from the liquid quantity in the storage tank. In the three experiments using 5.0 M sodium hydroxide solutions, all curves exhibited a sharp increase as soon as the sodium hydroxide solution reached the reactor at approximately 30 s. Similar to Figure 2 (a), no significant difference was observed among the three curves. On the contrary, for the 1.0 M sodium hydroxide solution case, effect of water pressure was dominant, when 1.0 M sodium hydroxide solution and water pressure of 10 MPa was used, the reaction was very slow showing an induration period of 100 s. A comparison with solutions of 5.0 and 1.0 M, represented by open and closed circles, respectively, indicates that 5.0 M sodium hydroxide solution was very effective for obtaining a rapid and large reaction degree because there was no duration time and the reaction curve was accorded with the theoretical one even at the relatively lower pressure of 10 MPa. In addition, a remarkable effect of pressure on the reaction curves using 1.0 M sodium hydroxide solution in Figure 2 (b) was shown. The slope angle was about four times that of the theoretical value. Compressing the water to a pressure of 30 MPa resulted in a very rapid reaction. The increased reaction degree when using 1.0 M sodium hydroxide solution and 30 MPa water pressure can probably be explained on the basis of the structure of the subcritical water. It is well-known that subcritical water can easily oxidize due to small clusters of water [3].

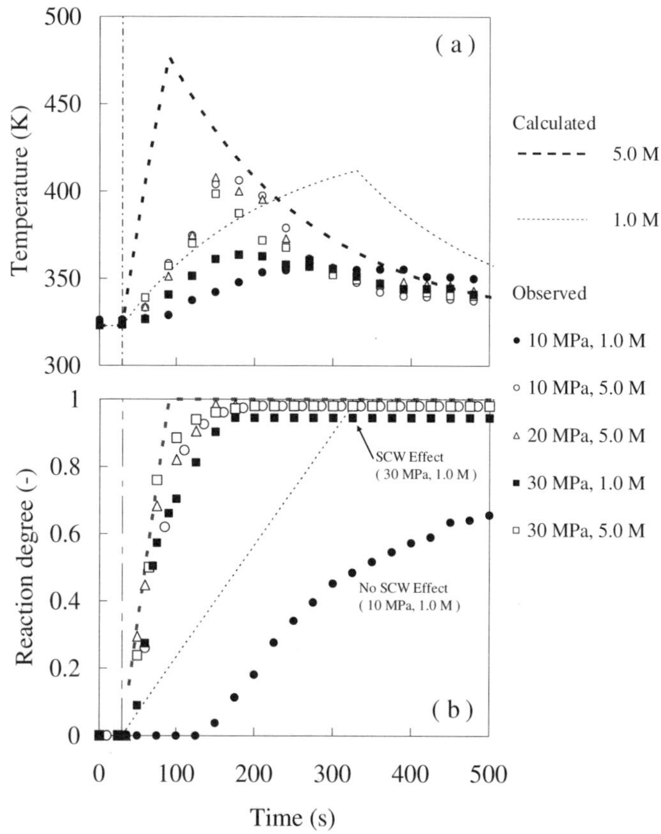

Figure 2. Changes in temperature (a) and reaction degree (b) under different experimental conditions of water pressure and sodium hydroxide solution concentration. SCW; Suv-critical water

3.2. Exergetic-LCA

Figure 3 shows the energy flow diagram of the proposed system in which the process is represented as a black trapezoid because the energy is always lost despite the conservation of enthalpy between the input and output materials. It should be noted that the input energy of waste aluminum in this system is significantly larger than that of the other materials such as water and fuel. This implies that waste aluminum is a type of hidden resource, although it contains only 15 mass% metallic aluminum. Furthermore, NaOH, whose energy is 36 MJ, goes through a cycle in the system because of its recovery during the production process of aluminum hydroxide.

Figure 4 summarizes the EXLs in the proposed and conventional systems during the co-production of 1 kg of hydrogen at 30 MPa and 26 kg of aluminum hydroxide. The EXL in the proposed system is 150.9 MJ, while that in the conventional system is 337.7 MJ. This implies that the energy loss in the proposed system is 45% of that in the conventional system, and the former can conserve an energy of 186.8 MJ. This is probably due to its advantage of the co-production of hydrogen and aluminum hydroxide in the same process. Moreover, it does not require the gas compressor because hydrogen produced in the closed reactor is self-compressive. The EXL in the hydrogen compression process is as much as 13% of that in the conventional system.

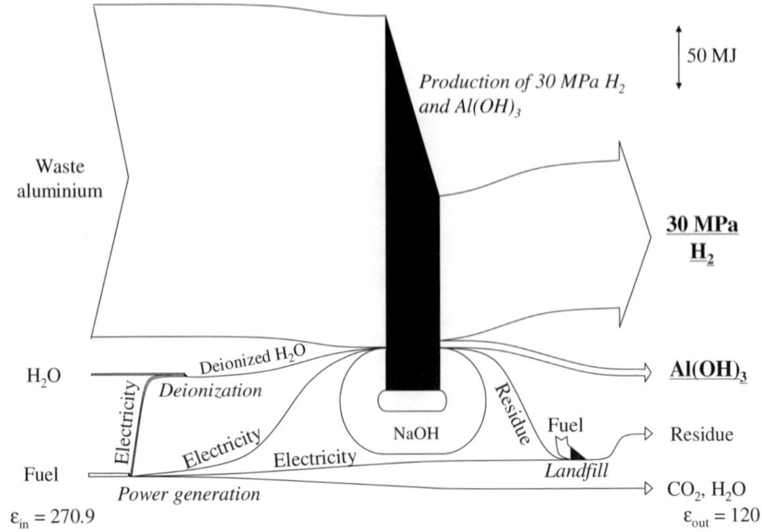

Figure 3. Energy flow diagram of the proposed system for producing 1 kg of hydrogen at 30 MPa and 26 kg of aluminum hydroxide from waste aluminum containing 15 mass% metallic aluminum.

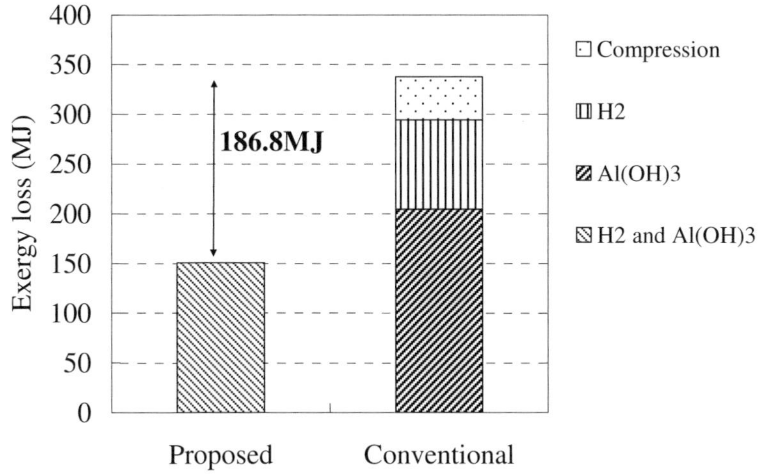

Figure 4. Comparison of total energy loss in the proposed and conventional systems during the production of 1 kg of hydrogen at 30 MPa and 26 kg of aluminum hydroxide.

Acknowledgments

This study was supported by the project 'The Model of Green-Hydrogen Community in Honjo-Waseda area' of the Ministry of the Environment, Japan, and a Grant-in-Aid for Scientific Research (Grant No. B-17360365) by the Japan Society for the Promotion of Science (JSPS). Technical support and fruitful discussions provided by the staff of ITEC Co., Ltd. was greatly appreciated.

References

1. T. Nobusawa, Energy Nyuumon; Ohmu Co., Ltd.: Tokyo (1980).
2. T. Hiraki, M. Takeuchi, M. Hisa and T. Akiyama, *Mater. Trans., JIM.* **46**, 1052 (2005).
3. M. Sasaki, Z. Fang, Y. Fukushima, T. Adschiri and K. Arai, *Ind. Eng. Chem. Res.*, **39**, 2883, (2000).

HYDROGEN PRODUCTION FROM HYDROCARBONS BY USING OXYGEN PERMEABLE MEMBRANES

HITOSHI TAKAMURA

Department of Materials Science, Graduate School of Engineering, Tohoku University, 6-6-11-301-2 Aramaki Aza Aoba, Sendai, 980-8579, Japan

This paper describes the fabrication of a novel hydrogen production system based on an oxygen permeable membrane and its methane reforming properties. The planar-type membrane reformer module with dimensions of 6 cm x 6 cm was developed by using the composite-type membrane of $(Ce_{0.85}Sm_{0.15})O_2$-15vol%$MnFe_2O_4$ and ferric stainless-steel separators. For the reformer module, high CH_4 conversion and H_2 selectivity of 96% and 89% were achieved, respectively. Based on energy analysis, it can be found that ΔG of partial oxidation of methane was effectively used for oxygen separation and heat compensation. The durability of the membrane was confirmed for 450 h.

1. Introduction

Oxygen permeable membranes based on mixed oxide-ion and electronic conductors (MIECs) have been widely studied for use in partial oxidation (POX) of hydrocarbons, especially methane, to produce syngas and hydrogen [1]. To date, a number of oxygen permeable membranes have been developed for realizing the membrane POX reformer (MPOX reformer). In addition to perovskite-type MIECs such as (Ba, Sr)(Co, Fe)$O_{3-\delta}$ [2-7], composite-type membranes consisting of acceptor-doped CeO_2 and spinel-type ferrites also exhibit a high oxygen flux density of approximately 10 μmol/cm^2 s (13.4 sccm/cm^2) at 1000 °C [8-10]. Compared to perovskite-type oxides, the advantages of composite-type membranes are higher mechanical strength and lower thermal expansion coefficients. For example, a thermal expansion coefficient of the composite of $(Ce_{0.85}Sm_{0.15})O_2$-15vol% $MnFe_2O_4$ is approximately 12 x 10^{-6} / °C between room temperature and 1000 °C. This value is almost same as that of yttria-stabilized zirconia, suggesting that component materials such as a ferric stainless steel and a sealing glass developed for solid oxide fuel cells can be used for the fabrication of the MPOX reformer.

The advantage of the MPOX reformer can be emphasized in the context of efficient usage of energy. As shown in Fig. 1, the POX reaction of methane gives larger $|\Delta G|$ than $|\Delta H|$. This implies that the conventional POX reaction with exhaust heat to ambient temperature causes a large amount of energy loss shown as a downward vector. To minimize the energy loss, a process with $\Delta G>0$ and $\Delta S<0$ should be combined. Gas separation is a typical process having these requirements. In other words, as shown in Fig. 2, oxygen permeable membranes can use ΔG of the POX reaction for oxygen separation, and supply heat in the form of joule heat as a result of oxygen permeation, simultaneously.

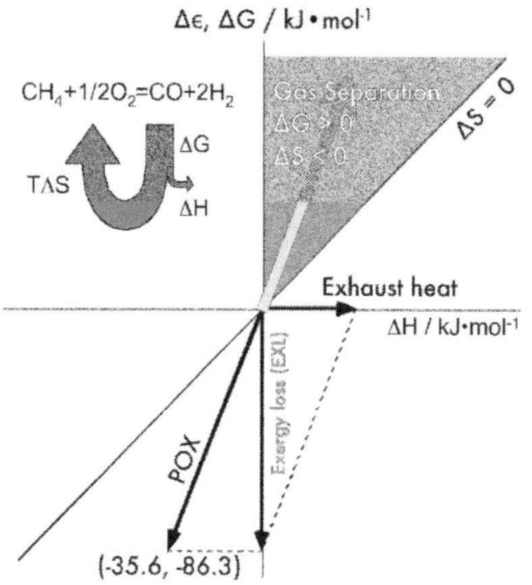

Figure 1. Thermodynamic compass of POX.

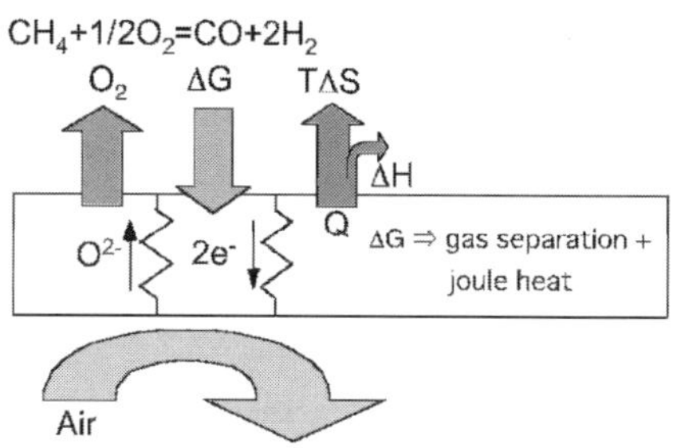

Figure 2. Schematic diagram of MPOX.

In this paper, the development of the planar-type MPOX reformer based on the oxygen permeable membrane of $(Ce_{0.85}Sm_{0.15})O_2$-15vol%$MnFe_2O_4$ and its reforming properties will be reviewed. In addition, the durability of the oxygen permeable membrane, which is one of the most important characteristics in practical use, will be examined.

2. Experimental

Samples of $(Ce_{0.85}Sm_{0.15})O_2$-15vol%$MnFe_2O_4$ (CSO-15MFO) were prepared by a conventional solid-state reaction. A tape-casting technique was used for the fabrication of membranes with dimensions of 3.6 cm x 3.6 cm; Additives and process parameters have been described elsewhere [9]. In this study, the membrane sintered had a thickness of 135 μm. The membrane was then attached to a ferric stainless steel frame by using a glass seal. At this point, it is important to manage thermal expansion coefficients (TECs) of materials. As mentioned above, the composite-type membrane has a TEC of 12 x 10^{-6} /°C. ZMG232® (Hitachi Metals, Ltd.) with a TEC of approximately 13 x 10^{-6} /°C was used as a support metal frame. To joint the ceramics membrane on the metal frame without causing cracks, slow heating and cooling rates of 2 °C/min were used; sealing was performed at 850 °C for 1 h. The jointed membrane module had an effective oxygen permeation area of 9 cm^2 (3 cm x 3 cm). A slurry of 10 mass% Ni supported on Pr-doped CeO_2 powders [8], which can work as a reforming catalyst, was adopted on one side of the membrane surfaces. The module plate can be stacked by rotating 90° to make gas flow channel as well as a heat exchanger. The schematic diagram of the fabrication process is shown in Fig. 3.

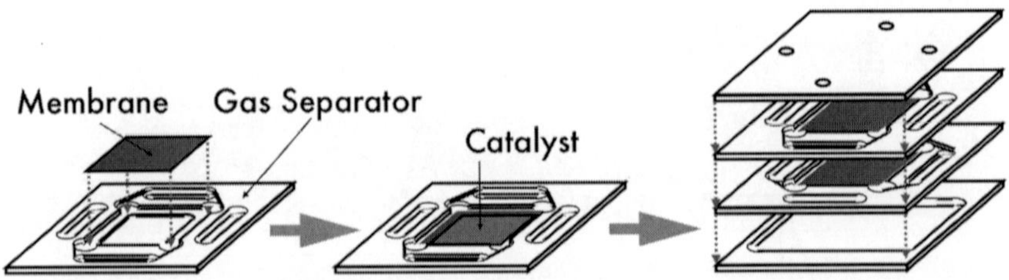

Figure 3. The schematic diagram of the fabrication process of the planar-type MPOX reformer.

100% CH_4 gas was fed at a rate of 150 sccm to the reformer at 780 °C. To avoid carbon deposition on the reforming catalyst and cool down the membrane through endothermic reaction of steam reforming of methane (SMR), steam was also fed to be a steam/carbon (S/C) ratio of 0.88. CH_4 conversion, CO and H_2 selectivity were calculated as follows, respectively:

$$CH_4 conversion(\%) = \frac{[CO]+[CO_2]}{[CH_4]+[CO]+[CO_2]} \times 100 \quad (1)$$

$$CO selectivity, S_{CO}(\%) = \frac{[CO]}{[CO]+[CO_2]} \times 100 \quad (2)$$

$$H_2 selectivity, S_{H2}(\%) = \frac{[H_2]}{[H_2]+[H_2O]} \times 100 \quad (3)$$

where, $[CH_4]$, $[CO]$, $[CO_2]$, $[H_2]$, $[H_2O]$ denote concentration in % of respective gases.

3. Results and Discussion

Figure 4 shows a top-view of the MPOX reformer modules (12 modules) with the CSO-15MFO composite-type membrane on the ZMG232® stainless steel frame with dimensions of 6 cm x 6 cm. As a result of TEC management, no cracks and exfoliation were found in the membrane and sealing parts.

Figure 4. MPOX reformer modules comprising of CSO-15MFO and ferric stainless steel frame.

The methane reforming properties were evaluated by using the single stack unit in this study. Table 1 summarizes the reforming properties and oxygen permeation flux (jO_2) of the single stack MPOX reformer. The carbon balance was found to hold within experimental errors (150 sccm of CH_4 input vs. 152 sccm of output gases including carbon). High CH_4 conversion, CO and H_2 selectivity of 96%, 84%, and 89% were achieved, respectively. Even though furnace temperature was set to be 780 °C, judging from the oxygen permeation flux (3.3 μmol/cm^2 s) [8], membrane temperature appears to be higher than the furnace temperature by approximately 100 °C, presumably due to joule heat caused by oxygen permeation. By using these reforming characteristics, the amount of CH_4 processed by POX and SMR reactions can be estimated as shown in Table 2, where heat values in watt were also calculated from flow rates.

Table 1. Reforming properties and oxygen permeation flux of the single-stack MPOX reformer.

CH_4 sccm	Air sccm	Temp. °C	S/C	jO_2 μmol/cm^{2}s	CH_4 conversion %	S_{CO} %	S_{H2} %
150	500	780	0.88	3.3	96	84	89

C-balance: 150 sccm (input), 152 sccm (output)

Table 2. Heat values for POX and SMR reactions estimated from H, C, and O balances.

	CH_4 sccm	Temp. °C	ΔG kJ/mol	ΔH kJ/mol	ΔG W	ΔH W	TΔS W
MPOX	80	900	-253.8	-21.8	-15.1	-1.3	13.8
SMR	70	900	-70.8	227.1	-3.7	11.8	15.5

The joule heat caused by the oxygen permeation in the MPOX reformer was then calculated based on the electrical conductivity data as follow:
The total electrical conductivity of CSO-15MFO at 900 °C can be simulated by the following equation [11]:

$$\sigma_{calc(total)} = \sigma_{calc(ionic)} + \sigma_{calc(electronic)} = \sigma_i + \sigma_n^0 P(O_2)^{-\frac{1}{6}} \tag{4}$$

where σ_i = 0.074(4) S/cm and σ_n^0 = 1.82(2) x 10^{-3} S/cm atm$^{1/6}$ give a good fitting result. By solving Wagner's equation with boundary conditions of jO_2 = *const.* and $pO_2(0) = p_0$ atm, one can obtain:

$$pO_2(x) = p_0\left(\frac{\sigma_n^0}{(\sigma_n^0 + \sigma_i p_0^{1/6})\exp\left(\frac{8jO_2F^2x}{3\sigma_i RT}\right) - \sigma_i p_0^{1/6}}\right)^6 \tag{5}$$

where void formation and/or cation demixing in the case of $\nabla jO_2 \neq 0$ are not taken into account for simplification. Now, joule heat in watt per unit volume caused at a given position, x, in the membrane is expressed as:

$$q(x) = 16jO_2^2F^2\left(\frac{1}{\sigma_n^0 pO_2(x)^{-1/6}} + \frac{1}{\sigma_i}\right) \tag{6}$$

Therefore, assuming an effective area size as *S* and a membrane thickness as *L*, respectively, the total joule heat value caused by oxygen permeation can be derive as:

$$Q = \int_0^L q(x)Sdx \tag{7}$$

Based on this treatment, joule heat caused in CSO-15MFO was estimated to be 8.4 W as shown in Table 3.

Table 3. Joule heat caused by oxygen permeation.

jO_2 μmol/cm²s	Current density A/cm²	Thickness mm	Joule heat (σ_i) W	Joule heat (σ_e) W	Area cm²	Joule heat (total) W
3.3	1.27	0.135	0.31	0.62	9	8.4

Even though further detailed energy analysis and measurements of actual temperature distribution in the membrane are required, the joule heat caused by oxygen permeation (8.4 W) appears to be originating from ΔG of reforming reactions and returns to the system to be used as a part of heat required for POX and SMR reactions [11].

The durability of the oxygen permeable membrane was also examined. Figure 5 shows the oxygen permeation flux, conversion, and selectivity of the CSO-15MFO membrane at 800 °C as a function of time. This durability test was performed for the same membrane as in the reformer; however, a diameter of the membrane was limited to 0.5 cm². Even though the oxygen permeation flux (jO_2) and conversion slightly decrease with increasing time, the membrane can be operated at 800 °C for 450 h. The durability test at higher temperatures and microstructural analysis after the operation will be required in future.

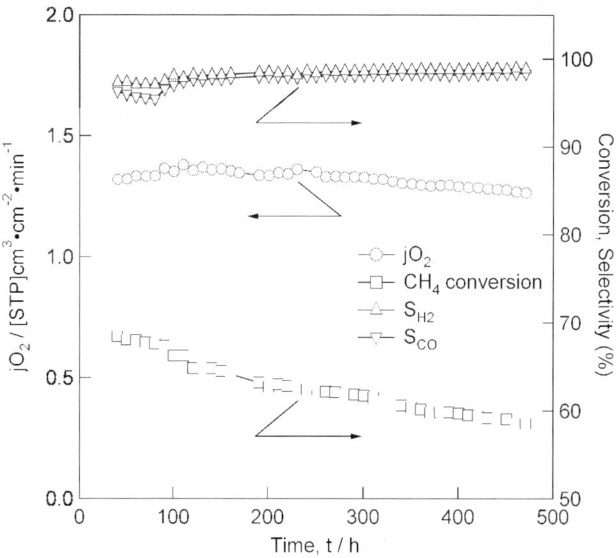

Figure 5. Oxygen permeation flux, conversion, and selectivity of CSO-15MFO at 800 °C as a function of time. The oxygen permeation flux of 1 μmol/cm²s corresponds to 1.34 sccm/cm².

4. Conclusions

The proto-type of MPOX reformer based on the composite-type $(Ce_{0.85}Sm_{0.15})O_2$-15vol% $MnFe_2O_4$ membrane has been prepared. By using ZMG232® ferric stainless steel with a comparable TEC value as a support, the membrane was successfully jointed on the support frame. As methane reforming properties, high CH_4 conversion, CO and H_2 selectivity of 96%, 84%, and 89% were achieved, respectively. Based on C, H, and O balances, the oxygen permeation flux was found to be 3.3 μmol/cm^{2}s. Joule heat caused by the oxygen permeation was estimated to be approximately 8.4 W. The membrane was found to operate at 800 °C for 450 h.

Acknowledgments

This work is supported in part by Industrial Technology Research Grant Program in 2005 from New Energy and Industrial Technology Development Organization (NEDO) of Japan, CREST, Japan Science and Technology Agency, and Ministry of Education, Science, Sports and Culture, Grant-in-Aid for Scientific Research (B) under contraction No. 18360326.

References

1. P. N. Dyer, R. E. Richards, S. L. Russek, and D. M. Taylor, *Solid State Ionics* **134**, 21 (2000).
2. Z. P. Shao, H. Dong, G. X. Xiong, Y. Gong, and W. S. Yang, *J. Membr. Sci.* **183**, 181 (2001).
3. Z. P. Shao and S. M. Haile, *Nature* **431**, 170 (2004).
4. T. Ishihara, Y. Tsuruta, T. Todaka, H. Nishiguchi, and Y. Takita, *Solid State Ionics* **152**, 709 (2002).
5. K. Brinkman, T. Iijima, and H. Takamura, *Jpn. J. Appl. Phys. Pt 2-Lett. & Exp. Lett.* **46**, L93 (2007).
6. H. Takamura, K. Enomoto, Y. Aizumi, A. Kamegawa, and M. Okada, *Solid State Ionics* **175**, 379 (2004).
7. H. Takamura, Y. Aizumi, A. Kamegawa, and M. Okada, *J. Fuel Cell Sci. Tech.* **3**, 175 (2006).
8. H. Takamura, K. Okumura, Y. Koshino, A. Kamegawa, and M. Okada, *J. Electroceram.* **13**, 613 (2004).
9. H. Takamura, T. Kobayashi, T. Kasahara, A. Kamegawa, and M. Okada, *J. Alloys Comp.* **408-412**, 1084 (2006).

10. H. Takamura, H. Sugai, M. Watanabe, T. Kasahara, A. Kamegawa, and M. Okada, *J. Electroceram.* **17**, 741 (2006).
11. H. Takamura, M. Ogawa, K. Suehiro, M.Okada, submitted to *Solid State Ionics*.

HYDROGEN PRODUCTION VIA WATER SPLITTING IN SOLAR REACTORS: THE HYDROSOL PROCESS

A. G. KONSTANDOPOULOS[a,b,*], C. SATTLER[c], P. STOBBE[d], A.M. STEELE[e]

[a]Aerosol & Particle Technology Laboratory, CERTH/CPERI, P.O. Box 361, Thermi, Thessaloniki 57001, Greece and [b] Department of Chemical Engineering, Aristotle University, P.O. Box 1517, 54006, Thessaloniki, Greece

[c]Deutsches Zentrum für Luft- und Raumfahrt e.V. (DLR), Institut für Technische Thermodynamik, Solarforschung, D-51170 Köln, Germany

[d]Stobbe Tech Ceramics, Vejlemosevej 60, DK-2840, Holte, Denmark

[e]Johnson Matthey Technology Centre, Sonning Common, RG4 9NH, Reading, UK

**E-mail: agk@cperi.certh.gr*

The present paper reviews recent work in the field of solar thermochemical hydrogen production via water splitting in monolithic reactors, also known as the Hydrosol process. The process employs a reactor concept, adapted from the well-known automotive emission control field, and consists of multichannel ceramic honeycombs, coated with active water-splitting materials, that are heated by concentrated solar radiation to the desired temperature.. When water vapor passes through the reactor, the coating material splits the water molecule by "trapping" its oxygen and leaving in the effluent gas stream pure hydrogen. In a next step, the oxygen "trapping" material is regenerated, by increasing the amount of solar heat absorbed by the reactor; hence a cyclic operation is established. Multi-cyclic solar thermo-chemical splitting of water was successfully demonstrated on a pilot solar reactor achieving constant hydrogen production exclusively at the expense of solar energy.

1. Introduction

The harnessing of the huge energy potential of solar radiation and its effective conversion to chemical energy carriers such as hydrogen is a subject of primary technological interest. One of the reactions with tremendous economical impact because of the low value of its reactants is the dissociation of water (water splitting) to oxygen and hydrogen. However because of unfavorable thermodynamics interesting yields can only be achieved at very high temperatures imposing therefore technological difficulties. The integration of solar energy concentration systems with systems capable to split water is of immense value and represents an important long term goal for hydrogen production with virtually zero CO_2 emissions [1-3]. The state of the art is focusing on two-step processes, based on redox materials that can act as effective water splitters at lower temperatures [4-6].

The HYDROSOL process employs water splitting materials coated on a monolithic honeycomb solar reactor, inspired from the well-known automobile catalytic converters [7], and it was recently introduced in [8]. The HYDROSOL reactor contains no moving parts and is constructed from special refractory (Silicon carbide) ceramic thin-wall, multi-channelled (honeycomb) monoliths, optimised to absorb solar radiation and develop sufficiently high temperatures. When steam passes through the solar reactor, the coating material splits water vapor by "trapping" its oxygen and leaving in the effluent gas stream pure hydrogen (Eq. 1), without any need for expensive and complicated gas separation post-processing steps. In a subsequent step (Eq. 2), the oxygen "trapping" material is regenerated (i.e. releases the oxygen absorbed), by increasing the amount of solar heat absorbed by the reactor and hence a cyclic operation is established:

$$MO_{red} + H_2O\ (g) \rightarrow MO_{ox} + H_2 \qquad \text{(exothermic)} \qquad ....(1)$$

$$MO_{ox} \rightarrow MO_{red} + \tfrac{1}{2}\, O_2 \qquad \text{(endothermic)} \qquad ...(2)$$

The inherent advantage of two-step thermochemical cycles is that the production of pure hydrogen and the removal of oxygen take place in separate steps, avoiding the need for high-temperature separation and the chance of explosive mixtures formation. In addition, with the HYDROSOL reactor configuration, with the active redox pair materials coated upon the substrate walls, the whole process can be carried out on a single solar energy converter, the process temperature can be significantly lowered compared to other thermo-chemical cycles and, last but not least, this reactor concept does not involve any circulation of (hot) solid reactants or products and therefore has no problems with the recovery of high temperature heat. Such redox-material-coated-honeycombs have achieved continuous solar operation water splitting – regeneration cycles in the temperature range 850–1200°C demonstrating the "proof-of-concept" of the proposed reactor design and producing the first solar hydrogen with monolithic honeycomb reactors [8,9]. The present work summarizes the work performed so far and highlights the current research efforts focussed on long-term material stability and scale-up of the solar reactor.

2. Redox Material Development

Four different routes were employed for the synthesis of iron-oxide-based redox water-splitting materials: Solid-State Synthesis (SSS), Self-Propagating High-Temperature Synthesis (SHS), Gel Combustion (GC) and Aerosol Spray Pyrolysis (ASP) [8]. These synthesis methods were chosen with the rationale to exploit particular characteristics of each one for the synthesis of products with "tunable" oxygen vacancies concentration. The synthesis details have been reported previously [8], therefore only the general reaction concepts are reported below, where A and B denote the bivalent dopant metals: Ni, Mn or Zn.

i) Solid state synthesis (SSS) involved powder mixing of the component oxides (or carbonates), pre-firing, milling, spray drying and calcination at high temperatures (≈1250°C). The products were doped spinel ferrites of the structure $(A_xB_yFe_z)Fe_2O_4$.

ii) Self-Propagating High-temperature synthesis (SHS or Combustion Synthesis) of the targeted materials is based on the heat released from the reaction of iron metal powder ("fuel") with oxygen ("oxidizer") in the presence of the dopant metal oxides and of Fe_2O_3 powder as a "thermal ballast/moderator" to control the synthesis temperature.

iii) Gel combustion (GC) is based on the reaction in aqueous solutions of nitrate salts $A_x(NO_3)_y$ ("oxidant") with amino-groups ("fuel") to form explosive ammonium nitrate; upon heating the solution is first transformed to a gel which is then combusted to produce a very fine powder of spinel structure $(A_xB_yFe_z)Fe_2O_4$.

iv) Aerosol spray pyrolysis (ASP) employs the atomisation of a metal precursor salts solution in a spray of fine droplets that is subsequently passed through a hot-wall reactor where it transforms within a very short time to ultra-fine, nanostructured spherical particles [10].

The first "screening" of the synthesized material compositions with respect to water splitting activity was performed in a laboratory unit described in detail previously [11] and subsequently in a scaled-up testing rig version consisting of a 20-mm-diameter quartz glass tubular reactor enclosed within a high-temperature programmable furnace capable of reaching temperatures of 1500°C. A bed of the redox material powder to be tested was placed in the middle of the reactor and subsequently heated under inert atmosphere (Nitrogen) to the water-splitting testing temperature. When this temperature was reached, steam was introduced to the reactor. The effluent, after passing through a water trap, was diverted to the analysis rig, consisting of a mass spectrometer (MS). The quantities of unconverted water and of produced hydrogen were calculated based on the areas of the corresponding MS peaks.

These experiments have shown that systems from all the synthesis routes were able to split water and generate hydrogen as the only product, at temperatures as low as 800°C and could be repeatedly regenerated under inert atmospheres at temperatures below 1200°C [8]. The "best" products from each synthesis route are compared with respect to water-to-hydrogen conversion and total hydrogen yield in Figs. 1a, 1b respectively. Both SHS and ASP materials exhibited very high water conversions (amount of water converted/total amount of injected water) at 800°C – 57% and 81% respectively. Overall the ASP materials exhibited both the highest water conversion and hydrogen yield.

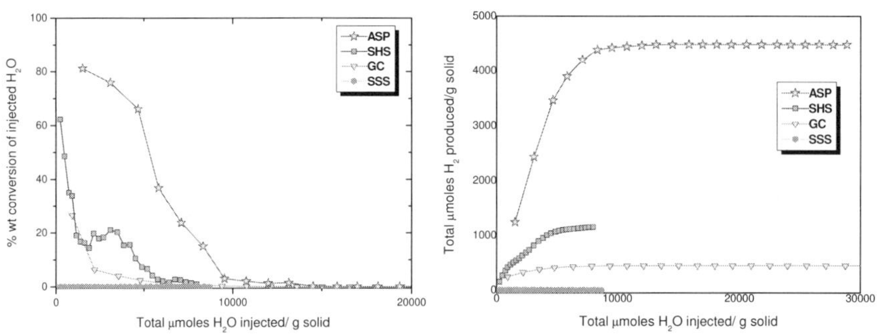

(a) (b)

Figure 1: Comparison of "best'' redox powders from each synthesis route with respect to: (a) water conversion and (b) total Hydrogen yield.

3. Manufacturing and Coating of Honeycomb Solar Receivers

Several series of small-scale (Ø25x50 mm) and large-scale (Ø144x200 mm) monolith extruded multi-cell SiC supports were produced (shown in Figs. 2a and 2b respectively) and coated with the synthesized materials via the "washcoating" technique employed for the coating of automotive catalysts, in which the porous supports are impregnated in a slurry of the coating powder [7]. With subsequent drying and firing at the temperature range 500-800°C, an adherent oxide layer is formed on the walls of the support. The coated honeycombs were employed for the experimental campaigns in a solar furnace to demonstrate the "proof-of-concept" of the proposed approach and to "screen" redox material formulations.

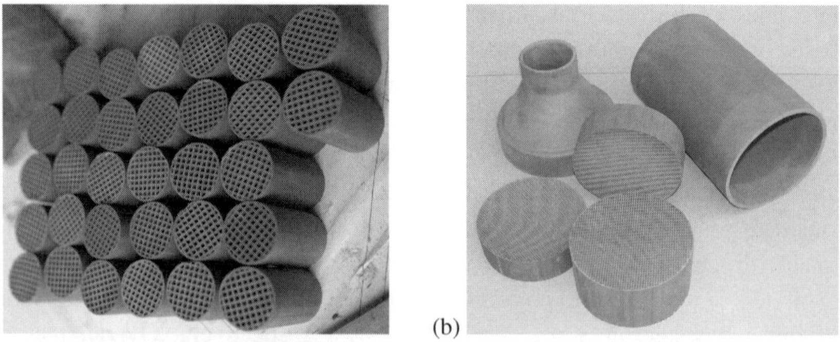

(a) (b)

Figure 2: (a) Small-size and (b) large-size, extruded SiC honeycombs and housing vessel.

4. Solar Reactors

Two reactors have been developed for different purposes. The first reactor built (Fig. 3a) was designed and constructed to monitor the performance and feasibility of solar chemical hydrogen production by the HYDROSOL process. This receiver-reactor is operated in the the Solar Furnace at the DLR facilities, in Cologne, Germany. This reactor was mainly designed for the investigation of the general feasibility of both steps of the process and for screening different "families" and "generations" of redox pairs coated on small-scale honeycombs as depicted in Fig. 3b. Both steps of the thermochemical cycle were successively performed in the same reactor.

The first solar campaign demonstrated the in-principle-feasibility of water splitting by the proposed method. The first solar hydrogen was successfully produced by irradiating a redox material coated on a SiC monolith at 800°C in a mixture of steam and nitrogen, whereas after completion of the water splitting step and by raising the operating temperature to 1200°C under flushing by pure nitrogen, a release of oxygen was initiated. The second campaign proved the feasibility of multi-cycling, i.e. a periodic and alternating performance of water splitting and regeneration of the redox system. Not only the reactor was capable for producing hydrogen from steam at the expense of solar energy alone, but multi-cyclic operation (water splitting and redox material regeneration) at the temperature range 800-1200°C was successfully demonstrated several times and for several of the redox materials synthesized [9].

The main objectives of the third campaign were on the one hand large monolith (Ø144 x200 mm) testing for the demonstration of continuous production of hydrogen and on the other hand, further improvement of the coated monoliths with respect to multi-cycling capability and the amount of hydrogen produced. In addition, the stability of coating/support assembly was examined. The results were quite encouraging. One of the samples was irradiated in a long-term operation over five days and it maintained its activity after 40 cycles. More cycles would have been possible if more testing time had been available but the solar furnace had to be used for other scheduled projects.

(a) (b)

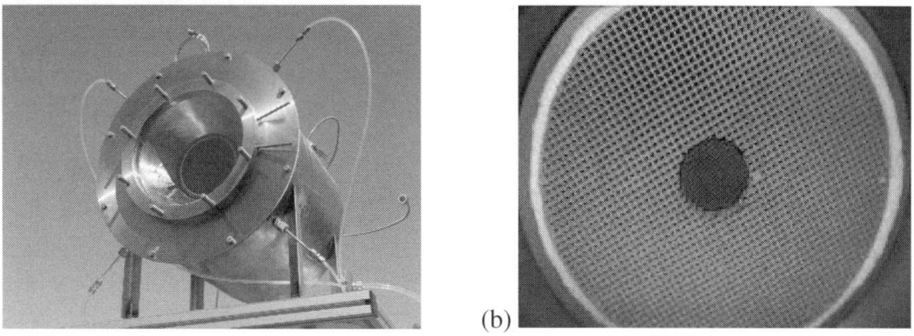

Figure 3: (a) Front view of the first solar water-splitting receiver-reactor, (b) Small-scale honeycomb coated with redox material (black) in the centre of the reactor ready for testing.

The second reactor constructed, was designed with the purpose to be capable for continuous hydrogen production [12]. Different approaches of receiver-reactors have been analysed and compared. The final decision was in favor of a multi-chamber reactor with fixed honeycomb absorber allowing a modular set-up. This is the so-called "conti reactor", where one module splits water while the other is being regenerated, shown in Fig. 4. The test programme aimed at exploring suitable operation conditions to verify the concept of a continuous hydrogen production in the "conti reactor".

Fig. 5 presents the quasi-continuous production of hydrogen in 13 subsequent cycles during the first day of solar testing of two large coated monoliths in the "conti reactor". Another 10 cycles were carried out during the following day. A subsequent campaign proved the long-term stability of the redox-coated honeycomb systems: 53 cycles of solar hydrogen generation with the same redox coating were performed during a 5-day campaign, proving the capability of the "conti-reactor" to reliably operate the HYDROSOL two-step water splitting process quasi-continuously.

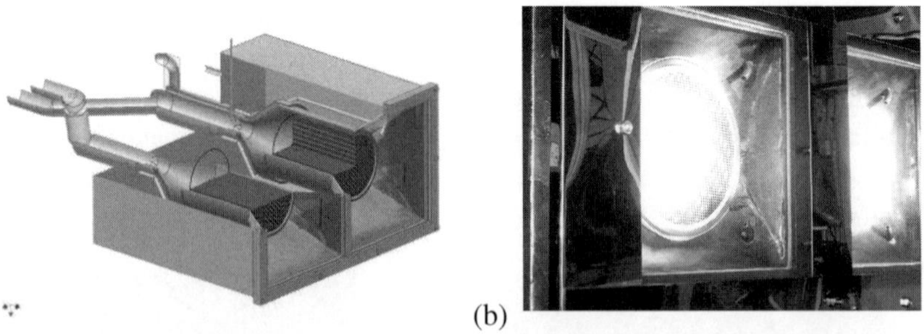

(a) (b)

Figure 4: The dual-chamber ("conti" reactor) for continuous solar hydrogen production: (a) vertical-horizontal cut, (b) front view of the reactor, in operation at the solar furnace facility.

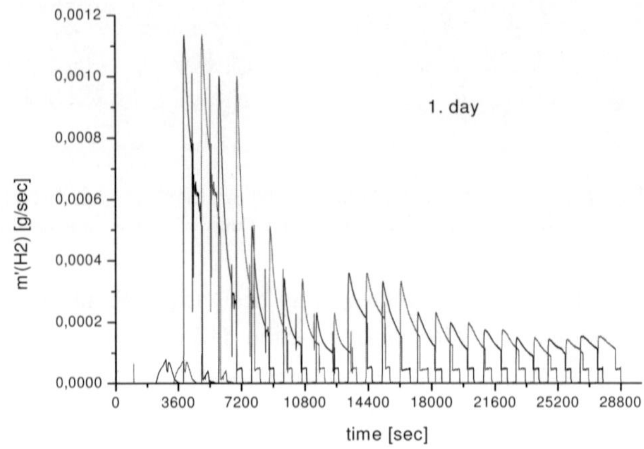

Figure 5: Campaign with the "conti" reactor; first quasi-continuous production of hydrogen: mass flow of hydrogen for 13 cycles during the first day of testing of two coated monoliths.

5. System Scale-up

The next steps involve the development and build of an optimized pilot plant (100 kWth) for solar Hydrogen production based on this novel reactor concept, involving further scale-up of the HYDROSOL technology and its effective coupling with solar platform concentration systems, in order to exploit and demonstrate all potential advantages. Specific challenging problems currently addressed include:

The first scaled-up version of the solar reactor/receiver currently under construction involves a dual-reactor unit, each part assembled from 9 square-shaped SiC honeycomb pieces with dimensions 146x146 mm. The unit is going to be installed on a solar tower and coupled with the heliostat field at the Plataforma Solar in Almeria, Spain for test operation in 2008.

6. Conclusions

An innovative technology has been developed for the production of hydrogen from the splitting of water by a two-step thermochemical cycle using solar energy. Highly active water splitting (redox) materials were produced via unconventional synthesis routes (combustion synthesis and aerosol spray pyrolysis). The HYDROSOL process was successfully put into practice in a pilot scale and the stability of the redox/support assemblies during multi-cyclic solar thermo-chemical splitting of water was successfully demonstrated: the reactor produces hydrogen by cyclic operation exclusively at the expense of solar energy. Up to 52 cycles of constant hydrogen production were operated in a row during the five-day campaign that the solar furnace was available to us.

The HYDROSOL process represents the world's first closed, solar-thermochemical cycle in operation that is capable of continuous, pure renewable hydrogen production. It is expected that deployment of the HYDROSOL process will proceed with the ongoing commercialization of solar thermal power plants.

Due to the fact that the HYDROSOL process employs entirely renewable and abundant energy sources and raw materials - solar energy and water respectively - it holds a significant potential for large-scale, emissions-free hydrogen production, particularly for regions of the world that lack indigenous resources but are endowed with ample solar energy.

Acknowledgements

Colleagues who have contributed to this research are from the APT Lab: C. Agrafiotis, S. Lorentzou, C. Pagkoura and from DLR: M. Roeb, M. Neises, P.M. Rietbrock, J.P. Säck. The authors would like to thank the European Commission for partial funding of this work within Projects HYDROSOL (ENK6-CT-2002-00629) and HYDROSOL-II (FP6-2002-Energy-1, 020030).

References

[1] Rubbia, C., *Hydrogen at crossroads between science and politics*. Conference on the Hydrogen Economy – A Bridge to Sustainable Energy, Brussels, June 16-17, 2003.

[2] Kodama, T., *High-temperature solar chemistry for converting solar heat to chemical fuels*. Progress in Energy and Combustion Science, 29 (6), 567-597, 2003.

[3] Steinfeld, A., *Solar thermochemical production of Hydrogen-a review*. Solar Energy, 78, 603-615, 2004.

[4] Tamaura, Y., Steinfeld, A., Kuhn, P., Ehrensberger, K., *Production of solar Hydrogen by a novel, 2-step, water-splitting thermochemical cycle*. Energy 20(4), 325-330, 1995.

[5] Ehrensberger, K., Frei, A., Kuhn, P., Oswald, H. R., Hug, P., *Comparative experimental investigations of the water-splitting reaction with iron oxide* $Fe_{1-y}O$ *and iron manganese Oxides* $(Fe_{1-x}Mnx)_{1-y}O$. Solid State Ionics, 78, 151-160, 1995.

[6] Steinfeld A., *Solar hydrogen production via a two-step water-splitting thermochemical cycle based on Zn/ZnO redox reactions*. International Journal of Hydrogen Energy 27, 611-619, 2002.

[7] Heck R. M., and Farrauto, R. J., *Catalytic Air Pollution Control-Commercial Technology*. Van Nostrand Reinhold, New York U.S.A, 1995.

[8] Agrafiotis, C., Roeb, M., Konstandopoulos, A.G., Nalbandian, L., Zaspalis, V.T., Sattler, C., Stobbe, P., Steele, A.M., *Solar Water Splitting for Hydrogen Production with Monolithic Reactors*. Solar Energy, 79(4), 409-421, 2005.

[9] Roeb, M., Sattler, C., Klüser, R., Monnerie, N., deOliveira, L., Konstandopoulos, A.G., Agrafiotis, C., Zaspalis, V.T., Nalbandian, L., Stobbe, P., Steele, A.M., *Solar hydrogen production by a two-step cycle based on mixed iron oxides*. Journal of Solar Energy Engineering - Transactions of the ASME, 128, 125-133, 2006.

[10] Lorentzou S., Karadimitra K., Agrafiotis C., Konstandopoulos A. G., *New Routes for Ferrite Powders Synthesis*. PARTEC International Conference for Particle Technology, March 16 –18, Nuremberg, Germany, 2004.

[11] Nalbandian, L., Zaspalis, V.T., Evdou, A., Agrafiotis, C., Konstandopoulos, A.G., *Redox materials for Hydrogen production from the water decomposition reaction*. Chemical Engineering Transactions, 4, 43-48, 2004.

[12] M. Roeb, M., Monnerie, N., Schmitz, M., Sattler, C., Konstandopoulos, A.G., Agrafiotis, C., Zaspalis, V.T., Nalbandian, L., Steele, A.M., Stobbe, P., *Thermo-chemical production of hydrogen from water by metal oxides fixed on ceramic substrates*. Proceedings of the 16th World Hydrogen Energy Conference, Lyon, France, 13-16 June, 2006.

Hydrogen Storage

H_2 BINDING AND REACTIVITY ON TRANSITION METAL COMPLEXES UNDERLYING BIOMIMETIC H_2 PRODUCTION AND NEW MATERIALS FOR H_2 STORAGE

GREGORY J. KUBAS

Chemistry Division, Los Alamos National Laboratory, MS J582
Los Alamos, NM 87545, USA

The H_2 molecule is held together by a very strong two-electron H–H bond but is only useful chemically when the two H's are split apart in controlled fashion. The reverse of this process is formation of H_2 from for example protons and electrons as performed elegantly and efficiently in Nature by hydrogenase enzymes. Both splitting of H_2 and formation of H_2 occurs on transition metal complexes via binding of molecular H_2 to the metal center, often observed as stable solids. Splitting of H_2 occurs by both homolytic (dihydride formation) and heterolytic (formation of metal hydride plus proton) pathways depending on the nature of the metal complex. The molecular chemistry and spectroscopic features of dihydrogen complexes will be the major topic of this talk. We are engaged in synthesizing catalysts for biomimetic photocatalytic hydrogen production consisting of first-row metals such as iron capable of binding and splitting/forming H_2. Hydrogen binds reversibly to a surprisingly large variety of both metal and main-group atoms, especially at low temperatures, and we are also studying such H_2 complexes for hydrogen storage.

1. Metal-Dihydrogen Complexes

1.1. *Structure and Bonding of Metal-Dihydrogen Complexes*

The H_2 molecule is married together by a very strong two-electron H–H bond but is only useful chemically when the two H's split apart in controlled fashion. However the mechanism by which this occurs was established only relatively recently because such electronically saturated molecules had never been caught in the act of chemically binding to a metal, usually the first step in breaking apart a strong bond. The discovery by Kubas and coworkers in 1984 of coordination of a nearly intact H_2 molecule to a metal complex (L_nM; L= ligand) thus led to a new paradigm in chemistry [1-4].

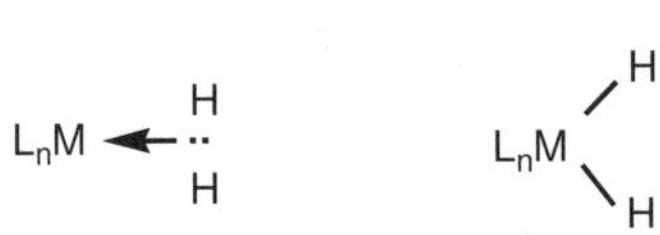

η^2-H_2 complex dihydride complex Eq. (1)

The H_2 binds side-on (η^2) to M primarily via donation of its two σ electrons to a vacant metal orbital to form a *stable* H_2 complex. It is remarkable that the electrons already strongly bonded can donate to a metal to form a nonclassical 2-electron, 3-center bond, first demonstrated in the complex $W(CO)_3(PR_3)_2(H_2)$ [1-3]. The H–H bond length in $W(CO)_3(P^iPr_3)_2(H_2)$ (0.89 Å) is lengthened ~20% over that in H_2 (0.74 Å), showing that H_2 is not physisorbed but rather chemisorbed, with the bond "activated" towards rupture. The vibrational modes for $M(\eta^2\text{-}H_2)$ are distinct from those for hydrides, which have only two fundamental modes: ν(MH) at 1700-2300 cm^{-1} and a M–H bending mode at 700-900 cm^{-1}. However H_2 complexes show *six* modes [5], including ν(HH) at 2200-3100 cm^{-1}, well over a 1000 cm^{-1} lower than in H_2 gas.

Molecular binding was proven by observation of a large HD one-bond coupling constant in the proton NMR of $W(CO)_3(P^iPr_3)_2(HD)$ (J_{HD} = 33.5 Hz, cf 43.2 Hz in HD gas), proving that the H–D bond was mostly intact. Observation of J_{HD} higher than that for a dihydride complex (>2 Hz) became the premier criterion for an H_2 complex. Over 600 H_2 complexes are known for nearly every transition metal and are the focus of >1500 publications. The 3-center metal-H_2 interaction complements classical Werner-type coordination complexes where a ligand donates electron density through its *nonbonding* electron pair(s) and π-complexes in which electrons are donated from π-electrons.

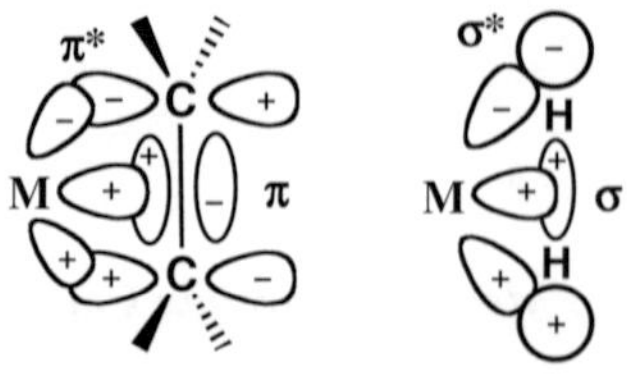

M–π bond M–σ bond

It is remarkable that the *bonding* electron pair in H_2 can interact with a metal center as strongly as a nonbonding pair. The resulting side-on bonding in M-H_2 is *nonclassical*, by analogy to the 3c-2e bonding in carbocations and diborane. The M center may be considered to be electronically equivalent to H^+ and CH_3^+ [6], mimicking carbocation chemistry; i. e. a complex such as M^+–CH_4 is related to CH_5^+, which is viewed as a highly dynamic H_2 complex of CH_3^+ [7]. H_2 is thus a weak Lewis base that can bind to strong electrophiles, but transition metals are unique in stabilizing H_2 complexes by *backdonation* of electrons from a filled metal d orbital to the σ^* antibonding orbital of H_2, an interaction unavailable to main group atoms [2-4, 8]. The backdonation is analogous to that for π-complexes, e.g. M-ethylene.

Backdonation of electrons from M to H_2 σ* is crucial not only in stabilizing the bonding but also in splitting the H-H bond. If it is too strong, the H-H bond cleaves to form a dihydride because of overpopulation of the H_2 σ* orbital. There is often a fine line between H_2 and dihydride coordination, and in some cases *equilibria* exist in solution for $W(CO)_3(PR_3)_2(H_2)$, showing that side-on coordination of H_2 is the first step in H–H cleavage [2, 3].

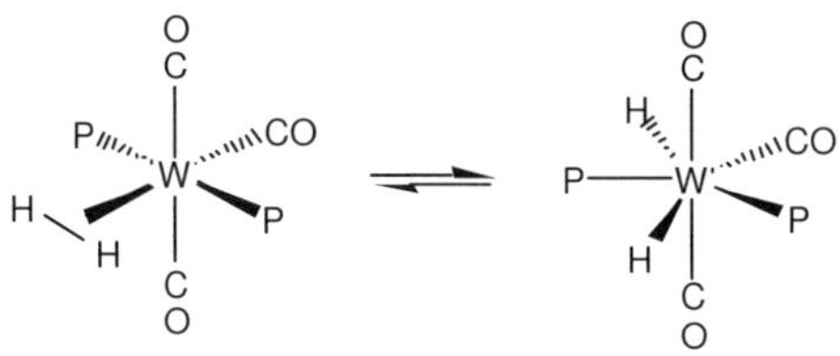

Eq. (2)

H_2 complexes are also stable with non-bulky co-ligands such as NH_3, in some cases with greatly elongated d_{HH} (1.3 Å for the Os complex) [9].

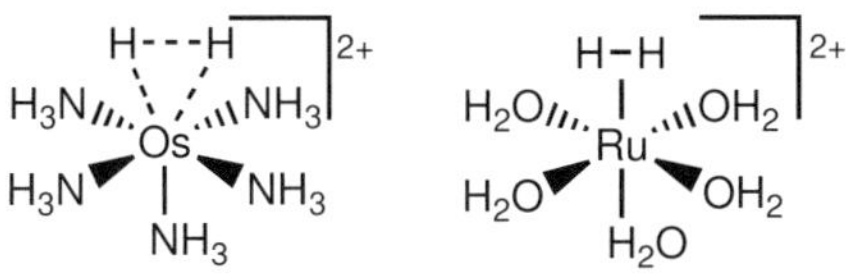

Variation of M, L, and other factors shows "arresting" of bond rupture along its entire reaction coordinate where d_{HH} varies enormously from 0.82 Å to 1.5 Å.

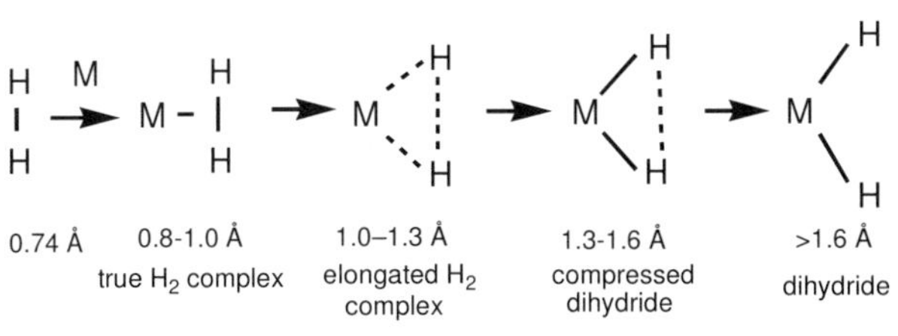

Eq. (3)

Although the d_{HH} ranges shown are arbitrary, each category of complexes has distinct properties. The d_{HH} is relatively short (0.8-1.0 Å) in "true" H_2 complexes best exemplified by $W(CO)_3(PR_3)_2(H_2)$, much as in physisorbed H_2 where d_{HH} is <0.8 Å. Importantly the H_2 binding is often completely *reversible* here, i.e. H_2 can be removed simply upon exposure to vacuum and re-added many times at room temperature. Elongated H_2 complexes and "compressed hydrides" are relative terms since a near *continuum* of d_{HH} has been observed, including in intermetallic rare-earth hydrides such as CeNiInH (d_{HH}= 1.48 Å.) [10]. First-row M, electron-withdrawing L, and positive charge (cationic complex) favor H_2 binding and shorten d_{HH}. The ligand trans to H_2 has a powerful influence: strongπ-acceptors such as CO greatly reduce backdonation and normally keep d_{HH}<0.9 Å.

1.2. *Heterolytic Cleavage of H_2 Complexes*

In addition to homolytic cleavage of H_2, *heterolytic cleavage* of bound H_2 can occur on electrophilic metal centers [11, 12].

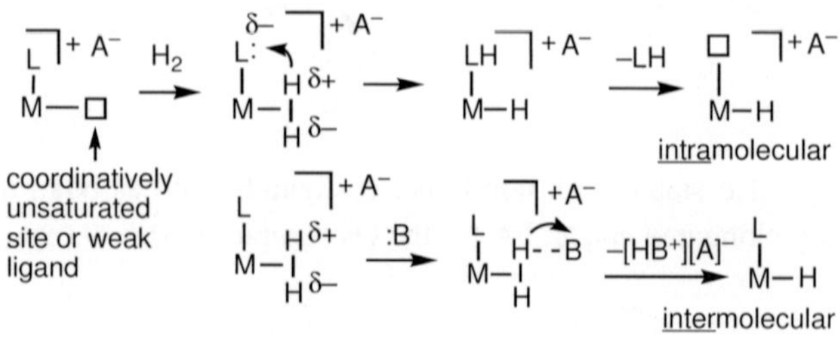

Eq. (4)

The H_2 ligand is deprotonated and the remaining hydrogen ligates to the metal as a hydride. The formal *oxidation state of M does not change* on binding of H_2, whereas formation of a dihydride formally increases the metal oxidation state by two. H_2 ligands can have far greater thermodynamic and kinetic acidity than hydrides. H_2 gas can be turned into a very strong acid: free H_2 is an extremely weak acid (pK_a ~35 in THF), but binding it to an electrophilic cationic metal increases the acidity *up to 40 orders of magnitude* (pK_a can be as low as –6). Heterolysis of H_2 is a crucial step in many industrial and biological processes, including the function of hydrogenase enzymes being modeled for H_2 production. H_2 can heterolyze in two ways (Eq. 4). *Intramolecular* heterolysis is extremely facile for proton transfer to a cis ligand L (e. g. H or Cl). *Intermolecular* heterolysis involves protonation of an external base B to give a metal hydride and HB+. This is the reverse of protonation of a metal hydride that is often used to synthesize H_2 complexes (reactions in Eq. 4 are reversible).

1.3. *Binding of H_2 to Surfaces and Non-Metals*

Molecular binding and heterolysis of H_2 on metal surfaces and small metal clusters is rarely observed since formation of hydrides is favored. H_2 binding to a stepped Ni(510) surface containing unsaturated sites was seen by electron energy loss spectroscopy [13] and is the first step in hydriding other surfaces [14]. $RuO_2(110)$ has also been found to bind H_2 at 85 K [15], and IR data suggests that, as above, the binding of H_2 is side-on similar to that in organometallic complexes. H_2 also ligates at low T in small clusters such as $Cu_3(H_2)$ [16], $Pd(H_2)$ [17], and similar species [18]. Oxides adsorb and activate H_2, including Cr_2O_3, MgO, and ZnO even at 25 °C; some of these could involve molecular interaction with oxide (X) via electron donation to H_2 σ^* orbitals.

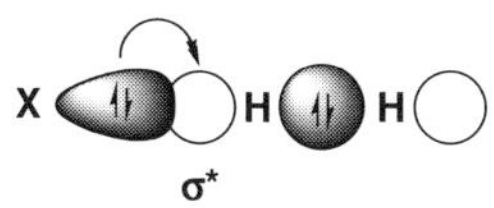

Other nonmetal systems such as fullerenes bind H_2 weakly at low T, possible via similar interaction where X is a C=C bond.

The first example of *reversible* splitting of H_2 on a *nonmetal center* has been found [19]. The phosphine borane in Eq. (5) has a strong Lewis acidic center (boron) linked to a Lewis basic site (phosphorus).

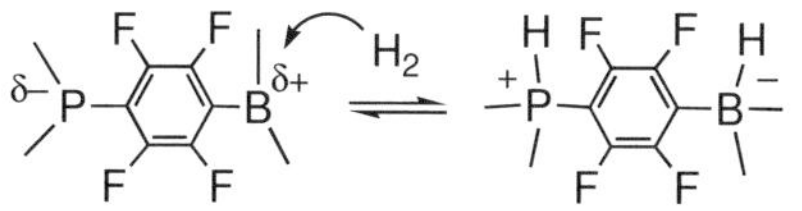

Eq. (5)

It is likely that H_2 heterolysis takes place at boron. Such main group systems could be useful for H_2 storage/production.

2. Activation of H_2 in Hydrogenases and Biomimetic H_2 Production

Hydrogenases are billion-year old redox enzymes in microorganisms that catalyze Eq. (6) to either utilize H_2 as an energy source or dispose of excess electrons as H_2 [20-22].

$$H_2 \rightleftharpoons 2H^+ + 2e^- \qquad \text{Eq. (6)}$$

Biologically unprecedented CO and CN ligands are present in dinuclear active sites [23] that are remarkably organometallic-like and have been extensively modeled for biomimetic H_2 production [21,22, 24-27].

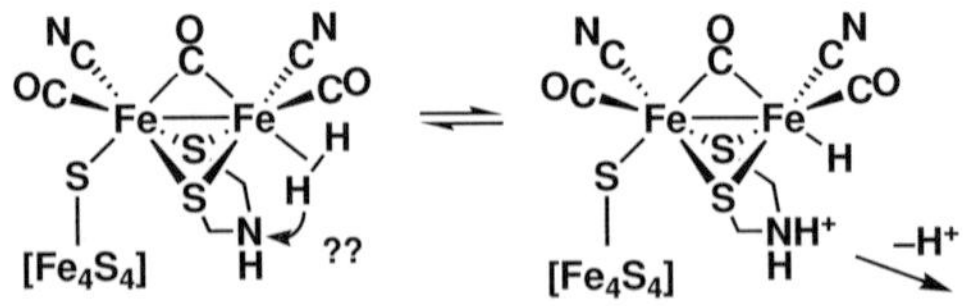

active site: low spin Fe

This complex presumably transiently binds and heterolytically splits H_2, most likely at a site trans to bridging CO, where a proton transfers to a basic ligand site such as amine [22]. Electron transfer and further deprotonation completes the splitting of H_2, a catalytic cycle which can be reversed in some hydrogenases to *produce* H_2 from protons and electrons. Production of H_2 fuel from water via solar energy is of high interest [28]. Catalysis may involve H_2 complexes at least as intermediates, and H_2 complexes have been implicated in solar energy conversion schemes based on photoreduction of water [29]. Biomimetic H_2 production, particularly solar driven (photocatalysis), is desirable and may take a cue from models of the active site of hydrogenase and photosystems [24-27]. Formation of H–H bonds from protons and electrons, the microscopic reverse of H_2 heterolysis, will be crucial in production of H_2 and is very rapid at the Fe sites in hydrogenases. Coupling model catalysts with photochemical water splitting is being investigated by us in a modular approach.

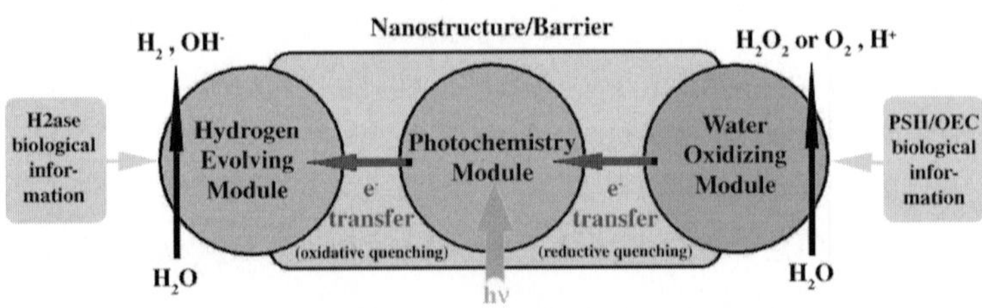

Water would be oxidized in the right module in a molecular system mimicking biological photosystem II and electrons transferred to a hydrogen-evolving module mimicking hydrogenase. Most of the efforts on modeling hydrogenase activity have been on bimetallic systems, but we are studying *monometallic* iron complexes with octahedral geometry (six ligands surrounding Fe) and divalent oxidation state (Fe^{II}). The ligands could include two CO and one CN^- as in the enzymes, and a SR^- group, although this and a sixth ligand Y could be varied.

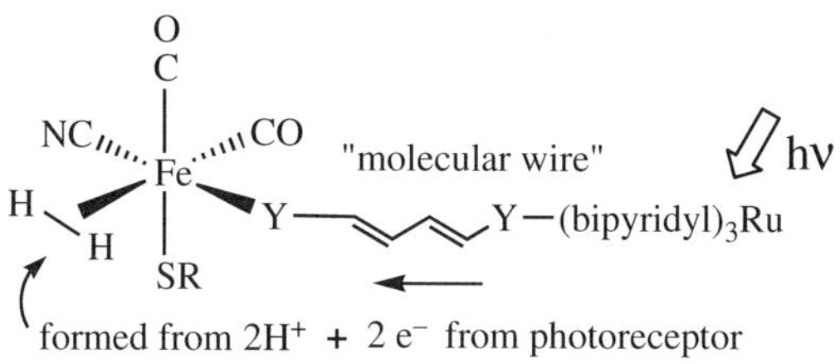

The Y group could be the conduit for the electrons supplied by the photochemical module, e.g. the well known Ru(bipyridyl)$_3$ type system, via a molecular wire linker, e.g. unsaturated C–C bonds or phenyl groups such as studied by Sun in dimetallic complexes [25]. We hope to coordinate an H_2 ligand and directly observe heterolytic cleavage of H_2 to simulate this function in hydrogenase, and then reverse this for H_2 production. The H_2 ligand would dissociate and be catalytically reformed from protons from water and electrons from the photoreceptor. We are synthesizing complexes of the type $FeCl(CO)_3(LL)$ that can be converted to e.g. $Fe(CN)(H_2)(CO)_2(LL)$ (LL = NH_2CRHCH_2S) and derivatives that can be linked to a photoreceptor via the R group on the chelating sulfido-amine ligand.

3. H_2 Complexes Relevant to H_2 Storage

Materials for H_2 storage are difficult to design because they must contain >6% by weight H_2, reducing prospects for known ideal reversible systems such as metal-H_2 or hydride complexes. Materials such as metal-organic frameworks (MOFs) [30-32] are now being examined for H_2 storage and have huge surface area capable of binding large numbers of H_2 molecules. Here neutron scattering (INS) studies by a collaborator, Juergen Eckert, are critical in determining whether H_2 binds to unsaturated metal centers and/or is physisorbed in the framework. Metal-doped zeolites were shown to bind H_2 side-on as in metal complexes, particularly strongly to the Cu^+ in Cu-ZSM-5 even at RT [33-35]. We have found that even light oxides such as nanoporous MgO bind the equivalent of 2.5 H_2 monolayers at 77 K and 13 atm, probably to O (see Section 1.3). Calculations indicate that metal complexes with multiple H_2, i.e. $Cr(H_2)_6$ may be stable [36], and species such as $[M(H_2)_n]^+$ have a fleeting gas-phase existence [37], but isolation in condensed phases is problematic. We are investigating synthesis of multi-H_2 complexes via protonation of metal polyhydrides such as $[FeH_6]^{4-}$[38].

$$[FeH_6]^{4-} + H^+ \longrightarrow [FeH_5(H_2)]^{3-} \quad [FeH_2(H_2)_4] \text{ on further protonation?} \qquad \text{Eq. (7)}$$

Reactions with acids at low T have yielded products that are being characterized. If the H_2 complexes are unstable, it may be possible to embed such H_2-rich species into nanoporous media (zeolites, MOFs) for reversible H_2 storage.

Acknowledgments

GJK is grateful to the U.S. Department of Energy, Basic Energy Sciences, Chemical Sciences, and Los Alamos National Laboratory for funding.

References

1. G. J. Kubas, R. R. Ryan, B. I. Swanson, P. J. Vergamini and H. J. Wasserman, *J. Am. Chem. Soc.* **106**, 451 (1984).
2. G. J. Kubas, *Chem. Rev.* **107**, 4152 (2007).
3. *Metal Dihydrogen and σ Bond Complexes*; G. J. Kubas, Kluwer Academic/Plenum Publishers: New York, NY, **2001**.
4. G. J. Kubas, *J. Organometal. Chem.* **635**, 37 (2001).
5. B. R. Bender, G. J. Kubas, L. H. Jones, B. I. Swanson, J. Eckert, K. B. Capps, and C. D. Hoff, *J. Am. Chem. Soc.* **119**, 9179 (1997).
6. M.Elian, M. M. L. Chen, D. M. P. Mingos, and R. Hoffmann, *Inorg. Chem.* **15**, 1148 (1976).
7. K.C. Thompson, D.L. Crittenden, and M.J.T. Jordan, *J. Am. Chem. Soc.* **127**, 4954 (2005).
8. F. Maseras, A. Lledós, E. Clot, and O. Eisenstein, *Chem. Rev.* **100**, 601 (2000).
9. (a) T. Hasegawa, Z. Li, S. Parkin, H. Hope, R. K. McMullan, T. F. Koetzle, and H. Taube, *J. Am. Chem. Soc.* **116**, 4352 (1994). (b) N. Aebischer, U. Frey, and A. E. Merbach, *Chem. Comm.*, 2303 (1998).
10. K. Ghoshray, B. Bandyopadhyay, M. Sen, A. Ghoshray, and N. Chatterjee, *Phys. Rev. B* **47**, 8277 (1993).
11. R.H. Morris in: *Recent Advances in Hydride Chemistry*, eds. M. Peruzzini and R. Poli, Elsevier Science B.V.: Amsterdam, **2001**, pp 1-38.
12. G. J. Kubas, *Adv. Inorg. Chem.* **56**, 127 (2004).
13. A.-S. Martensson, C. Nyberg, and S. Andersson, *Phys. Rev. Lett.* **57**, 2045 (1986).
14. Kresse, G. *Phys. Rev. B* **2000**, 62, 8295.

15. J. Wang, C.Y. Fan, Q. Sun, K. Reuter, K. Jacobi, M. Scheffler, and G. Ertl, *Angew. Chem. Int. Ed. Engl.* **42**, 2151 (2003).
16. R. H. Hauge, J. L. Margrave, and Z. H. Kafafi, *NATO ASI Ser., Ser. B* **158**, 787 (1987).
17. Ozin, G. A.; Garcia-Prieto, J. *J. Am. Chem. Soc.* **1986**, 108, 3099.
18. L. Andrews, *Chem. Soc. Rev.* **33**, 123 (2004).
19. G. C. Welch, R. R. San Juan, J. D. Masuda, and D. W. Stephan, *Science* **314**, 1124 (2006).
20. A. Volbeda, and J. C. Fonticella-Camps, *Coord. Chem. Rev.* **249**, 1609 (2005).
21. X. Liu, S. K. Ibrahim, C. Tard, and C. J. Pickett, *Coord. Chem. Rev.* **249**, 1641(2005).
22. M. Y. Darensbourg, E. J. Lyon, Z. Zhao, and I. P. Georgakaki, *PNAS* **100**, 3683 (2003).
23. J. W. Peters, W. N. Lanzilotta, B. J. Lemon, and L. C. Seefeldt, *Science* **282**, 1853 (1998).
24. J.-F. Capon, F. Gloagen, P. Schollhammer, and J. Talarmin, *Coord. Chem. Rev.* **249**, 1664 (2005).
25. L. Sun, B. Akermark, and S. Ott, *Coord. Chem. Rev.* **249**, 1653 (2005).
26. J. Alper, *Science* **299**, 1686 (2003).
27. R. M. Henry, R. K. Shoemaker, R. H. Newell, G. M. Jacobsen, D. L. DuBois, and M. Rakowski DuBois, *Organometallics* **24**, 2481 (2005).
28. N. S. Lewis, and D. G. Nocera, *PNAS* **103**, 15729 (2006).
29. N. Sutin, C. Creutz, and E.Fujita, *Comments Inorg. Chem.* **19**, 67 (1997).
30. Rosi, N.L.; Eckert, J.; Eddaoudi, M.; Vodak, D.T.; Kim, J.; O'Keeffe, M.; Yaghi, O.M. *Science* 2003, *300*, 1127.
31. J. L. C. Rowsell, J. Eckert, and O. M. Yaghi, *J. Am. Chem. Soc.* **127**, 14904 (2005).
32. P. M. Forster, J. Eckert, B. D. Heiken, J. B. Parise, J. W. Yon, S. W. Jhung, J.-S. Chang, and A. K. Cheetham, *J. Am. Chem. Soc.* **128**, 16846 (2006).
33. B.L. Mojet, J. Eckert, R.A. van Santen, A. Albinati, and R. E. Lechner, *J. Am. Chem. Soc.* **123**, 8147 (2001).
34. J. Eckert, J. M. Nicol, J. Howard, and F. R. Trouw, *J. Phys. Chem.* **100**, 10646 (1996).
35. P.A. Georgiev, A. Albinati, B.L. Mojet, J. Ollivier, and J. Eckert, *J. Am. Chem. Soc.* **129**, 8086 (2007).
36. L. Gagliardi, and P. Pyykko, *J. Am. Chem. Soc.* **126**, 15014 (2004).
37. J. C. Weisshaar, *Acc. Chem. Res.* **26**, 213 (1993).
38. D. E. Linn, Jr., S. G. Gibbins, *J. Organomet. Chem.* **554**, 171 (1998).

NANOSTRUCTURING IMPACT ON THE ENTHALPY OF FORMATION OF METAL HYDRIDES

VINCENT BERUBE AND M. S. DRESSELHAUS

Physics Department, Massachusetts Institute of Technology, 77 Massachusetts Avenue Cambridge, MA 02139, USA

GANG CHEN

Mechanical Engineering Department, Massachusetts Institute of Technology, 77 Massachusetts Avenue, Cambridge, MA 02139, USA

Metal and complex hydrides offer very promising prospects for hydrogen storage that reach the DOE targets for storage capacity for 2015. However, the slow sorption kinetics and the high release temperature must be addressed to make automotive applications feasible. Reducing the enthalpy of formation by destabilizing the hydride reduces the heat released during the hydrogenation phase and conversely allows desorption at a lower temperature. High-energy ball milling has been shown to decrease the release temperature, increase the reaction kinetics and lower the enthalpy of formation in certain cases. Increased surface and grain boundary energy could play a role in reducing the enthalpy of formation, but the predicted magnitude is too small to account for experimental observations. Since the particle and grain sizes are reduced considerably under high-energy treatments, structural defects and deformations are introduced. These deformed regions can be characterized by an excess volume due to deformations in the lattice structure, and these deformations have a significant effect on the physical properties of the hydride. We propose the use of two thermodynamic models to characterize the excess energy present in the deformed regions. The equations of state (EOS) provided by the models are used to explain the change in physical properties of metal hydrides. Particularly, the EOSs can predict which hydrides will be the most destabilized (if destabilized at all) by the introduction of excess volume regions.

1. Introduction

Hydrogen is considered a good energy carrier candidate for future automotive applications that could be part of a carbon-free cycle [1-2]. Metal hydrides are often preferred over pressurized gas and other hydrogen storage methods because of their gravimetric and volumetric storage capacities at safe operating pressures and non-cryogenic temperatures. The strong chemical bonds that form between hydrogen and metals during chemisorption explain the high storage capacity of hydrides and their stability at room temperature, but these strong bonds are also responsible for the following engineering hurdles that must be addressed before hydrogen storage in metal hydrides can be practically used for on-board automotive applications [3]:

1. Slow diffusion of hydrogen in the hydride.
2. The high hydrogen release temperature, which must be brought down to a level closer to the waste heat temperature of the operating fuel cells.
3. Low storage efficiency, resulting from the wasted energy needed to overcome the high enthalpy of hydride formation and elevated energy barriers to hydrogen release.
4. Thermal management during the highly exothermic hydriding reaction to prevent the high temperatures which may stall the hydriding reaction.

Even if progress has been made in the recent years to solve these bottlenecks, no technology satisfies all of the DOE's hydrogen storage requirements for on board automotive applications. A systems based approach to developing viable hydrogen technologies should aim at solving all those problems simultaneously. It has been shown that reducing[†] the enthalpy of formation of the metal hydride could positively impact the four bottlenecks mentioned above by reducing the energy barrier associated with the release of hydrogen [3]. Nanotechnology offers new ways to reduce the enthalpy of formation of metal hydrides by taking advantage of the distinctive chemical and physical properties available in nanostructures [4-6]. In this paper, we quantitatively looked at the contribution to enthalpy reduction from the surface energy, grain boundary energy, and lattice deformations. We compare the predicted change in enthalpy reduction to experimental data obtained for different size distribution of MgH_2 particles. The results show that only lattice deformations can account for the magnitude of the enthalpy reduction that has been observed in ball milled hydrides.

1.1. *Surface Area*

A decreased particle size leads to an increased surface to volume ratio. Creating surfaces has an energy cost that will ultimately reduce the enthalpy of formation of the system if the surface energy of the hydrided phase is higher than that of the non-hydrided phase. Compared to the large enthalpy of formation for metal hydrides (75 kJ/mol for MgH_2, for example), the surface energies are usually

[†] The enthalpy of formation of most hydrides is negative because the reaction is exothermic. By reduction of the enthalpy of formation we refer to a reduction of the magnitude of the enthalpy that results in a destabilization of the hydride.

negligible. However, for particles of sufficiently small size, the surface energy term cannot be ignored, and the molar free energy of reaction then becomes:

$$\Delta G(r) = \Delta G_0(r) + RT\ln\left(\frac{a_{\mathrm{MH_2}}}{a_{\mathrm{M}} P_{\mathrm{H_2}}}\right) + \frac{3V_{\mathrm{M}}\Delta_{\mathrm{M}\to\mathrm{MH_2}}(\gamma, r)}{r} \tag{1}$$

where r is the radius of a spherical particle and the volume-adjusted surface energy difference is [3]

$$\Delta_{\mathrm{M}\to\mathrm{MH_2}}(\gamma, r) = \left[\gamma_{\mathrm{MH_2}}(r)\left(\frac{V_{\mathrm{MH_2}}}{V_{\mathrm{M}}}\right)^{2/3} - \gamma_{\mathrm{M}}(r)\right] \tag{2}$$

Here, V_i denotes the molar volume of each phase (subscript M refers to the metal and MH_2 refers to the metal hydride) that accommodates the 10-30% volume increase that is usually observed in metal hydrides upon hydriding, γ is the surface energy of each phase, a_i is the activity coefficient of the phases [3] and P is the pressure of the gas.

Inclusion of the surface energy terms gives a new van't Hoff relation (3, 4), showing that size reduction lowers the enthalpy of hydride formation ($\Delta H'$) for the nanostructured hydride as long as $\Delta_{\mathrm{M}\to\mathrm{MH_2}}$ is positive. As seen in Eq. (2), $\Delta_{\mathrm{M}\to\mathrm{MH_2}}$ will be positive if the surface tension of the hydride is larger than that of the metal since a volume expansion usually accompanies the hydriding reaction:

$$\ln P_{\mathrm{H_2}}^{eq} = \frac{\Delta H'}{RT} - \frac{\Delta S_0}{R} \tag{3}$$

$$\Delta H' = \Delta H_0 + \frac{3V_{\mathrm{M}}\Delta_{\mathrm{M}\to\mathrm{MH_2}}}{r} \tag{4}$$

where ΔH_0 is a negative quantity corresponding to the enthalpy of formation of the crystalline hydride at standard pressure and temperature [3]. In the case of Mg, the bulk value of 1.76J/m^2 [13] for $\Delta_{\mathrm{M}\to\mathrm{MH_2}}$ predicts that $\Delta H'$ could be 10% smaller then ΔH_0 for hydride particles with radii smaller than 4 nm. Unfortunately, it would be hard to implement this size-dependent effect in Mg because of the limitation on the nanoparticle size that can be achieved. Calculations based on the repulsive energy between dislocations predict that a minimum particle radius size of 15 nm is achievable through ball milling for Mg [8]. At this size, the reduction in the enthalpy of formation predicted by Eq. (4) would only represent a 2-3% reduction in the enthalpy of formation of the hydride. This means that the additional surface energy created in nanopowders cannot explain the reduction of the enthalpy of formation observed in some experiments (above 20% in for MgH_2 in [7]) and that other mechanisms must be responsible for the large magnitude of the enthalpy reduction.

1.2. *Grain Boundaries*

High-energy mechanical treatments like ball milling also reduce the grain size of polycrystalline materials and introduce many grain boundaries. These grain boundaries are the result of mismatched crystal plane orientations and give rise to excess energy that in turn lead to an excess enthalpy using an approach similar to that found in Eqs. (1-4). A simple estimate of the maximum contribution of grain boundary can be obtained by equating the grain boundary energy difference to the surface energy difference between the metal hydrides and the metal, and by assuming that all grains have the smallest grain size in the sample (7-9nm for the samples we consider in the relevant experimental study [7]). This naturally overestimates the contribution since a grain boundary has a lower energy than the two surfaces creating it, due to the binding between the two planes. But the estimate thus made still provides an order of magnitude estimate for the enthalpy change. As in the case of the surface energy, the potential contribution to the enthalpy reduction is strongly limited by the relatively weak energies involved for grain boundaries (of the order of 1 J/m^2 as shown in Figure 3) compared to those of hydride formation: our calculation for MgH_2 shows a maximum reduction in the enthalpy of formation of only 3% for a grain size of the order of 10 nm. Those calculations were performed using experimental data from [7] where the smallest grain size measured was 8nm.

1.3. *Excess Volume in Deformed Regions*

The surface energy and grain boundary energy in nanostructured metal hydrides favor a reduction of the enthalpy of formation, although their combined contribution, at least for the MgH_2 example, is insufficient to explain the experimental observations [7]. Another mechanism must therefore be responsible for the observed reduction in the enthalpy of formation of certain metal hydrides. In heavily milled metal hydride samples, it is likely that non-crystalline regions will arise where the material is deformed. The resulting lattice distortions will change the energy content of the metal and hydride states and therefore could explain why the enthalpy of formation is changed, as discussed below.

A simple yet accurate way to explain the effect of lattice distortions on the energy of the crystal is by formulating an equation of state (EOS) that relates the energy of the crystal to its actual volume relative to the equilibrium volume. A dimensionless excess volume V_{exc} (defined as the ratio of the actual volume to the equilibrium volume) in deformed regions of metallic nanoparticles due to their longer atomic bonds was demonstrated to result in an excess energy, which can be related to a hydrostatic pressure that scales up with the atomic volume v [10, 11]:

$$p(V) = \frac{3B_0\left[\left(\frac{v}{v_0}\right)^{1/3} - 1\right]}{\left(\frac{v}{v_0}\right)^{2/3}} \exp\left[-a_*\left(1 - 0.15a_* + a_*^2\right)\right] \tag{5}$$

Here B_0 is the bulk modulus, v_0 is the equilibrium atomic volume and a_* is a scaling parameter for the excess energy specific to each material [12]. From this equation of state (Eq. (5)), the excess energy, enthalpy, entropy, and change in specific heat can be obtained from the usual thermodynamic relations. The excess enthalpy associated with the excess volume results in a reduction of the enthalpy of formation during hydrogenation if the non-hydrogenated phase has a smaller enthalpy increase. This can be understood from the fact that the enthalpy of formation is the difference between the enthalpy of the metal hydride and the enthalpy of the metal and the hydrogen gas together. If the enthalpy content of the metal is increase more than that of the metal hydride then the difference between the two enthalpies will increased instead of being reduced.

Another simple model that can explain how the excess volume in deformations can lead to a reduction of the enthalpy of formation in non-metallic materials is the Birch-Murnaghan (BM) equations of state that give the molar energy E as a function of the equilibrium molar energy E_0, the equilibrium molar volume V_0, the actual molar volume V, and the bulk modulus at equilibrium B_0 [12]. At 0K, the enthalpy of formation coincides with the molar energy of formation between the two deformed materials.

$$E = E_o + \frac{B_o(V - V_o)^2}{2V_o} \tag{6}$$

Figures 1 and 2 show a comparison between the two EOSs for MgH_2 and TiH_2. For MgH_2, both equations of state (Eqs. (5) and (6)) predict that an excess volume will lead to a substantial reduction of the enthalpy of formation. In the case of TiH_2, the bulk modulus of the hydrided phase is substantially smaller than that of titanium alone. This example shows that the hydride phase is less destabilized than the metal alone, which leads to an increase in the enthalpy of formation. A simple argument is thus capable of predicting which materials are favored by the presence of excess volume, namely that MgH_2 is favored and that TiH_2 is not. From the simple first order Birch-Murnaghan equation of state, we see that to first approximation, an excess volume will reduce the enthalpy of formation if the following condition is satisfied (see table 1).

$$\eta = V_o^{MH_2} B_o^{MH_2} - V_o^M B_o^M \geq 0 . \tag{7}$$

In Eq. (7), η gives an idea of the degree of destabilization of a hydride with respect to its metallic phase. A large positive η is favorable while a negative η indicates that the presence of highly deformed regions in the sample should be avoided for hydrogen storage applications.

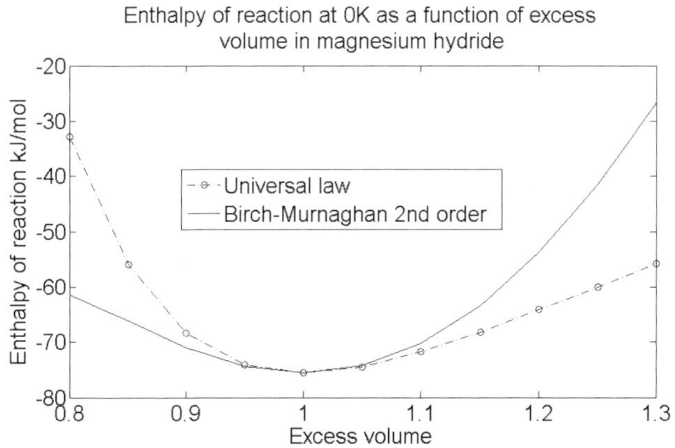

Figure 1. Predicted enthalpy of formation of MgH_2 as a function of the excess volume according to the universal law (Eq. (5)) and the Birch-Murnaghan 2nd order equation (Eq. (6) can be modified to include the change of the bulk modulus with regard to pressure. Incorporating this change leads to the Birch-Murnaghan 2nd order equation). Even if the two EOSs do not agree on the magnitude of the change of enthalpy, they both predict a reduction of the enthalpy of formation and the magnitude of the reduction in both cases is large enough to explain experimental data (see Figure 3). Future work using density functional theory will investigate the difference between the two EOSs.

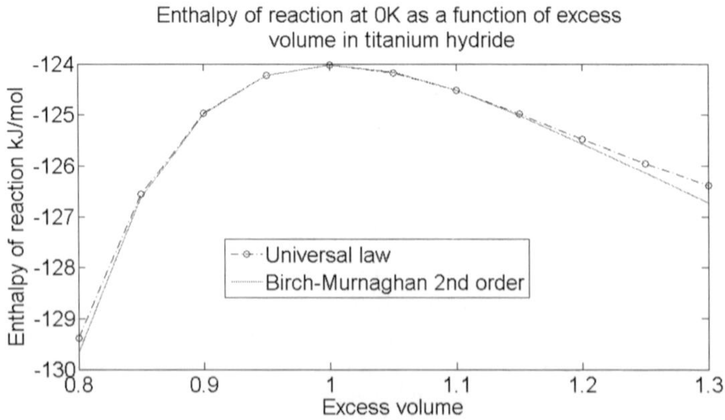

Figure 2. Predicted enthalpy of formation of TiH_2 from Ti + H_2 as a function of the excess volume according to the universal law and the Birch-Murnaghan 2nd order equation. The effect leads to an increase in enthalpy of formation for TiH_2 when such regions are introduced. The Ti hydriding reaction illustrates the fact that not all materials will benefit from the introduction of high energy structural defects (Eq. (7) and Table 1).

Table 1. Comparison of the destabilization parameter for different metal hydrides. A large positive value of η indicates that the enthalpy of formation of the hydride is reduced by the introduction of deformed regions while a negative value shows that introducing regions of excess volume would increase the heat released upon the hydriding reaction.

Material	**η [kJ/mol]**
Mg/MgH_2	805
Li/LiH	1330
Ti/TiH_2	-2270

A difference in the value of the excess volume will have a dramatic influence on the enthalpy of formation of the destabilized hydride. The fraction of the sample that is in a state of excess volume also has an important effect on the enthalpy reduction. The excess volume and the excess volume fraction will vary from sample to sample and even within distinct regions in the same sample. This is why a microscopic study of hydride samples in which enthalpy reductions have been observed is necessary to fit the theoretical models and to determine if crystal deformations can be responsible for the reduction in the enthalpy of formation. This can be done by investigating the crystal structure of heavily milled materials with a TEM to determine the concentration of extended regions and determine if any appreciable excess volume is present.

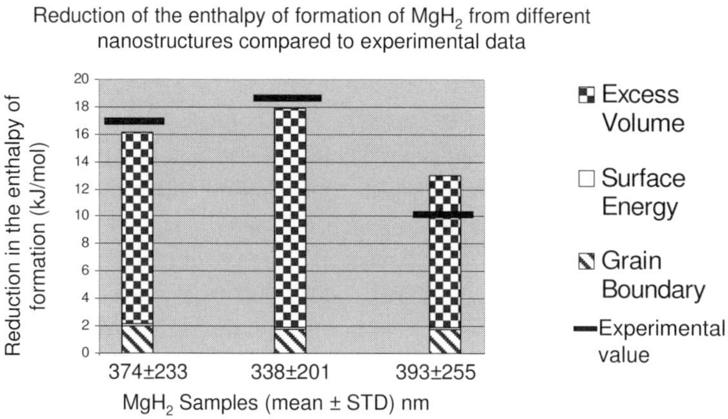

Figure 3. Comparison of the reduction of the enthalpy of formation in MgH_2 predicted by the three mechanisms presented in this paper with the experimental data obtained by Varin [7] (listed as black lines). Each column represents a different size distribution of nanoparticles obtained by high energy ball milling (see [7] for more details on the exact process used to produce the particles). Calculations show that surface energy and the grain boundary energy can't explain the observed values of enthalpy. The presence of deformed regions containing excess volume, on the other hand, can account for the order of magnitude reduction in the enthalpy of formation seen experimentally. For the calculations plotted on this graph, we chose a dimensionless excess volume V_{exc} of 1.3 over 35% of the sample.

Figure 3 shows the enthalpy reduction experimentally observed in three different powders made with MgH_2 nanoparticles [7]. It also shows the relative importance of the three different nanostructures studied in this paper. It is clear from the data that the particle size and size distribution are not enough to predict the enthalpy of formation and that deformed regions with excess enthalpy can explain the experimental observation. Figure 3 also illustrates that surface and grain boundary effects do not reduce the enthalpy sufficiently and that the excess volume effect is needed to explain the observed reduction in the enthalpy of formation.

2. Conclusions

Metal hydrides possess high storage capacity but there remain many issues that must be addressed before automotive applications are feasible. These are:

1. Slow sorption kinetics
2. High hydrogen release temperature
3. Low storage efficiency
4. Thermal management

Reducing the enthalpy of formation through the introduction of nanostructures is a strategy to address these issues because it reduces the energy barrier for hydrogen release. That in turns reduces the temperature needed for hydrogen extraction which indirectly addresses the other bottlenecks mentioned above. The introduction of surfaces and grain boundaries can lead to a reduction of the enthalpy of formation of a few percent, but cannot explain the experimental observations where reductions of more then 20% have been observed [7]. Only when considering the presence of deformed regions with excess volume were we able to reproduce the experimental data. TEM measurements and density functional theory calculations will help us confirm the validity of our EOS and to independently study the effect of the different nanostructures on release temperature, hydrogen diffusion and reaction kinetics. The TEM experiments will also help us understand the distribution of regions containing excess volume that can be expected in samples produced by ball milling.

Acknowledgments

The authors gratefully acknowledge support from the DOE through grant No. DE-FG02-05ER46241.

References

1. Hoffmann P. 1981. *The Forever Fuel*. Westview Press: Boulder.
2. Dresselhaus MS, Crabtree GW, Buchanan MV. *Basic research needs for the hydrogen economy*. Office of Basic Energy Sciences, US Department of Energy. http://www.sc.doe.gov/bes/reports/abstracts.html#NHE.
3. V. Bérubé, G. Radtke, M. Dresselhaus, G. Chen. 2007. Size effects on the hydrogen storage properties of nanostructured metal hydrides: A review, *Int J Energy Research*, 31:637-663..
4. Zaluska A, Zaluski L, Ström-Olsen JO. 2001. Structure, catalysis and atomic reactions on the nano-scale: a systematic approach to metal hydrides for hydrogen storage. *Applied Physics A*. **72**(2): 157–165.
5. Zaluska A, Zaluski L, Strom-Olsen JO. 1999. Nanocrystalline magnesium for hydrogen storage. *Journal of Alloys and Compounds*. **288**(1-2): 217-225.
6. Zaluski L, Zaluska A, Strom-Olsen JO. 1997. Nanocrystalline metal hydrides. *Journal of Alloys and Compounds*. **253-254**(1-2): 70-79.
7. Varin RA, Czujko T, Chiu CH and Wronski Z. 2006. Particle size effects on the desorption properties of nanostructured magnesium dihydride (MgH_2) synthesized by controlled reactive mechanical milling (CRMM) *JAC* **424**:356–364.

8. Eckert J. 1995. Relationships governing the grain size of nanocrystalline metals and alloys. *Nanostructured Materials*. **6**(1-4): 413-416.
9. Rose JH, Smith JR, Guinea F, Ferrante J. 1984. Universal features of the equation of state of metals. *Physical Review B*. **29**: 2963-2969.
10. Wolf D. 1989. Correlation between the energy and structure of grain boundaries in b.c.c metals. I. Symmetrical boundaries on the (110) and (100) planes. *Philosophical Magazine B*. **59**(6): 667-680.
11. Ferrante J, Smith JR. 1985. Theory of the bimetallic interface. *Physical Review B*. **31**(6): 3427–3434.
12. Murnaghan FD. 1944. The Compressibility of Media under Extreme Pressures. *Proceedings of the National Academy of Sciences.*. **30**: 244-247.
13. Johnson K. and Dai B. at the University of Pittsburgh (private communication).

DEHYDROGENATION MECHANISM FROM TITANIUM-ACTIVATED SODIUM ALANATE

SA LI AND P. JENA

Department of Physics, Virginia Commonwealth University, Richmond, VA 23284, USA

Addition of a small amount of Ti precursors to sodium alanate has recently been found to substantially improve the kinetics and thermodynamics of hydrogen sorption. In spite of several attempts, a fundamental understanding of how the catalyst works has remained unattainable. Using first principles calculations we have investigated the mechanisms for hydrogen desorption in this material by substituting Ti at various sites as well as creating a variety of vacancies. The lowest energy cost is when Ti replaces an AlH pair. Following this replacement, Ti attracts neighboring H atoms. Hydrogen desorption from this Ti neighborhood is much more efficient than from AlH_4 complex in pristine $NaAlH_4$. The formation of the AlH_3 vacancy, even though it is the easiest among all vacancy formation, yields higher creation energy than when Ti replaces the AlH pair. These results provide important new insight into the design of future catalysts for hydrogen storage materials.

1. Introduction

The increasing world-wide demand on fossil fuels as the primary energy source for the transportation sector and its dwindling supply have made it necessary to look for alternate energy sources that are safe, secure, abundant, renewable, cost effective, and environmentally friendly. Hydrogen is the third most abundant element on earth, is clean when it burns, and packs the highest energy per unit mass among all the elements in the periodic table. While considerable difficulties remain in the production of hydrogen and its use in fuel cells, storing hydrogen is the greatest of all challenges. Materials capable of storing hydrogen with high gravimetric and volumetric density, fast kinetics, and favorable thermodynamics are considered to be critical to a new hydrogen economy. To store hydrogen at about 10 wt % gravimetric density, which is the system target set for mobile applications,[1] hydrogen has to be stored in light hosts such as Li, B, N, C, Na, Mg, and Al. However, hydrogen in these materials is held by strong covalent or ionic bonds. Consequently, the hydrogen desorption temperatures are high and the kinetics are slow. Ideally hydrogen should be stored in such a way that it is neither easy (as would be the case if they are molecularly physisorbed) nor difficult (as would be the case if hydrogen is held in strong covalent or ionic bonds) for it to desorb at moderate temperatures. The central challenge then is to find materials that can store hydrogen like methane but whose kinetics and thermodynamics mimic that of intermetallics.

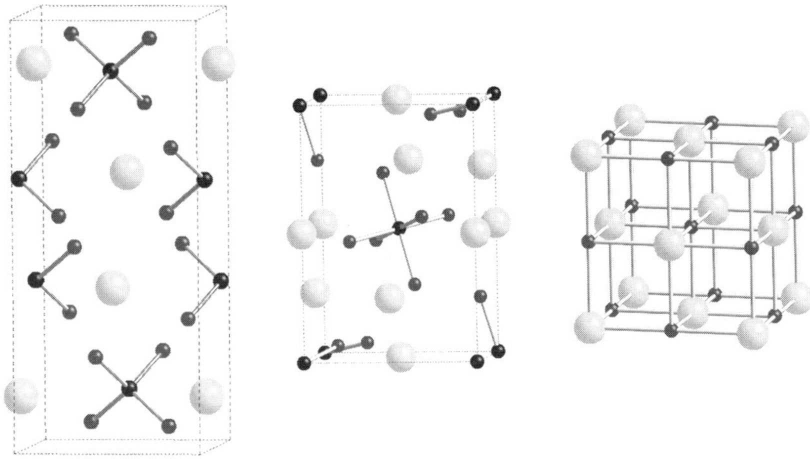

(a) (b) (c)

Figure 1. Crystal structure of (a) $NaAlH_4$, (b) Na_3AlH_6 and (c) NaH.

In the recent years, a great deal of attention has been focused on complex light-metal hydrides and in particular on sodium alanate, due to their high hydrogen content. In $NaAlH_4$, the four hydrogen atoms form a tetrahedron that encapsulates an Al atom much as in methane and the AlH_4 unit is stabilized by the transfer of one electron from the Na atom. The four hydrogen atoms in AlH_4 are covalently bonded to Al while the AlH_4^- unit is bonded to Na^+ by an ionic bond. The dehydrogenation and rehydrogenation in $NaAlH_4$ takes place in the following three steps as shown in Figure 1:

$$3NaAlH_4 \leftrightarrow Na_3AlH_6 + 2Al + 3H_2, \tag{1}$$

$$Na_3AlH_6 \leftrightarrow 3NaH + Al + 1.5H_2, \tag{2}$$

$$NaH \leftrightarrow Na + 1/2H_2. \tag{3}$$

The first decomposition reaction occurs at 353 K, releasing 3.7 wt % of hydrogen. In the second step, which occurs at 423 K, 1.9 wt % of hydrogen is released. The remaining 1.9 wt % hydrogen, released at 698 K does not have much practical value as the temperature is too high for on-board applications. Hence the hydrogen-storage capacity of sodium alanate is considered to be

5.6 wt%, i.e. the sum of the first two steps. The restoration of $NaAlH_4$ and Na_3AlH_6, i.e. the reverse reactions of Eqs. (1) and (2), can be accomplished through hydrogenation of NaH-Al conglomerates. These reversible reactions can only be completed under certain conditions and the recovery is only partial. For reversible hydrogen storage, the reactions must proceed rapidly under acceptable conditions during dehydrogenation-rehydrogenation cycles.

In 1997 Bogdannovic and Schwickardi showed that reactions in Eqs. (1) and (2) can be accelerated by adding a few mol% of selected Ti compounds, such as β-$TiCl_3$, $Ti(OBu)_4$ and $Ti(O\text{-}n\text{-}C_4H_9)_4$. The addition of small amount of Ti compounds to sodium-alanate was found not only to accelerate the adsorption and desorption process, but also markedly lower the hydrogen desorption temperature[2]. Six years later, the use of Ti nanocomposite[3,4] was also reported to improve hydrogen exchange kinetics. Doping with nanosized Ti brought hydrogenation times close to that required for practical applications, combined with high capacity (4.5 wt% H_2).

These discoveries have revitalized research into complex light metal hydrides as potential hydrogen storage materials. Moreover, the role of catalysts has been highlighted[5]. In spite of large amount of research work (both experiment and theory) a full understanding of where Ti resides and how it helps to lower the hydrogen desorption temperature remains elusive.

1.1. *Experiment Findings*

Numbers of experiments have been recently carried out to understand the mechanisms for hydrogen desorption and the role of Ti precursors play in the process. There are mainly three explanations:

1.1.1. *Formation of $TiAl_3$*

The reaction product $TiAl_3$ has been commonly observed in a large number of experiments.[6-12] The high catalytic activity of $TiCl_3$ has been attributed to microcrystalline intermetallic $TiAl_3$, which rapidly forms *in situ* from $TiCl_3$ and $NaAlH_4$ during mechanical processing and then acts as a heterogeneous dehydrogenation catalyst.[6] Graetz *et al.*[8] reported that the decomposition of $NaAlH_4$ during mechanical milling liberates Al and H_2, either of which may form a compound with Ti. Even though the liberated H_2 is considerably more mobile and therefore more likely to react with Ti and form TiH_{2-x}, TiHx is less stable than $TiAl_3$. The local environment around Ti is nearly invariant during the hydrogenation cycle [13]. They exclude the possibility of Ti bulk substitution and conclude that Ti catalyst is present on the surface in the form of amorphous $TiAl_3$.

1.1.2. *Ti Hydrides*

A number of research groups found that mechanical milling of a NaH/Al mixture or $NaAlH_4$ with metallic Ti powder resulted in the formation of nanocrystalline Ti hydrides[14]. The variation of the preparation conditions during the doping process leads to a slight composition variation of Ti hydrides. The catalytic enhancement arising upon doping the hydride with commercial TiH_2 was quite similar to that achieved in the hydrides doped with metallic Ti. Moreover, the cycling stability that was previously established in metallic Ti-doped hydrides was also observed in the hydrides doped with TiH_2. These results clearly demonstrate that the *in situ* formed Ti hydrides act as active species to catalyze the reversible dehydrogenation of $NaAlH_4$. At the same time, the catalytic effect of TiH_2 on the decomposition of $LiAlH_4$ and $NaAlH_4$ is well-known.[15] Most likely, the presence of a titanium hydride phase in the catalyst is responsible for the catalytic effect of Ti-additives.[16]

1.1.3. *Defects*

XANES and EXAFS data indicates that $TiCl_3$ is reduced to Ti^0 during the ball-milling process and stays in this state during desorption and adsorption of hydrogen. The experimental observations and correlations support a mechanism where the number of defects created by a partial substitution of Al by Ti determines the rates of transformation of the alanate material.[17] A systematic study[18,19] of the dehydrogenation process indicates that the most likely process involves a defect of type AlH_x ($x < 6$) which gives rise to local vacancy dynamics. The formation of defects in Na_3AlH_6 during dehydrogenation takes place at lower temperatures in Ti-doped samples than in undoped samples. The results show that not all the hydrogen released during the decomposition reactions evolves out of the samples as gas, but part of it remains in the lattice.

1.2. *Theoretical Findings*

Even though large experimental works have been done, their results are not conclusive as they do not provide an atomistic understanding of where Ti atoms reside and how they influence the bonding between hydrogen and metal atoms.

Theoretical calculations have been carried out to solve this enigma. Some of theoretical investigations are listed in the following:

1.2.1. *Ti substitution at the Na or Al site*

The site preferred by Ti has been controversial. Iniguez *et al.* reported that substitutional Ti doping is energetically possible and Ti prefers to substitute the Na site. Ti is a powerful hydrogen attractor that facilitates multiple Al-H bond breaking.[20] Their later calculations on surface suggest that Ti would occupy the surface Na site and the most likely product of the Ti doping is the formation of $TiAl_n$ ($n > 1$) compounds on the surface.[21] Lovvik and Opalka[22] have done calculations for bulk and surface and they argue that Ti doping is unstable in $NaAlH_4$ and the least unfavorable location of Ti is on the sub-surface layer, replacing Al in the host lattice. This difference on energetics and preferable sites arises due to the reference energies one uses. Iniguez *et al.* have chosen this reference to be isolated atoms while Lovvik and Opalka have used bulk cohesive energies of Ti, Al, and Na as reference. The use of the cohesive energies of Al, Na, and Ti leads to the Al site being the least unfavorable one. We found out that in both cases the bonding of hydrogen to Al metal atoms is weakened and the energy necessary to remove a hydrogen atom is consistently lower than that from pristine sodium alanate irrespective of whether Ti occupies the Na or the Al site.[23]

1.2.2. *$TiAl_3$ Cluster Formation On Ti-Doped $NaAlH_4$ Surface*

Liu *et al.*[24] carried out a calculation on $NaAlH_4$ (001) plane. In the simulation, Ti was substituted at the Na site and was found to bind to three Al atoms nearby forming a $TiAl_3H_{12}$ cluster. The complex structure may play important role in the reversible hydrogen release/uptake in Ti-doped $NaAlH_4$. Their calculations showed that desorption of hydrogen within the cluster can cause hydrogen atoms of the neighboring $(AlH_4)^-$ units to migrate to the Al atoms of the complex. The migration of hydrogen during relaxation after desorbing hydrogen indicates that the barrier for hydrogen diffusion across different AlH_4 units is small. The authors did not compare the substitution energy of Ti placed at different sites.

1.2.3. *Vacancy Mediated Hydrogen Desorption*

Besides the Ti substitution, the presence of Na vacancies is shown to play an even larger role. The energy need to remove a hydrogen atom is not only an order of magnitude smaller than that from Ti-doped sodium-alanate, but the removal of hydrogen associated with a Na vacancy is exothermic with respect to formation of H_2 molecule. The authors[25] assigned the diminished value of the hydrogen-removal energy to unusual stabilization of the magic AlH_3 cluster in the vacancy containing sodium-alanate. Later on, the possibility of forming NaH and AlH_3 vacancies was reported. AlH_3 vacancy is reported to be easier to form and diffuse than NaH vacancy. In the above paper, the authors[26] claim that bulk substitution of Ti yields higher formation energy and is accompanied with large volume change. However, this judgment was based on calculations made using different methods and at different temperatures.

2. Calculation Methods

We note that all the above calculations, even though they used first principles methods, were performed either at the 0 K or used different reference energies, In addition, none of these calculations have studied all the above substitutions using the same approach. Note that 0 K relaxation will very likely lead to some local minimum in energy surface. For example, in Lovvik and Opalka's calculation, 0 K relaxations for the case of Ti→Al led to two coordination H spheres around Ti. In the first sphere, four H atoms were bonded to Ti at a distance of 1.81 Å. In the second sphere of four nearest neighbor H's were 2.34 Å away from Ti. On the other hand, molecular dynamics (MD) calculation at 300 K yields a configuration where Ti is bound to eight hydrogen atoms with bond distance in the rage of 1.79-1.94 Å. This configuration is lower in energy by 0.33 eV. Thus the results Lovvik and Opalka obtained at 0 K only correspond to local energy minima. The creation of an Al vacancy is most exceptional. When an Al vacancy is created, the initial optimization at 0 K leads to four hydrogen atoms forming a square structure with the H-H distance of 1.24 Å (Fig. 2(a)). However, molecular dynamics simulation at 5 K shows this structure to be dynamically unstable and the four hydrogen atoms combine to form two hydrogen molecules with bond distance 0.78 Å (see Fig. 2(c)). Higher temperature MD simulations, on the contrary, lead to higher energy configuration with H_2 dissociated and bind to the nearby AlH_4 complexes. The Al vacancy formation energy listed in Fig. 3 corresponds to configuration Fig. 2(c).

Thus the question arises: Among all the above scenarios which provides the correct picture for hydrogen desorption? Using molecular dynamic simulations, we have compared the substitution energies for all possibilities discussed above for bulk sodium alanate. We first constructed a (2x2x1) super cell consisting of

96 atoms ($Na_{16}Al_{16}H_{64}$).[23,30] We have calculated the total energies by allowing full geometry optimization for the following cases: (a) Na, NaH, Al, AlH, AlH_2 and AlH_3 atoms were substituted by Ti., (b) Na, NaH, Al and AlH_3 vacancies were created. The calculations are carried out using generalized gradient approximation (GGA)[27] in the spin polarized density functional theory (DFT)[28] and the projector augmented wave (PAW)[29] method. The PAW potentials with the valence states 3p, 3d and 4s for Ti, 2p and 3s for Na, 3s and 3p for Al and 1s for H were used as prescribed in the Vienna *ab initio* simulation package (VASP).[30] *Ab initio* molecular dynamics simulations were carried out at 300 K for all the above systems. Two thousand time steps, each 1 fs long, were chosen for the equilibration. The velocities were scaled at each time step. The structures obtained from these molecular dynamics simulations were further relaxed.

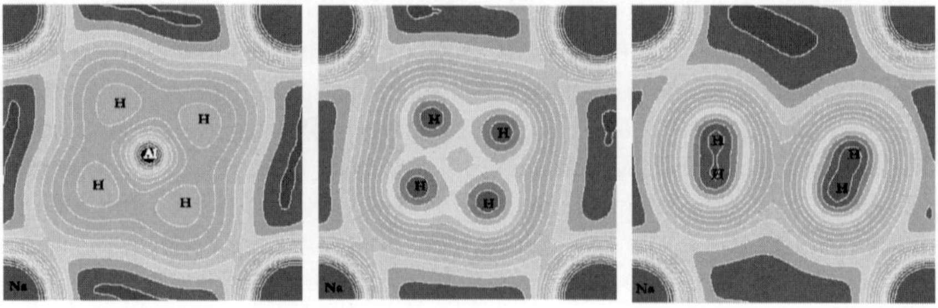

(a) (b) (c)

Figure 2. Charge density plot of AlH_4, H_4 units in part of the (001) plane of (a) $Na_{16}Al_{16}H_{64}$ (b) $Na_{16}Al_{15}H_{64}$ at 0K and (c) $Na_{16}Al_{15}H_{64}$ at 5 K, respectively. Deep red and blue colors correspond to highest and lowest charge densities.

3. Results and Discussions

In the following we present the results of these comprehensive calculations. We have used the cohesive energies of hcp Ti, bcc Na, fcc Al, fcc NaH and the binding energy of the H_2 molecule as reference energies. These energies are found to be 5.508 eV/atom, 1.102 eV/atom, 3.498 eV/atom, 3.814 eV/formula unit and 4.511 eV/H_2 respectively from our spin polarized calculations. We should emphasize that spin polarized atomic energy correction is important especially for the isolated Ti and H atom. For example, without the spin polarized energy correction, cohesive energy of H_2 molecule is as large as 6.737 eV. This is in agreement with 6.781 eV calculated by Ke *et al.*[31]

We define the energy cost in replacing Na, NaH, Al, AlH, AlH_2, and AlH_3 atoms by Ti and energy cost in creating Na, NaH, Al and AlH_3 vacancies in terms of the cohesive energies of bulk Ti, Na, Al and NaH as:

$$\Delta E_{Ti\rightarrow Na} = E[(TiNa_{15})Al_{16}H_{64}]_{coh}+E(Na_{bcc})_{coh}-E(Ti_{hcp})_{coh}-E[Na_{16}Al_{16}H_{64}]_{coh}$$

$$\Delta E_{Ti\rightarrow NaH} = E[(TiNa_{15})Al_{16}H_{63}]_{coh}+E(NaH_{fcc})_{coh}-E(Ti_{hcp})_{coh}-E[Na_{16}Al_{16}H_{64}]_{coh}$$

$$\Delta E_{Ti\rightarrow Al} = E[Na_{16}(TiAl_{15})H_{64}]_{coh}+E(Al_{fcc})_{coh}-E(Ti_{hcp})_{coh}-E[Na_{16}Al_{16}H_{64}]_{coh}$$

$$\Delta E_{Ti\rightarrow AlHx} = E[Na_{16}(TiAl_{15})H_{64-x}]_{coh}+E(Al_{fcc})_{coh}+x/2E(H_2)-E(Ti_{hcp})_{coh}-E[Na_{16}Al_{16}H_{64}]_{coh}$$

$$\Delta E^{H}_{Nav} = E\,[Na_{15}Al_{16}H_{63}]_{coh} + E(Na_{bcc})_{coh} - E\,[Na_{165}Al_{16}H_{64}]_{coh}$$

$$\Delta E^{H}_{NaHv} = E\,[Na_{15}Al_{16}H_{63}]_{coh} + E[NaH_{fcc}]_{coh} - E\,[Na_{16}Al_{16}H_{64}]_{coh}$$

$$\Delta E^{H}_{Alv} = E\,[Na_{16}Al_{15}H_{64}]_{coh} + E(Al_{fcc})_{coh} - E\,[Na_{16}Al_{16}H_{64}]_{coh}$$

$$\Delta E^{H}_{AlH3v} = E\,[Na_{16}Al_{15}H_{64}]_{coh} + E(Al_{fcc})_{coh} + 3/2E(H_2) - E\,[Na_{16}Al_{16}H_{64}]_{coh}$$

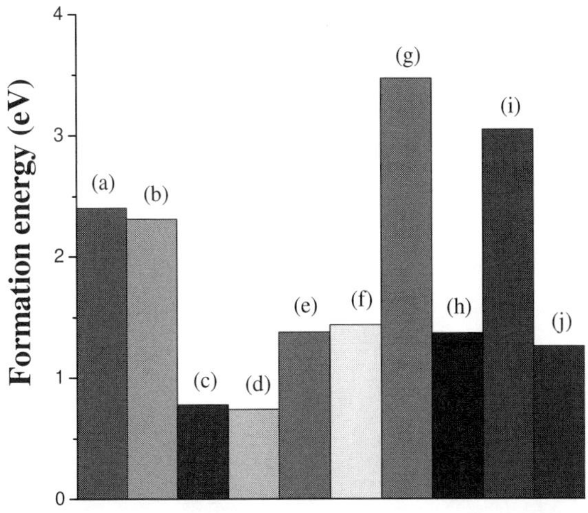

Figure 3. The formation energy of (a) Ti substituted at the Na site (b) Ti substituted after NaH pair is removed. (c) Ti substituted at the Al site, (d) Ti substituted after AlH pair is removed. (e) Ti substituted after AlH_2 is removed.(f) Ti substituted after AlH_3 is removed. (g) Na vacancy, (h) NaH vacancy, (i) Al vacancy and (j) AlH_3 vacancy.

In Figure 3, the energies required to substitute Ti after removing Na, NaH, Al, AlH, AlH_2 and AlH_3 atoms in $NaAlH_4$ as well as energies necessary to create a Na, NaH, Al or AlH_3 vacancy are shown. We can see that Ti substituted after AlH atoms are removed is energetically most preferable. Ti substituted by removing AlH is 0.04 eV lower in energy than Ti substituted at the Al site, which suggests the first hydrogen desorption following Ti substitution at the Al site does not cost energy and can even be mildly exothermic. The formation of the AlH_3 vacancy follows next. It is 0.52 eV higher in energy than Ti substitution following AlH removal.

3.1. *Ti Substitution at the Al Site*

When Ti is substituted at the Al site, besides the four hydrogen atoms originally bonded to Al, Ti is either edge shares or corner shares with the four hydrogen atoms belonging to the nearest AlH_4 complex. Ti is thus bonded to eight hydrogen atoms at distances between 1.79 to 1.94 Å and four Al atoms at distances between 2.8 to 2.97 Å (Fig. 4(a)). Upon removal of one hydrogen atom at 1.94 Å, the H atom which was at distance of 3.62 Å from Ti migrated and bound to Ti at a distance of 1.93 Å. Ti is found to bind to four Al atoms lying within a range of 2.68 to 2.98 Å and eight hydrogen atoms at the distances between 1.78-1.93 Å (Fig. 4(b)). The energy cost for removal of second and third hydrogen atoms is 0.64 eV/H and 0.06 eV/H, respectively. The higher energy cost to remove the second hydrogen is due the decrease in the hydrogen coordination number from 8 to 7. The number of hydrogen atoms bound to Ti is seven when Ti is substituted following the removal of both AlH_2 and AH_3 atoms. Note that even though Ti substitution after AlH_3 removal is slightly higher (0.18 eV) in energy than the formation energy of an AlH_3 vacancy, hydrogen can desorb step by step following Ti substitution. Instead of direct

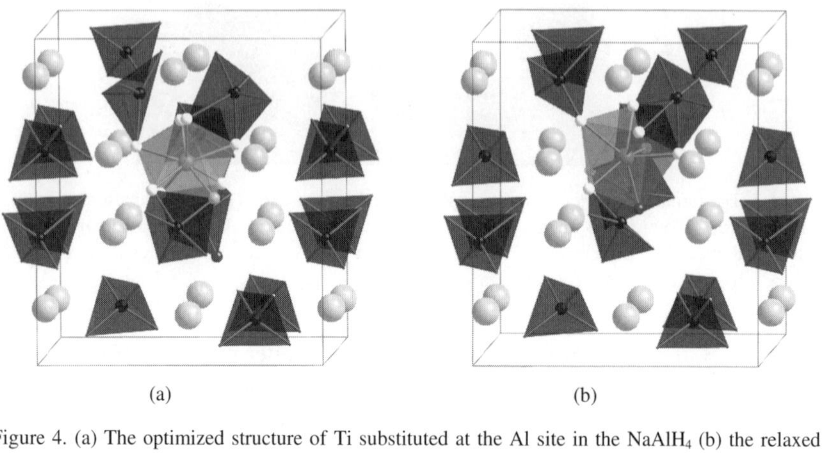

(a) (b)

Figure 4. (a) The optimized structure of Ti substituted at the Al site in the $NaAlH_4$ (b) the relaxed structure when Ti replaces AlH.

formation of AlH_3 vacancy, Ti can first replace the Al or AlH and then desorb H atoms one after the other at moderate temperature. This is a likely pathway of how Ti catalyst works. Beyond that, we found out that low percentage of Ti substitution do not lead to big volume change. Replacing an Al atom with a Ti atom in a 96 atom unit cell (5 wt% of Ti) results only in a 0.3% volume change. This is equivalent to the volume change due to an AlH_3 vacancy. Therefore, small amount, as low as 0.02%, of Ti substitution should not even result in any observable lattice change.

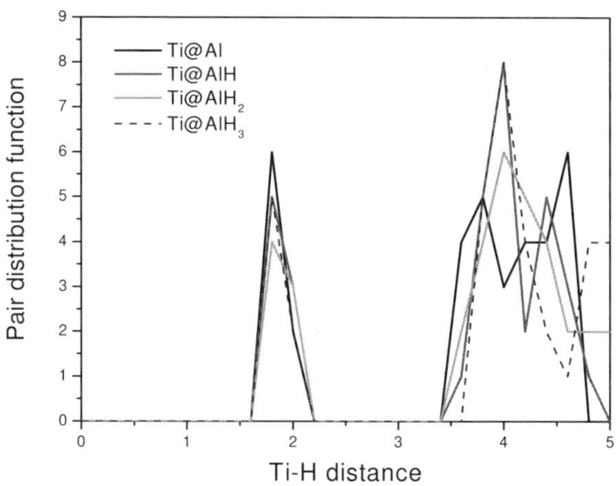

Figure 5. The Ti-H pair distribution function for Ti replacing Al, AlH, AlH_2 and AlH_3 sites.

We have used the pair distribution function (PDF) (Figure 5) to study the number of hydrogen atoms within the distance of 5 Å from the Ti center in the case of Ti substituting the Al site. There are 34 H atoms within a distance of less than 5 Å from the Ti atom when one Al atom is replaced. Eight H atoms are in the first neighbor (1.8-2.2 Å) and 26 H atoms are in the second neighbor (3.6-4.8 Å). When one, two and three hydrogen atoms in a AlH_4 complex are removed, the number of hydrogen atom within the sphere with radius of 5 Å is 33, 34 and 35. The number of hydrogen atoms in the first neighbor is 8, 7 and 7. Instead of getting less upon the removal of hydrogen, the number of hydrogen atoms inside the 5 Å sphere increases. Thus, one can visualize the Ti atom to serve as a magnet that continues to attract nearby H atoms as the nearest ones are successively desorbed

3.2. *Ti Substitution at the Na Site*

Next, we studied the effect of Ti substitution at the Na site. In pristine $NaAlH_4$, each Na is coordinated to eight H atoms at the distance of 2.4 Å and eight Al atoms at distances ranging between 3.52 and 3.72 Å. When a Na atom is replaced by a Ti atom, Ti is bonded to seven hydrogen atoms between at distances of 1.82-1.92 Å and six Al atoms at the distances of 2.72-3.18 Å. The shortening of Ti-Al distance indicates the formation of TiAlx complexes. When one hydrogen atom in the vicinity of the Ti atom is removed, Ti can still bind to seven H atoms at distances varying from 1.72 to 1.99 Å after optimization. As shown in Figure 6, the first nearest neighbors of Ti (the first peak) are

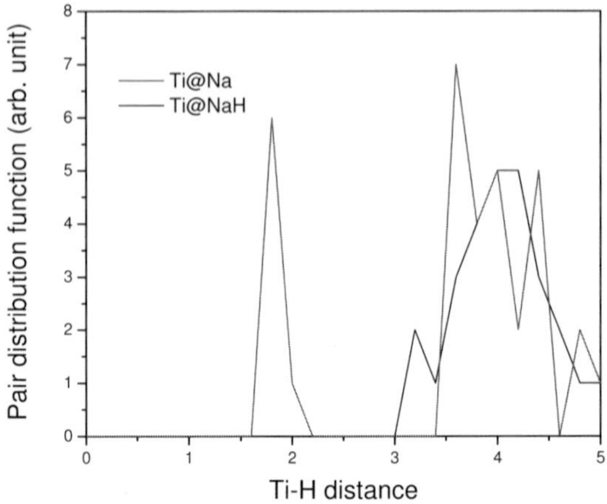

Figure 6. The Ti-H pair distribution function for Ti replaces Na and NaH.

essentially unchanged upon the removal of a H atom in the vicinity of the Ti atom. Similar to the situation when Ti substituted an Al site, one H atom migrates from the second nearest neighbor to the first nearest neighbor upon hydrogen removal.

3.3. *Formation of a NaH and AlH_3 vacancy*

From Figure 3 we can see that the formation energies of a NaH and AlH_3 vacancy are much lower than that for a Na and Al vacancy, respectively. Introduction of NaH and AlH_3 vacancies has been discussed by Gunaydin *et al*[26]. The creation of a NaH vacancy results in the formation of $(Al_2H_7)^-$, i.e. one H atom sharing two tetrahedral (Fig. 7(a)). For the case of AlH_3 vacancy (Fig. 7(b)), the extra hydrogen atom recombines with nearest AlH_4 complex to form $(AlH_5)^{2-}$. The Na and AlH_3 vacancy are reported to diffuse together with a shared H atom and the $(AlH_5)^{2-}$, respectively, as a result of Coulomb interaction. Our calculated NaH and AlH_3 vacancy creation energy (per vacancy) of 132 KJ/mol and 122 KJ/mol agrees well with 144 KJ/mol and 116 KJ/mol, respectively given by Gunaydin *et al*[26]. The good agreement shows that our calculated formation energies are reliable.

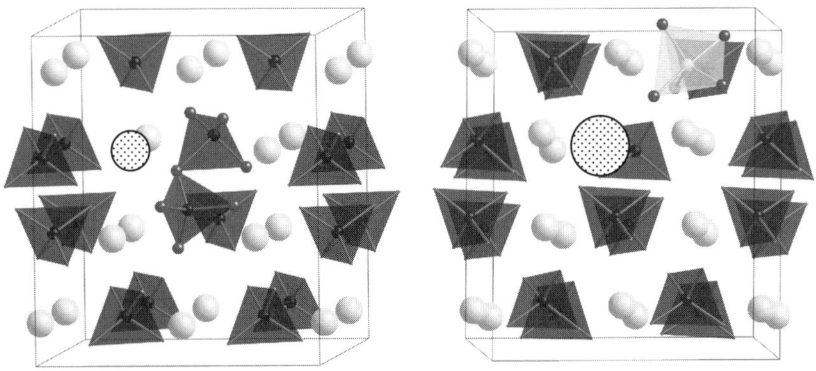

(a) (b)

Figure 7. The optimized structure of (a) NaH and (b) AlH_3 vacancies in $NaAlH_4$.

Based on the above calculations, we conclude that replacing AlH pair with Ti atom is energetically most favorable. The small amount of Ti substitution does not introduce large lattice distortion. The Ti atom serves as a magnet that continues to attract nearby H atoms as the nearest ones are successively desorbed. The Al atoms near to the Ti site remain at four upon hydrogen desorption when Ti is substituted at the Al site. Considerable amount of work still needs to be done to understand Ti-catalyzed dehydrogenation-rehydrogenation process.

Acknowledgments

This research used resources of the National Energy Research Scientific Computing Center, which is supported by the Office of Science of the U.S. Department of Energy. Partial support of this work by the Department of Energy is also acknowledged.

References

1. L. Schlapbach and A. Zuttel, Nature **414,** 353 (2001).
2. B. Bogdanovic and M. Schwickardi, J. Alloys Compd. **253,** 1 (1997).
3. M. Fichtner, O. Fuhr, O. Kircher, and J. Rothe, Nanotechnology **14,** 778 (2003).
4. B. Bogdanovic, M. Felderhoff, S. Kaskel, A. Pommerin, K. Schlichte, and F. Schuth, Advan. Mater. **15,** 1012 (2003).
5. K. J. Gross, G. J. Thomas, and C. M. Jensen, J. Alloys Compd. **330,** 683 (2002).
6. V. P. Balema, J. W. Wiench, K. W. Dennis, M. Pruski, and V. K. Pecharsky, J. Alloys Compd. **329,** 108 (2001).
7. E. H. Majzoub and K. J. Gross, J. of Alloys Compd. **356,** 363 (2003).
8. J. Graetz, J. J. Reilly, J. Johnson, A. Y. Ignatov, and T. A. Tyson, Appl. Phys. Lett. **85,** 500 (2004).
9. E. H. Majzoub, J. L. Herberg, R. Stumpf, S. Spangler, and R. S. Maxwell, J. Alloys Compd. **394,** 265 (2005).
10. J. H. Shim, G. J. Lee, and Y. W. Cho, J. Alloys Compd. **417,** 69 (2006).
11. A. Leon, O. Kircher, M. Fichtner, J. Rothe, and D. Schild, J. Phys. Chem. B **110,** 1192 (2006).
12. C. P. Balde, H. A. Stil, A. M. J. van der Eerden, K. P. de Jong, and J. H. Bitter, J. Phys. Chem. C **111,** 2797 (2007).
13. B. Bogdanovic, M. Felderhoff, M. Germann, M. Hartel, A. Pommerin, F. Schuth, C. Weidenthaler, and B. Zibrowius, J. Alloys Compd. **350,** 246 (2003).
14. P. Wang, X. D. Kang, and H. M. Cheng, J. Phys. Chem. B **109,** 20131 (2005).
15. K. J. Gross, E. H. Majzoub, and S. W. Spangler, J. Alloys Compd. **356,** 423 (2003).

16. V. P. Balema and L. Balema, Phys. Chem. Chem. Phys. **7,** 1310 (2005).
17. M. Fichtner, P. Canton, O. Kircher, and A. Leon, J. Alloys Compd. **404,** 732 (2005).
18. O. Palumbo, A. Paolone, R. Cantelli, C. M. Jensen, and R. Ayabe, Mater. Sci.& Eng. A: Structural Materials: Properties, Microstructure. and Processing 442**,** 75 (2006).
19. O. Palumbo, R. Cantelli, A. Paolone, C. M. Jensen, and S. S. Srinivasan, J. Phys. Chem. B **109,** 1168 (2005).
20. J. Iniguez, T. Yildirim, T. J. Udovic, M. Sulic, and C. M. Jensen, Phys. Rev. B **70,** 060101(R) (2004).
21. J. Iniguez and T. Yildirim, Appl. Phys. Lett. **86,** 103109 (2005).
22. O. M. Lovvik and S. M. Opalka, Phys. Rev. B **71,** 054103 (2005).
23. C. Moyses Araujo, R. Ahuja, P. Jena, and J. M. Osorio Guillen, Appl. Phys. Lett. **86,** 251913 (2005).
24. J. J. Liu and Q. F. Ge, Chem. Commun.**,** 1822 (2006).
25. C. Moyses Araujo, S. Li, R. Ahuja, and P. Jena, Phys. Rev. B **72,** 165101 (2005).
26. H. Gunaydin, K. N. Houk, and V. Ozolins, PNAS **105,** 3673 (2008).
27. J. P. Perdew, K. Burke, and M. Ernzerhof, Phys. Rev. Lett. **77,** 3865 (1996).
28. W. Kohn and L. J. Sham, Phys. Rev. **140,** A1133 (1965).
29. P. E. Blochl, Phys. Rev. B **50,** 17953 (1994).
30. G. Kresse and J. Furthmuller, Phys. Rev. B **54,** 11169 (1996).
31. X. Z. Ke and I. Tanaka, Phys. Rev. B **71,** 024117 (2005).

COMPARISON OF THE DEHYDROGENATION CHEMISTRY OF CARBORANE AND DECABORANE ON THE Pt(111) SURFACE

AASHANI TILLEKARATNE, MICHAEL TRENARY

Department of Chemistry, University of Illinois at Chicago, 845 W. Taylor Street, Chicago, IL 60607-7061, USA

The surface chemistry of carborane ($C_2B_{10}H_{12}$) and decaborane ($B_{10}H_{14}$) on Pt(111) has been studied with reflection absorption infrared spectroscopy (RAIRS), temperature programmed desorption (TPD), and X-ray photoelectron spectroscopy (XPS). It is found that the Pt surface catalyzes the release of hydrogen from both molecules at temperatures much lower than their thermal decomposition temperatures. The thermal degradation of these two molecules was found to occur in stages as shown by the TPD results. From XPS data, it was concluded that boron remains on the surface up to very high temperatures.

1. Introduction

Boranes and carboranes constitute a vast class of molecules with unique structures and properties and there is an extensive literature on the structure, bonding and reactivity of these molecules.[1-7] In addition, the interaction of boranes and carboranes with metal atoms is another area of research that has been widely explored. Although fascinating structures and interesting chemistry of a large number of metalloboranes and metallocarboranes are now known[1, 2, 8-13], the interaction of boranes and carboranes with metal surfaces has received little attention. However, understanding these surface interactions is important to subjects such as the use of carboranes[14-17] in the growth by chemical vapor deposition (CVD) of thin films of boron carbide and the use of catalysts to promote the release of hydrogen from boron-containing hydrogen storage materials.

Surface science studies of boranes and carboranes are rare, with the notable exception of the work of Dowben and coworkers.[14-17] Decaborane ($B_{10}H_{14}$) has also been used widely as a source for the deposition of boron on surfaces.[18-22] Avouris, *et al.*, found that at low temperatures, B adsorbs as an adatom on a T_4 site on a Si(111) surface when $B_{10}H_{14}$ was used as the boron precursor.[20] Dujardin, *et al.*, used STM to select a particular adsorbed decaborane molecule on the Si(111)-(7x7) surface, probe its electronic structure, dissociate the molecule by using the electrons from the STM tip, and examine the dissociation products.[22] However, with STM, they observed only the overall shape of the molecule; individual B-H bonds were not resolved.

The structures of decaborane and carborane are shown in Figure 1. Both molecules feature a three-dimensional cage involving delocalized boron-boron bonding with hydrogen atoms decorating the exterior of the cage. Whereas carborane has a closed icosahedral structure with only terminal B-H or C-H bonds, decaborane has an open structure with both terminal and bridging B-H bonds. In both cases, the cage structures resemble the structures found in elemental boron and in boron carbide.

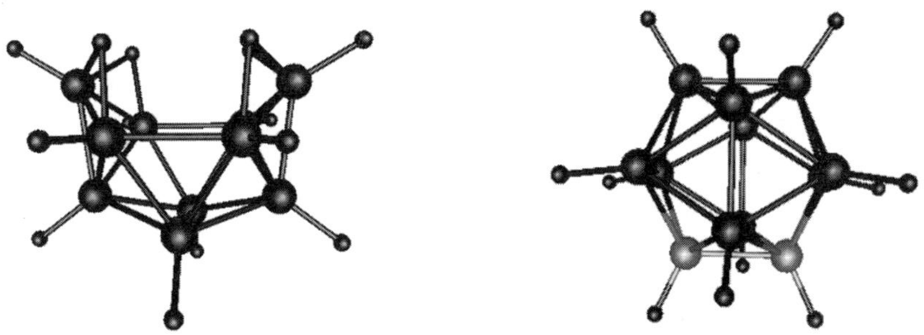

Figure 1. The structures of $B_{10}H_{14}$ (left) and $C_2B_{10}H_{12}$ (right).

2. Experimental

The experiments were performed in two different ultra high vacuum (UHV) chambers using two different Pt(111) single crystals. The X-ray photoelectron spectra were obtained in a chamber with a base pressure of ~ $1x10^{-10}$ Torr. The system has been described in detail elsewhere.[23] In brief, the UHV chamber is equipped with low energy electron diffraction (LEED), an X-ray photoelectron spectrometer (XPS), a quadrupole mass spectrometer (QMS) for temperature programmed desorption (TPD), and a Fourier transform infrared spectrometer (FTIR) for reflection absorption infrared spectroscopy (RAIRS). All RAIRS and TPD experiments were performed in a second chamber with a base pressure of ~2×10^{-10} Torr. The system has been described in detail elsewhere.[24] In brief, the UHV chamber is equipped for LEED, Auger electron spectroscopy (AES) and TPD experiments with a QMS. The chamber is coupled to a commercial FTIR spectrometer, a Bruker IFS 66v/S. To achieve maximum sensitivity, an

InSb detector was used with a tungsten source for the B-H stretch region, which contains the only peaks of appreciable intensity for both molecules. A resolution of 4 cm^{-1} was used. For the TPD results, signal from the QMS was recorded for each mass using a linear heating rate of 2 K/sec. The Pt(111) surfaces were cleaned and judged free of impurities by a standard procedure described earlier.[25] Before exposing to carborane ($C_2B_{10}H_{12}$) or to decaborane ($B_{10}H_{14}$), the crystal was flashed to ~1200 K and cooled down to 85 K. The carborane was purchased from Fisher Scientific and decaborane from Alfa Aesar with quoted purities of 99%.

3. Results

Figure 2 shows RAIR spectra in the B-H stretch region as a function of annealing temperature following 2.0 L exposures of $C_2B_{10}H_{12}$ and $B_{10}H_{14}$ to the Pt(111) surface at 85 K. The crystal was held at each temperature above 85 K for 30 seconds and then cooled back down to 85 K before acquiring a spectrum. All background spectra were also acquired at 85 K. In the case of $C_2B_{10}H_{12}$, a 200 K anneal does not change the peak positions or the total peak area, which indicates that the molecule remains undissociated up to this temperature. The first sign of dissociation of $C_2B_{10}H_{12}$ is seen at 250 K (not shown), where a new peak at 2499 cm^{-1} appears. This peak dominates the spectrum obtained after a 300 K anneal, and shifts to 2507 cm^{-1} at 350 K. Because the frequency of this new peak is significantly different from the B-H stretches of the parent carborane, it is assumed to belong to a stable surface intermediate containing B-H bond.

In contrast to this, significant changes in the B-H stretch region occur for $B_{10}H_{14}$ after annealing to only 200 K. Whereas the most intense peak at 85 K is at 2605 cm^{-1}, after the 200 K anneal the most intense B-H stretch is at 2551 cm^{-1}. This peak is still the most intense one up to 300 K, although the anneal at this temperature results in the appearance of another peak at 2563 cm^{-1}. At 350 K, the spectrum changes significantly with the dominant peak now at 2565 cm^{-1}. The RAIR spectra are featureless for annealing temperatures of 400 K and above for both $C_2B_{10}H_{12}$ and $B_{10}H_{14}$, suggesting that either all B-H bonds have been dissociated or that surface species have been formed with B-H stretch vibrations that are too weak to be observed.

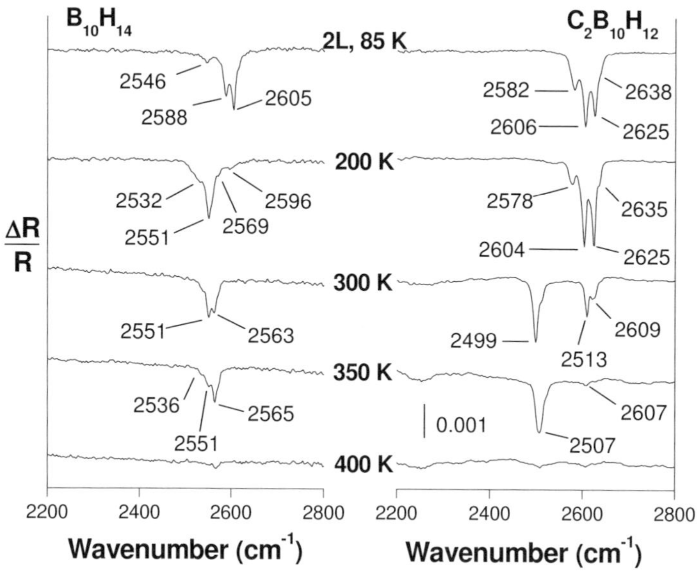

Figure 2. RAIR spectra following a 2 L exposure of $C_2B_{10}H_{12}$ and $B_{10}H_{14}$ to the Pt(111) surface at 85 K and annealing to the indicated temperatures.

The dehydrogenation chemistry of both $C_2B_{10}H_{12}$ and $B_{10}H_{14}$ was also studied with TPD. Figure 3 compares desorption of H_2 (m/e = 2) for a series of $C_2B_{10}H_{12}$ and $B_{10}H_{14}$ exposures. For the 0.5 L case, there is a large contribution from H_2 that adsorbed from the background. For $C_2B_{10}H_{12}$, the growth of the peak at ~ 300 K with increasing carborane exposure confirms that for the 1.0, 2.0, and 3.0 L cases dehydrogenation of carborane occurs at or below ~ 300 K. The 5.0 L exposure shows a more complex peak shape in the ~ 300-350 K range, with a main peak at 300 K and a resolved component at 346 K. For $B_{10}H_{14}$, the pattern is different in that a peak at ~400 K is clearly observed at an exposure of 1.0 L that shifts to ~335 K for 2.0, 3.0, and 5.0 L exposures. These desorption peaks are from the dissociation of B-H bonds of chemisorbed $B_{10}H_{14}$ molecules that are in contact with the platinum surface. The other major desorption peak at ~222 K is due to the desorption of molecular $B_{10}H_{14}$ from a multilayer. This was established by monitoring m/e = 124 (not shown), which corresponds to molecular decaborane ($B_{10}H_{14}$). This desorption occurs at ~224 K, giving rise to the fragment peak at ~222 K peak in the H_2 TPD results.

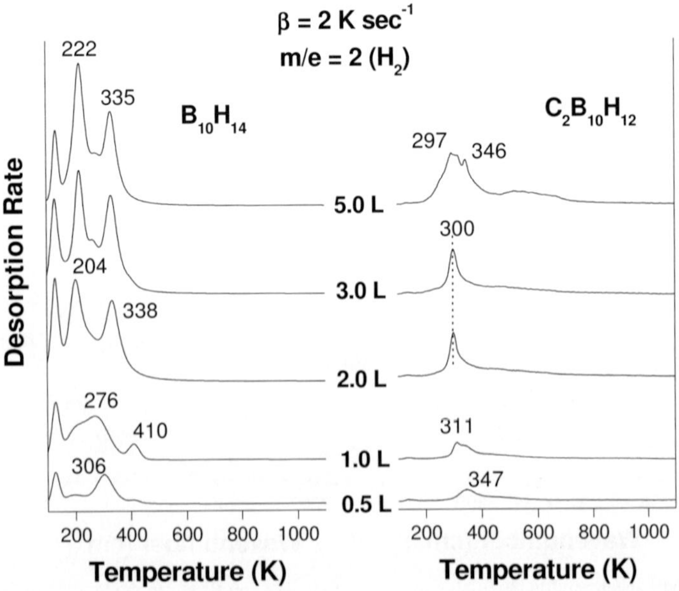

Figure 3. Temperature Programmed Desorption of $C_2B_{10}H_{12}$ and $B_{10}H_{14}$ to the Pt(111) surface at 85 K as a function of exposure.

Figure 4 compares the B 1s region in X-ray photoelectron spectra following 10.0 L exposures to $C_2B_{10}H_{12}$ and $B_{10}H_{14}$ at 85 K and after annealing to the indicated temperatures. The spectra reveal that boron remains on the surface after annealing to temperatures as high as 700 K, which is well above the point at which B-H stretch vibrations are no longer observed in the RAIR spectra. There is a steady and continuous decrease in B 1s peak intensity, indicating that boron is removed from the surface as the temperature is increased.

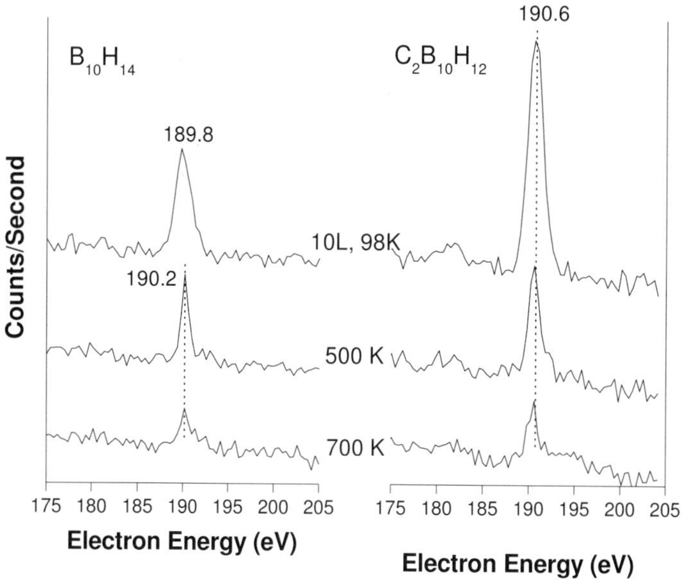

Figure 4. X-Ray photoelectron spectra of $C_2B_{10}H_{12}$ and $B_{10}H_{14}$ on Pt(111) as a function of annealing temperature following a 10.0 L exposure at 98 K.

4. Discussion

Carborane and decaborane show similar behavior in that both molecules adsorb molecularly on Pt(111) at 85 K, undergo some structural changes at low temperatures, and lose hydrogen in stages when the temperature is raised, as determined by both RAIRS and TPD. The RAIR spectra show remarkably sharp B-H stretch peaks. The fact that the individual B-H stretch peaks can be resolved here is in marked contrast to experimental IR spectra of solid carborane[26], which show only a broad feature about 100 cm^{-1} wide centered at ~ 2620 cm^{-1}. Both $C_2B_{10}H_{12}$ and $B_{10}H_{14}$ are assumed to be adsorbed molecularly on Pt(111) at 85 K, because the RAIR spectra for submonolayer and multilayer coverages do not differ that much from each other as far as the B-H stretch positions are concerned.

The behavior of the two molecules on Pt(111) is quite different after annealing to 200 K. $C_2B_{10}H_{12}$ seems to be stable after this anneal, whereas $B_{10}H_{14}$ undergoes dissociation producing a new feature at 2551 cm^{-1}. The first new dissociation product for $C_2B_{10}H_{12}$ is observed after an anneal to 300 K. In the case of $B_{10}H_{14}$, the new peak appearing at 2551 cm^{-1} may be due to the enhancement of an already existing peak for the parent molecule. It lies within the range of the B-H stretch vibrations of the parent molecule and may be enhanced by a change in the molecular orientation on the surface. However, the B-H stretch vibration of the carborane intermediate at 300 K, which appears at ~2500 cm^{-1}, is significantly different from those of the parent $C_2B_{10}H_{12}$ molecule and is clearly due to a new surface intermediate. In her extensive survey of boranes and carboranes,[26] Leites has shown that the B-H stretch vibrations of a series of closo-borane anions are centered below 2500 cm^{-1}. This is further supported by the results Brint and coworkers[27] obtained for borane anions of the type $B_nH_n^{2-}$. Comparison of our results with theirs suggests that the surface intermediate formed from $C_2B_{10}H_{12}$ is in the form of a $B_nH_n^{2-}$ anion.

The amount of hydrogen desorbed from $B_{10}H_{14}$ is higher than that from $C_2B_{10}H_{12}$ as indicated by TPD. However, XPS results indicate that the amount of boron on the surface following a given exposure to $C_2B_{10}H_{12}$ is higher than the amount of surface boron obtained following the same exposure of $B_{10}H_{14}$. These differences can be attributed to a difference in the sticking probabilities of the two molecules.

5. Conclusions

Carborane ($C_2B_{10}H_{12}$) and decaborane ($B_{10}H_{14}$) adsorb molecularly on the Pt(111) surface at 85 K. Both molecules undergo thermal dissociation as the surface is annealed to higher temperatures, as indicated by RAIR spectra and by TPD. These results show that the platinum surface catalyses the low temperature release of hydrogen from both molecules. In both cases, boron remains on the surface up to temperatures as high as 700 K.

Acknowledgments

This work is supported by the Department of Energy under grant DE-FG02-05ER15726.

References

1. F. A. Cotton, G. Wilkinson, *Advanced inorganic chemistry: a comprehensive text* (Wiley, New York, ed. 4th, 1980).
2. R. H. Crabtree, *The organometallic chemistry of the transition metals* (Wiley, New York 1988).
3. N. N. Greenwood, *Chemistry of the elements / N.N. Greenwood and A. Earnshaw* (Butterworth-Heinemann Oxford ; Boston ed. 2nd, 1997).

4. R. N. Grimes, *Carboranes*. P. M. Maitlis, F. G. A. Stone, R. West, Eds., Organometallic Chemistry (Academic Press, New York, 1970).
5. W. N. Lipscomb, *Boron hydrides* (W. A. Benjamin, New York, 1963).
6. E. L. Muetterties, *Polyhedral boranes* (M. Dekker, New York, 1968).
7. E. L. Muetterties, *Boron hydride chemistry* (Academic Press, New York, 1975).
8. W. J. Evans, G. B. Dunks, F. M. Hawthorne, *J. Am. Chem. Soc.* **95**, 4565 (1973).
9. D. F. Gaines, G. A. Steehler, *J. Chem. Soc., Chem. Commun.* **2**, 122 (1982).
10. N. N. Greenwood, *Chem. Soc. Rev.* **3**, 231 (1974).
11. N. N. Greenwood, *Pure & Appl. Chem.* **55**, 1415 (1983).
12. R. N. Grimes, *Metal interactions with boron clusters*. J. P. F. Jr., Ed., Modern inorganic chemistry (Plenum Press, New York, 1982).
13. C. G. Salentine, F. M. Hawthorne, *Inorg. Chem.* **15**, 2872 (1976).
14. D. Byun *et al.*, *Jpn. J. Appl. Phys.* **34**, L941 (1995).
15. A. N. Caruso *et al.*, *Appl. Phys. Lett.* **84**, 1302 (2004).
16. A. N. Caruso *et al.*, *Mater. Sci. Eng., B* **135**, 129 (2006).
17. P. Lunca-Popa *et al.*, *J. Phys. D: Appl. Phys.* **38**, 1248 (2005).
18. F. K. Perkins, R. A. Rosenberg, L. Sunwoo, P. A. Dowben, *J. Appl. Phys.* **69**, 4103 (1991).
19. S. P. Alex, K. L. William, E. R. William, 2002.
20. A. Ph, L. In-Whan, F. Bozso, E. Kaxiras, *J. Vac. Sci. Technol. A* **8**, 3405 (1990).
21. I. W. Lyo, E. Kaxiras, P. Avouris, *Phys. Rev. Lett.* **63**, 1261 (1989).
22. G. Dujardin, R. E. Walkup, P. H. Avouris, *Science* **255**, 1232 (March 6, 1992).
23. D. H. Kang, M. Trenary, *Surf. Sci.* **470**, L13 (2000).
24. M. E. Brubaker, M. Trenary, *J. Chem. Phys.* **85**, 6100 (1986).
25. D. Jentz, H. Celio, P. Mills, M. Trenary, *Surf. Sci.* **341**, 1 (1995).
26. L. A. Leites, *Chem. Rev.* **92**, 279 (1992).
27. P. Brint, B. Sangchakr, P. W. Fowler, V. J. Weldon, *J. Chem. Soc. Dalton Trans.*, 2253 (1989).

SINGLE- AND DOUBLE-CATIONS BOROHYDRIDES FOR HYDROGEN STORAGE APPLICATIONS

SHIN-ICHI ORIMO, YUKO NAKAMORI, HAI-WEN LI, MOTOAKI MATSUO, TOYOTO SATO

Institute for Materials Research, Tohoku University, Sendai 980-8577, Japan

NOBUKO OHBA, KAZUTOSHI MIWA, SHIN-ICHI TOWATA

Toyota Central R&D Labs., Nagakute, Aichi 480-1192, Japan

The thermal desorption temperature T_d of single-cation borohydrides was found to decrease with increasing the value of the Pauling electronegativity χ_P of the cation (metal). We examined T_d of double-cation borohydrides $ZrLi_{n-4}(BH_4)_n$, and then the correlation between T_d and χ_P determined for single-cation borohydrides is extended to double-cation ones. T_d for $ZrLi_{n-4}(BH_4)_n$ with the composition n from 4 to 6 continuously increases from 440 K to 650 K, and approaches that of $LiBH_4$, 740 K: T_d correlates with the averaged χ_P calculated from n. The extended correlation might lead to precise adjustments in the thermodynamical stabilities of borohydrides.

1. Introduction

Experimental and theoretical researches on complex hydrides are important in order to develop solid-state hydrogen storage materials with high gravimetric hydrogen densities [1,2]. Candidates for the materials are metal borohydrides such as $LiBH_4$, $Mg(BH_4)_2$, $Ca(BH_4)_2$, $Y(BH_4)_3$ [3-11]; including the possible intermediate compounds of $LiBH_4$ such as LiBH and $Li_2B_{12}H_{12}$ [12-15].

Recently, the thermodynamical stabilities of a series of single-cation borohydrides ($M(BH_4)_n$ with M = Li, Na, K, Cu, Mg, Zn, Sc, Zr, and Hf; n = 1-4) were systematically investigated by using both the first-principles studies and thermal desorption measurements [16]. The former indicated that the charge transfer from the cation M^{n+} to the complex anion $[BH_4]^-$ is a key feature for the stability of $M(BH_4)_n$ [17-18], and also that there exists a linear relationship between the calculated heat of formation ΔH of $M(BH_4)_n$ and the Pauling electronegativity χ_P of M. Experimentally, $M(BH_4)_n$ was synthesized [16, 19] by mechanical milling on the basis of the following reaction:

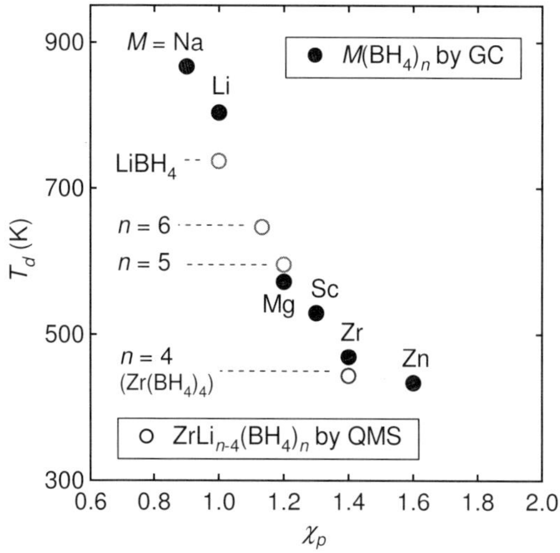

Figure 1. Thermal desorption temperature T_d as a function of the Pauling electronegativity χ_P [20]. Closed and open circles indicate the series of single-cation borohydrides $M(BH_4)_n$ (M = Li, Na, Mg, Zn, Sc, and Zr; n = 1-4) and double-cation ones $ZrLi_{n-4}(BH_4)_n$ (n = 4 ($Zr(BH_4)_4$), 5 and 6), examined by gas chromatography (GC) and quadrupole mass spectroscopy (QMS), respectively. There is a difference between T_d of $LiBH_4$/$ZrBH_4$ determined using GC (approximately 800 K) and that determined using QMS (approximately 740 K), owing to a longer distance between the detector and sample, and also to a lower gas flow rate, in GC.

$$MCl_n + nLiBH_4 \rightarrow M(BH_4)_n + nLiCl. \qquad (1)$$

The thermal desorption temperature T_d of $M(BH_4)_n$ determined using gas chromatography was also closely correlated with χ_P [20], as shown in Fig. 1. T_d (closed circles) decreases with an increase in the value of χ_P. Therefore, we conclude that the value of χ_P of the cation (metal) is an indicator that assists in the estimation of the thermodynamical stabilities of single-cation borohydrides $M(BH_4)_n$ with the corresponding value of T_d in various temperature ranges.

In this study, we examine the thermodynamical stabilities of double-cation borohydrides $MM'(BH_4)_n$, and then we verify whether the correlation between T_d and χ_P determined for single-cation borohydrides can be reasonably extended also to double-cation ones. (The possible "extended" correlation might lead to precise adjustments of the thermodynamical stabilities of borohydrides, which is considered to be difficult in case of single-cation borohydrides due to the discrete value of χ_P of each cation (metal).) For the abovementioned purpose, we propose Zr^{4+} (χ_P = 1.4, T_d of $Zr(BH_4)_4$ is approximately 440 K) and Li^+

(χ_P = 1.0, T_d of $LiBH_4$ is 740-800 K) to be a feasible combination of cations, and a nominal composition is $ZrLi_{n-4}(BH_4)_n$.

2. Experimental

The series of $ZrLi_{n-4}(BH_4)_n$ with n = 4 ($Zr(BH_4)_4$), 5, and 6 was synthesized by mechanical milling on the basis of the following reaction:

$$ZrCl_4 + nLiBH_4 \rightarrow ZrLi_{n-4}(BH_4)_n + 4LiCl. \quad (2)$$

The starting materials $ZrCl_4$ and $LiBH_4$ were purchased from Aldrich Co. Ltd. They were premixed manually using an agate mortar and pestle, and then mechanically milled by planetary ball milling with 20 steel balls in a hardened steel vial for 5 h under 0.1 MPa argon. The milling process was paused every 15 min to avoid an increase in the temperature of the sample. The samples prepared were subsequently examined by powder X-ray diffraction measurement (Cu-$K\alpha$), and laser Raman spectroscopy, and quadrupole mass spectroscopy (helium flow of 150 ml/min and heating at 5 K/min).

3. Results and discussion

In the X-ray diffraction profiles of $ZrLi_{n-4}(BH_4)_n$, that is, the mechanically milled $ZrCl_4$ + $nLiBH_4$ with n = 4, 5, and 6, no diffraction peaks of $ZrCl_4$ and $LiBH_4$ are observed in the milled samples, indicating the progress of the thermodynamically favorable reaction, Eq. (2). LiCl is observed as a by-product in all the diffraction profiles, no peaks of $ZrLi_{n-4}(BH_4)_n$ are detected. This is probably due to a lack of any long range ordering of the structure in $ZrLi_{n-4}(BH_4)_n$ synthesized by mechanical milling. The lack of the long range ordering was also reported in the other borohydrides.

The Raman spectra were examined to obtain the information on B-H bonding of $ZrLi_{n-4}(BH_4)_n$. Both the B-H bending and stretching modes around 1300 cm^{-1} and 2300 cm^{-1}, respectively, are detected in $LiBH_4$ as a reference. The sample with n = 4 ($Zr(BH_4)_4$) provides characteristic stretching modes in the range of 2150–2580 cm^{-1}, while the Raman shifts of n = 5 and 6 are similar to that of $LiBH_4$ mentioned above. Thus, so far, we have been unable to distinguish the local atomistic structures among $ZrLi_{n-4}(BH_4)_n$ with n = 5 and 6 and $LiBH_4$.

However, it is noteworthy that the thermal desorption reactions shown in Fig. 2 do not indicate the evident disproportionation of $ZrLi_{n-4}(BH_4)_n$ into the

$Zr(BH_4)_4$- and $LiBH_4$-based phases upon heating. Thus, the series of $ZrLi_{n-4}(BH_4)_n$ is regarded to be appropriate for experimentally clarifying the existence of the extended correlation.

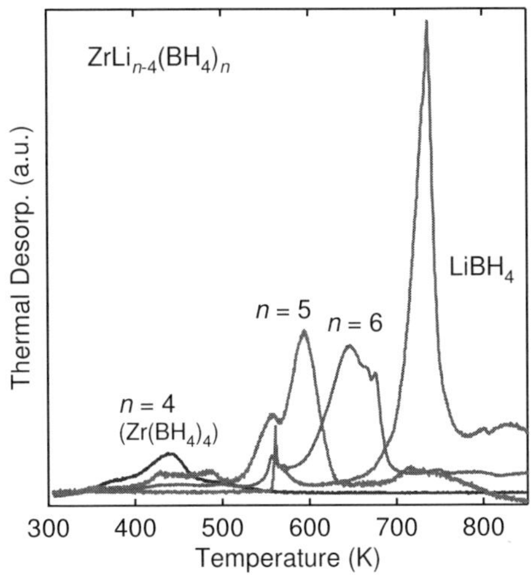

Figure 2. Thermal desorption reactions of $ZrLi_{n-4}(BH_4)_n$ and $LiBH_4$ for reference, examined by quadrupole mass spectroscopy (QMS) [20]. All the reactions originate only from $ZrLi_{n-4}(BH_4)_n$, because coexisting LiCl decomposes at temperature higher than 878 K. The desorption temperature T_d, defined as the peak temperature of the main desorption reaction, is summarized in Fig. 1 with open circles.

The value of T_d —defined as the peak temperature in the main desorption reaction shown in Fig. 2— are 440 K ($n = 4$), 595 K ($n = 5$), and 650 K ($n = 6$); and it continuously approaches toward 740 K which is T_d for $LiBH_4$. We found that T_d is uncorrelated with the smaller ($\chi_P = 1.0$) and larger ($\chi_P = 1.4$) values of χ_P; however, it is clearly correlated with the averaged value of χ_P of the cations (metals), as is shown also in Fig. 1 (open circles). Here, the averaged value of χ_P depends on n and it is simply calculated [20]. The extended correlation of T_d of double-cation borohydrides with the averaged value of χ_P of the cations (metals) suggests that the thermodynamical stabilities of borohydrides might be precisely adjusted by combinations of appropriate cations.

4. Conclusions

The thermal desorption temperature T_d of $ZrLi_{n-4}(BH_4)_n$ increases from 440 K to 650 K as the value of composition n increases, and continuously approaches toward 740 K —T_d of $LiBH_4$. The correlation between T_d and the Pauling electronegativity χ_P determined for single-cation borohydrides can be reasonably extended to double-cation ones; T_d correlates with the averaged value of χ_P calculated from n. The extended correlation is important to precisely adjust the thermodynamical stabilities of borohydrides employed as candidates of solid-state hydrogen storage materials with high gravimetric hydrogen densities. Further studies of double(multi)-cation borohydrides, focusing on the syntheses of well-crystallized single phases, the local atomistic/electronic structures, and the dehydring/rehydriding processes, are in progress.

Acknowledgements

This study was partially supported by NEDO, JSPS, MEXT.

References

1. S. Orimo, Y. Nakamori, J.R. Eliseo, A. Züttel and C.M. Jensen, *Chem. Rev.* **107**, 4111 (2007).
2. A. Züttel, A. Borgschulte and S. Orimo, *Scripta Mater.* **56**, 823 (2007).
3. A. Züttel, S. Rentsch, P. Fisher, P. Wenger, P. Sudan, Ph. Mauron and Ch. Emmenegger, *J. Alloys Compd.* **356-357**, 515 (2003).
4. Y. Nakamori and S. Orimo, *J. Alloys Compd.* **370**, 271 (2004).
5. R.A. Kumar and A.L. Cornelius, *Appl. Phys. Lett.*, **87**, 261916 (2005).
6. K. Miwa, M. Aoki, T. Noritake, N. Ohba, Y. Nakamori, S. Towata, A. Züttel, and S. Orimo, *Phys. Rev. B* **74**, 155122 (2006).
7. Y. Nakamori, S. Orimo and T. Tsutaoka, *Appl. Phys. Lett.* **88**, 112104 (2006).
8. M. Matsuo, Y. Nakamori, K. Yamada and S. Orimo, *Appl. Phys. Lett.* **90**, 232907 (2007).
9. H.-W. Li, K. Kikuchi, Y. Nakamori, K. Miwa, S. Towata and S. Orimo, *Scripta Mater.* **57**, 679 (2007).
10. T. Matsunaga, F. Buchter, P. Mauron, M. Bielman, Y. Nakamori, S. Orimo, N. Ohba, K. Miwa, S. Towata, K. Miwa and A. Züttel, *J. Alloys Compd.*, in press.
11. T. Sato, K. Miwa, Y. Nakamori, K. Ohoyama, H-W Li, T. Noritake, M. Aoki, S. Towata and S. Orimo, *Phys. Rev. B*, communicated.

12. J.K. Kang, S.Y. Kim, Y.S. Han, R.P. Muller and W.A. Goddard III, *Appl. Phys. Lett.* **87**, 111904 (2005).
13. S. Orimo, Y. Nakamori, N. Ohba, K. Miwa, M. Aoki, S. Towata and A. Züttel, *Appl. Phys. Lett.* **87**, 021920 (2006).
14. N. Ohba, K. Miwa, M. Aoki, T. Noritake, S. Towata, Y. Nakamori, S. Orimo and A. Züttel, *Phys. Rev. B* **74**, 075110 (2006).
15. H.-W. Li, K. Kikuchi, Y. Nakamori, K. Miwa, S. Towata and S. Orimo, submitted.
16. Y. Nakamori, K. Miwa, A. Ninomiya, H.-W. Li, N. Ohba, S. Towata, A. Züttel and S. Orimo, *Phys. Rev. B* **74**, 045126 (2006).
17. K. Miwa, N. Ohba, S. Towata, Y. Nakamori and S. Orimo, *Phys. Rev. B* **69**, 245120 (2004).
18. K. Miwa, N. Ohba, S. Towata, Y. Nakamori and S. Orimo, *J. Alloys Compd.* **404-406**, 140 (2005).
19. Y. Nakamori, H.-W. Li, K. Miwa, S. Towata, and S. Orimo, *Mater. Trans.* **47**, 1898 (2006).
20. H.-W. Li, S. Orimo, Y. Nakamori, K. Miwa, N. Ohba, S. Towata and A. Züttel, *J. Alloys Compd.* **446–447**, 315 (2007).

LOW TEMPERATURE TRANSMISSION IR SPECTRA OF SODIUM AND LITHIUM BOROHYDRIDE

PANCHATAPA JASH, MICHAEL TRENARY

Department of Chemistry, University Illinois at Chicago, 845 W Taylor Street Chicago, IL 60607, USA

Metal borohydrides of the general formula $M(BH_4)_x$ are attractive materials for use in the storage of hydrogen. In order to realize the potential of these compounds for such applications, more information is needed on the temperature-dependent hydrogen-loss mechanism. We have constructed a new apparatus that permits infrared spectra to be obtained on hydrogen storage materials over a wide range of temperatures and we demonstrate the capabilities of this method with spectra of $LiBH_4$ and $NaBH_4$ at room temperature and at 148 K. These compounds yield similar IR spectra and for $NaBH_4$ strong IR absorption peaks are observed at room temperature in the B-H stretch region at 2224, 2295, and 2385 cm^{-1} and in the BH^-_4 deformation region at 1127 cm^{-1}. Cooling the sample leads to a sharpening of all of the peaks, which allows new features to be resolved that were not observable at room temperature.

1. Introduction

The tetrahydroborate (BH_4^-) anion is the simplest of the many borane anions that are known. The BH_4^- ligand interacts more or less covalently with metal ions through bridging H atoms. This results in three-center two-electron M-H-B bonds. The metal tetrahydroborates are stable salts ($LiBH_4$ has a melting point of 541 K) that undergo thermal decomposition only at elevated temperatures, sometimes above their melting points. They are potential hydrogen storage materials because of their high gravimetric and volumetric hydrogen densities.[1] For example, $LiBH_4$ is over 18% hydrogen by weight and has almost twice the volumetric hydrogen density of liquid H_2. Consequently, its hydrogen storage properties have been extensively investigated. Ge and other theoreticians have carried out detailed DFT calculations of the bulk structure, the structure of different surfaces of $LiBH_4$, and of the energetic and structural changes that accompany the creation of hydrogen vacancies at the surfaces[2-5]. From thermal desorption experiments, Zuttel, *et al.*[6] established that the major H_2 release of 9 weight % occurs for pure $LiBH_4$ at temperatures above 673 K but that the addition of a SiO_2 catalyst lowers the release by 200 K.

In 1940, Schlesinger and Brown synthesized the complex borohydride $LiBH_4$ for the first time by a reaction between LiH and diborane in an ether medium.[7] Since then its properties have been extensively studied with various methods.[8-11] However, almost nothing is known about the relationship between the changes in structure of either $NaBH_4$ or $LiBH_4$ as a function of temperature and their vibrational spectra. Therefore, we have developed a new apparatus that permits us to acquire infrared spectra of hydrogen storage materials such $LiBH_4$ and $NaBH_4$ over a wide range of temperatures from 150 to over 1000 K.

In isolation, the BH^-_4 ion is tetrahedral, and consequently only two fundamentals, the asymmetric BH stretch (ν_3) and asymmetric BH_4 deformation (ν_4) are IR active for the isolated ion, whereas all four fundamentals are Raman active. The Raman active fundamentals were characterized in liquid ammonia solutions[12], whereas IR spectra of thin films of $NaBH_4$ on alkali halide crystals[13] or diluted in an alkali halide host crystal have been reported[14]. Raman and IR studies focused specifically on $NaBH_4$ and $LiBH_4$ have also been reported.[15, 16] The vibrational modes in borohydrides are of three distinct types: librational (below 1000 cm^{-1}), B-H bending (1127 cm^{-1}) and B-H stretching (2200-2400 cm^{-1}). The overtone of the deformation mode ($2\nu_4$) occurs around 2228 cm^{-1}.

There are two main purposes for this study: First, earlier IR studies of $NaBH_4$ by Schutte[13] in 1960 and by Harvey and McQuaker[15] in 1971 differ in several details. Harvey and McQuaker noted that some decomposition occurs and that a hydrated form of $NaBH_4$ develops upon atmospheric exposure. In light of these earlier differences, we have sought to obtain our IR spectra while minimizing exposure to the atmosphere. Immediately after preparation, the samples were transferred into an evacuable cell and the spectra were acquired under vacuum. Second, a thorough understanding of the IR spectra of these compounds is needed in order to lay the groundwork for later studies of their decomposition at higher temperatures.

2. Experimental

A new apparatus for transmission IR measurement has been built based on the design of Basu and Yates[17]. The IR cell consists of a stainless steel sphere with six 2.75 inch conflate flange ports. The IR beam enters and exits the cell through two differentially pumped KBr windows. The cell is pumped with a roughing pump to maintain a base pressure $\leq 2.0 - 5.0 \times 10^{-2}$ Torr. The front part of the IR cell is connected to a six-way cross with ports for gas dosing, pumping, and pressure measurements. Gases are introduced into the cell with a leak valve. The pressure in the cell is measured by a convectron gauge in the range of 5.0 – 1.0 × 10^{-3} Torr and in the pumping port by a thermocouple gauge from atmospheric pressure to 1 mTorr. The sample is pressed into a tungsten grid stretched between two Ni clamps, which are attached to 3 mm diameter Cu rods. The Cu rods are in good thermal contact with a liquid N_2 reservoir but are electrically isolated from it by BeO insulators. The grid can be heated resistively to ~1000 K and cooled to ~103 K. The sample temperature is measured by a type K (chromel-alumel) thermocouple spot-welded to the top of the grid. The sample can be moved vertically by up to 2 inches within the cell with a linear translator. The IR spectra were obtained with a Mattson (Cygnus 25) FTIR spectrometer. The samples were prepared following the procedures used by Ballinger, *et al.*[18, 19] Infrared grade KBr powder was obtained from Fluka and $LiBH_4$ (95%) and $NaBH_4$ (98%) were obtained from Sigma Aldrich and were used without further purification. Borohydrides were ground in a motor and pestle together with KBr powder and were pressed into a tungsten grid (0.002″ thickness) using a hydraulic press with an applied pressure of about 12,000 lb/square inch.

3. Results

3.1. *IR Spectra of $LiBH_4$ and $NaBH_4$ at Room Temperature*

Room temperature spectra of $LiBH_4$ and $NaBH_4$ are compared in Figure 1. The $NaBH_4$ spectra were collected from 800 to 3600 cm^{-1} whereas a wider range of 400 to 3600 cm^{-1} was used for the $LiBH_4$ spectra. Consequently, features due to the lower frequency vibrational modes of $NaBH_4$ were not recorded. Both spectra feature a single large peak due to the fundamental of the ν_4 deformation mode, and three peaks in the BH stretching region. The latter peaks are assigned to the ν_3 fundamental, the $2\nu_4$ overtone, and the $\nu_2+\nu_4$ combination band. These assignments follow those of the previous studies of these two compounds and are summarized in Table 1.

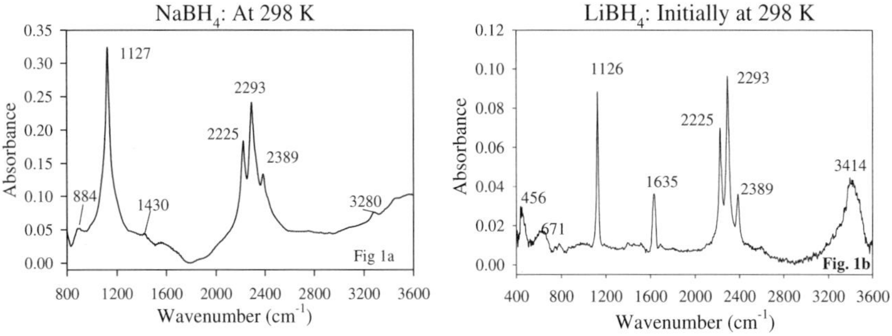

Figure 1. Transmission IR spectra at the room temperature

Table 1. Comparison of BH_4 stretching and bending frequencies in $NaBH_4$ and $LiBH_4$ at 298 K and at the low temperature

	ν_4		ν_3		$2\nu_4$		$\nu_2+\nu_4$	
	$NaBH_4$	$LiBH_4$	$NaBH_4$	$LiBH_4$	$NaBH_4$	$LiBH_4$	$NaBH_4$	$LiBH_4$
Room Temperature	1130	1127	2291	2295	2224	2227	2385	2388
At the low Temperature	1130	1127 (1137 sh)	2295 (2307 sh)	2294 (2307 sh)	2228 (2241 sh)	2224 (2241 sh)	2390 (2426 sh)	2389

Several other features are also apparent in the spectra. The peak at 884 cm^{-1} for $NaBH_4$ was assigned to an external lattice vibration by Harvey and McQuaker[16] but Maiti[20] notes that a peak at 880 cm^{-1} could be due to CO_3^{2-} impurities. The peak at 884 cm^{-1} remains unchanged after cooling to 148 K. Following Price's assignments[21], the peak at 3280 cm^{-1} in the $NaBH_4$ spectrum is attributed to the $\nu_3+\nu_4$ combination band. Water is apparent in the $LiBH_4$ sample and gives rise to the peaks at 3414 and 1635 cm^{-1}. The peaks at 671 and 456 cm^{-1} in the $LiBH_4$ spectra are assigned to impurities. Both the water peaks and the impurity peaks disappear from the $LiBH_4$ spectra after heating the sample to 413 K (Figure 4), which also results in the appearance of weak bands at 3107 and 3456 cm^{-1}. The fact that water is present in the $LiBH_4$ sample but not in the $NaBH_4$ sample reflects the fact that the former salt is more hygroscopic than the latter.

3.2. *IR Spectra of $LiBH_4$ and $NaBH_4$ at Low Temperature*

It is interesting to compare spectra of $LiBH_4$ and $NaBH_4$ at low temperature as the latter undergoes a phase transition from a face centered cubic structure to a tetragonal structure as it is cooled below ~ 190 K, whereas no phase transition for $LiBH_4$ occurs below room temperature. However, $LiBH_4$ changes from an orthorhombic to a hexagonal structure when it is heated above 411 K. For both compounds the IR peaks are much sharper at low temperature, which permits, in some cases, resolution of separate peaks due to the presence of ^{11}B and ^{10}B isotopes in the natural abundance ratio of 4:1. The satellite peaks due to $^{10}BH_4$ occur 10-20 cm^{-1} higher than the more intense peaks due to $^{11}BH_4$. The peak widths (full width half maxima (FWHM)) of the ν_3 and ν_4 fundamentals of $NaBH_4$ and $LiBH_4$ at room and at low temperature are given in Table 2. Although the widths are less at the lower temperature for both compounds, the widths for $NaBH_4$ are generally about twice those of $LiBH_4$ regardless of temperature. This difference is presumably due to the different crystal structure of the two compounds.

Table 2. Comparison of FWHM in $NaBH_4$ and $LiBH_4$ at 298 K and at the low temperatures

FWHM	ν_4		ν_3	
	$NaBH_4$	$LiBH_4$	$NaBH_4$	$LiBH_4$
Room Temperature	58	18	40	25
At the low Temperature	34	14	35	14

The peak positions found here are compared with literature values in Table 3. Although we observe most of the major peaks reported previously, we also see quite a few additional peaks. For example, in none of the previous studies of $NaBH_4$ reported the peaks at 1290, 2426 and 2630 cm^{-1} seen here in Figures 2 and 3, which are associated with the transition to the tetragonal structure. In addition, the increased sharpness of the peaks in the BH stretch region permits resolution of distinct new peaks. The weak band, which is observed at 3280 cm^{-1} in the room temperature spectrum of $NaBH_4$ shifts to 3295 cm^{-1} at the lower temperature, presumably due to the transition to the tetragonal structure.

Table 3. Comparison of $NaBH_4$ spectra with previous works (sh shoulder, v very, w week, m medium, s strong)

Assignment	Harvey	Schutte	This work
ν_4 (E)	1122	1123	1127
$^{10}\nu_4$ (E)	1134	1135 (vw)	1137
$3\nu_L$	1148 (sh)	--	1147
ν_4 (B_2)	1153	1152 (vw)	--
	--	2197	--
	--	2223 s	2224 s
$2\nu_4$ (E)	2236	2238	2239 w
$2^{10}\nu_4$ (E)	2256	2256 m	--
ν_3 (B_2)	--	2287 vs	2295 s
	2272 (sh)		
ν_3 (E)	2303	2305	2307 w
$^{10}\nu_3$ (E)	2334 (sh)	--	2329
$\nu_2 + \nu_4$ (E)	2404	2404 w	2389

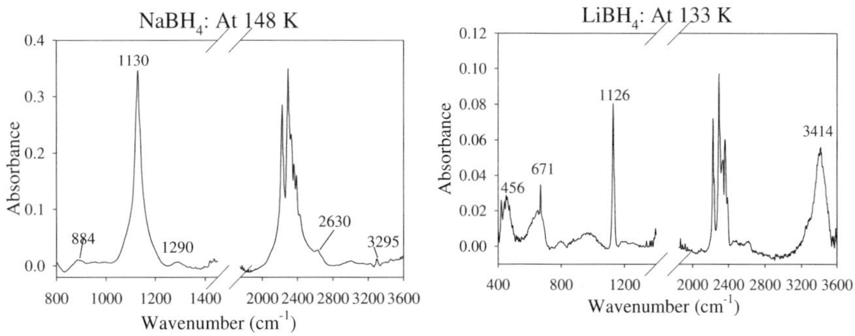

Figure 2. Transmission IR spectra at the low temperature

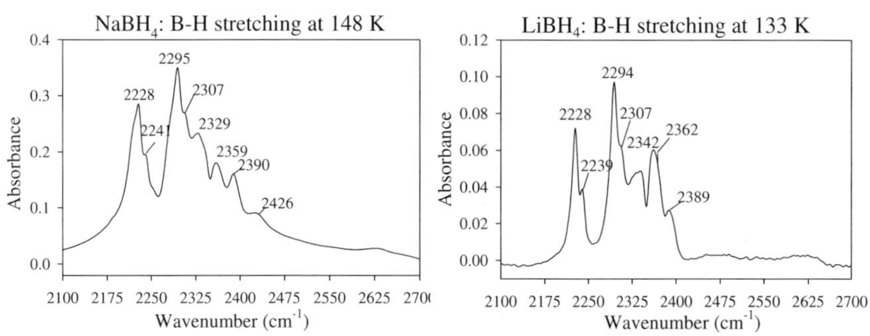

Figure 3. B-H stretching region of at the low temperature

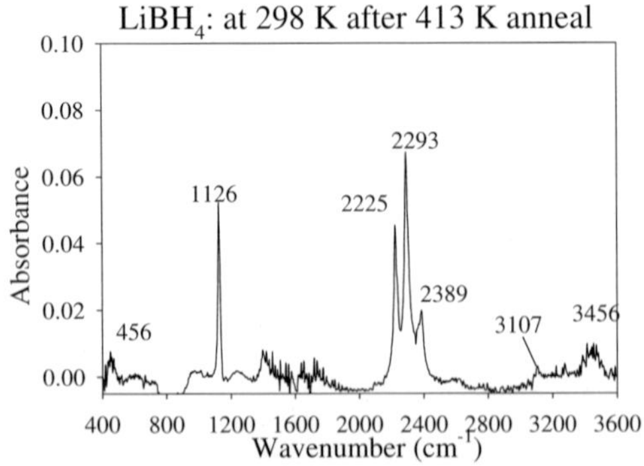

Figure 4. IR spectra of $LiBH_4$ at 298 K

4. Conclusions

The capabilities of a newly constructed apparatus for transmission IR studies of hydrogen storage materials has been demonstrated with spectra of $LiBH_4$ and $NaBH_4$ obtained at both room temperature and at low temperature. The narrower line widths in the low temperature spectra permit many additional features to be resolved that are not observable at room temperature. The ability to obtain spectra over a wide temperature range should prove especially useful in identifying intermediates formed during the process of dehydrogenation of potential hydrogen storage materials such as $LiBH_4$ and $NaBH_4$.

Acknowledgment

This work is supported by a grant from the Department of Energy (DE-FG02-05ER15726).

References

1. S. i. Orimo, Y. Nakamori, J. R. Eliseo, A. Zuttel, C. M. Jensen, *Chem. Rev.* **107**, 4111 (2007).
2. Q. Ge, *J. Phys. Chem. A* **108**, 8682 (2004).
3. Y. Nakamori *et al.*, *Phys. Rev. B* **74**, 045126 (2006).
4. N. Ohba *et al.*, *Phys. Rev. B* **74**, 075110 (2006).
5. T. J. Frankcombe, G.-J. Kroes, A. Zuttel, *Chem. Phys. Lett.* **405**, 73 (2005).
6. A. Zuttel *et al.*, *J. Alloys Compd.* **356-357**, 515 (2003).
7. H. I. Schlesinger, H. C. Brown, *J. Am. Chem. Soc.* **62**, 3429 (1940).

8. E. M. Fedneva, V. L. Alpatova, V. I. Mikheeva, *Russ. J. Inorg. Chem.* **9**, 826 (1964).
9. N. C. Hallett, H. L. Johnston, *J. Am. Chem. Soc.* **75**, 1496 (1953).
10. T. Tsang, T. C. Farrar, *J. Chem. Phys.* **50**, 3498 (1969).
11. J. P. Soulie, G. Renaudin, R. Cerny, K. Yvon, *J. Alloys Compd.* **346**, 200 (2002).
12. A. R. Emery, R. C. Taylor, *J. Chem. Phys.* **28**, 1029 (1958).
13. C. J. H. Schutte, *Spectrochim. Acta* **16**, 1054 (1960).
14. J. A. A. Ketelaar, C. J. H. Schutte, *Spectrochim. Acta* **17**, 1240 (1961).
15. K. B. Harvey, N. R. McQuaker, *Can. J. Chem.* **49**, 3272 (1971).
16. K. B. Harvey, N. R. McQuaker, *Can. J. Chem.* **49**, 3282 (1971).
17. P. Basu, T. H. Ballinger, J. T. Yates, Jr., *Rev. Sci. Instrum.* **59**, 1321 (1988).
18. T. H. Ballinger, J. C. S. Wong, J. T. Yates, Jr., *Langmuir* **8**, 1676 (1992).
19. T. H. Ballinger, J. T. Yates, Jr., *Langmuir* **7**, 3041 (1991).
20. G. C. Maiti, M. Baerns, *Thermochim. Acta* **261**, 69 (1995).
21. W. C. Price, *J. Chem. Phys.* **17**, 1044 (1949).

SYNTHESIS AND MODIFICATION OF LIGHT METAL AND COMPLEX HYDRIDES BY HIGH-ENERGY BALL MILLING

I. LLAMAS-JANSA, C. RONGEAT, S. DOPPIU, AND O. GUTFLEISCH

IFW Dresden, Institute for Metallic Materials, P.O. Box 270016, D-01171 Dresden, Germany

High-energy ball milling was used for the synthesis and modification of a variety of light metal hydrides (MgH_2) and complex hydrides ($LiAlH_4$, $NaAlH_4$, and $LiBH_4$) by changing the milling conditions and the added catalyst (Ti, $TiCl_3$, $ScCl_3$, and $MgCl_2$). The combination of the milling technique with a specially designed vial allowed the in-situ monitoring of the reaction pressure and vial temperature variations taking place during the milling process. This provided immediate and valuable information about the efficiency of the reactions before removing the products from the vial. The dependence of the powder composition with the milling conditions and the catalyst was carried out by X-ray diffraction and Raman spectroscopy at room temperature. The use of different catalysts and milling pressures was found to have a profound effect on the reaction efficiencies during the synthesis of doped $NaAlH_4$ from a mixture of NaH + Al + (4% mol) catalyst. Different milling pressures were also found to change the reactivity of the Mg99Ni1 alloy during the synthesis of MgH_2. The use of catalysts was observed to affect the milling of $LiAlH_4$ and $LiAlH_4$ + $TiCl_3$ and the synthesis of Mg $(BH_4)_2$ from a mixture of $LiBH_4$ and $MgCl_2$.

1. Introduction

The synthesis and modification of light metal and complex hydrides by high-energy ball milling has gained great interest in the hydrogen storage community [1–6] because it improves hydrogen absorption by decreasing the crystallite size and increasing the number of structural defects and chemical disorder [7–10]. The process also provides an effective mixing of the precursors in the nanoscale facilitating chemical reactions to occur.

On the other hand, the use of high-pressure reactive atmospheres at room temperature has been shown to improve the reactions occurring during milling in such a way that sintering is not anymore necessary for the synthesis of complex hydrides such as Mg_2FeH_6 [11]. Recently, pressure effects during the high-pressure ball milling of MgNi to obtain MgH_2 were studied by Doppiu et al. [12]. These authors also showed that the combination of the milling technique with a specially designed vial allows the in-situ monitoring of the gas pressure and vial temperature variations taking place during the synthesis process. With this combined system, information about the efficiency of the reactions can be obtained before removing the powder from the vial. Similarly, the effect of the milling parameters on the synthesis of $NaAlH_4$ and the decomposition of $LiAlH_4$ and $LiBH_4$ were studied by Rongeat et al. [13] and Llamas et al. [14] under different conditions.

In this paper, a summary of these previous results on the synthesis and modification of metal and complex hydrides by ball milling is presented. More information about the applied methods and the properties of the materials can be found in the original publications.

2. Experimental

The synthesis and modification of nanocrystalline hydrides was carried out by high-energy ball milling at room temperature and under different pressures and atmospheres (from 1 bar Ar up to 150 bar H_2). In this manner, crystallite sizes lower than 100 nm [12] were achieved after ball milling for periods between 1 and 16 h. The time length was determined by the synthesis or decomposition of the corresponding hydride. The milling process was further improved by adding Ti and chlorides such as $TiCl_3$, $ScCl_3$, or $MgCl_2$ as catalysts. Milling was performed in specially constructed stainless steel vials working at pressures between 1 and 150 bar (produced by Evico Magnetics). In-situ monitoring of the vial temperature and gas pressure variations during milling was achieved by a designed gas-temperature system consisting of different sensors, a transmitter, and an external receiver connected to the computer. The powders were handled in argon atmosphere inside dedicated glove boxes (oxygen and water content less than 1 ppm). The initial characterisation of the as-synthesised samples was performed by X-ray diffraction (XRD) analysis and Raman spectroscopy at room temperature.

3. Results

3.1. *$LiAlH_4$*

The variation of the hydrogen pressure (initially 80 bar H_2) and the vial temperature during the milling of pure $LiAlH_4$ and $LiAlH_4$ doped with $TiCl_3$ was monitored and plotted (Figure. 1). The curves showed the effect of the dopant on the reactions taken place during the milling process. While no variations in pressure were detected in the case of the undoped sample, in the case of the doped system, the decomposition of the hydride was shown through an increase of the pressure due to hydrogen gas release [14]. The increase of pressure observed in the curves within the first three hours of milling was related to the thermal gas expansion due to the mechanochemical process and not to hydrogen release.

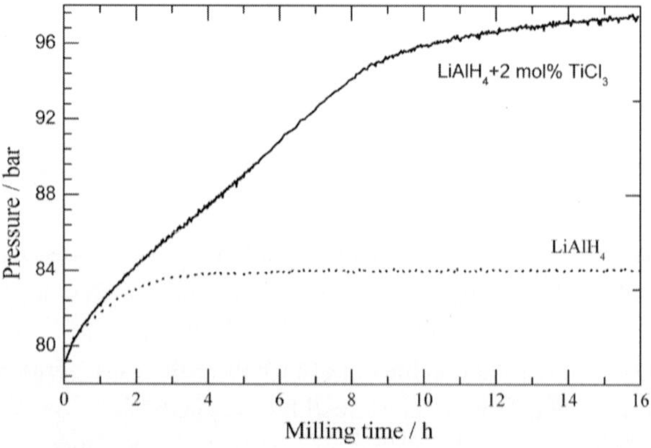

Figure 1: Monitored hydrogen pressure and vial temperature during the milling of $LiAlH_4$. The curves show an increase of the hydrogen pressure in the case of the doped sample (red line), whereas no changes are observed in the case of the undoped sample (black line) [data from: [14]].

3.2. *$NaAlH_4$*

A series of experiments on the synthesis of $NaAlH_4$ from NaH + Al + (4% mol) catalyst mixtures were performed using different high-pressure milling conditions under reactive atmospheres [13]. In particular, the efficiency of Ti, $TiCl_3$, and $ScCl_3$ as catalysts of the synthesis reaction was analyzed by monitoring vial temperature and hydrogen pressure variations during the milling process (Figure. 2). The comparison of the different traces showed that $TiCl_3$ is the most efficient catalyst followed by $ScCl_3$. In contrast, Ti appeared to lead to similar results than those observed for the sample prepared without catalyst. The inert material (in this case Al), showed a non-reactive behavior, with an increase of the temperature and of the hydrogen pressure beyond the 5 h of the experiment. The small bumps observed in the temperature curve corresponding to the $TiCl_3$-doped sample were related to slight thermal variations due to the synthesis of the Na_3AlH_6 and $NaAlH_4$ phases [13]. The comparison between the corresponding XRD patterns confirmed the results obtained by the monitoring of the reactions during milling.

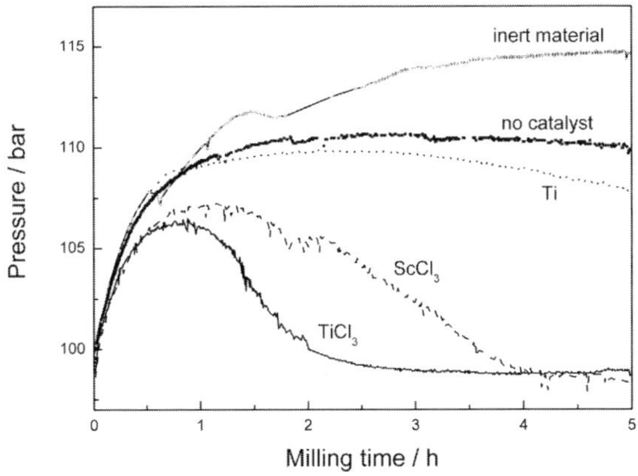

Figure 2: Monitored hydrogen pressure during the milling of a NaH + Al mixture with different catalysts. Curves corresponding to a sample milled without catalyst and an inert material (Al powder) are included for comparison [data from: [13]].

In this case, $TiCl_3$ and $ScCl_3$ led to clear $NaAlH_4$ features, with small amounts of NaCl and Al, while Ti led to strong NaH and Al features and to the formation of the intermediate phase $NaAlH_6$ indicating the incomplete reaction between the precursors [13].

3.3. *MgH_2/Mg99Ni1*

Different milling pressures (10, 40 and 90 bar H_2) were also found to change the reactivity of the Mg99Ni1 alloy during the synthesis of MgH_2 [12]. In particular, the reaction was found to be incomplete in the case of a 10 bar atmosphere, whereas higher milling pressures led to the complete transformation of Mg into the hydride. The results showed an increase of the incubation period for the samples synthesized at higher pressures. The XRD patterns of the same samples showed the presence of the Mg phase for the samples milled under 10 bar indicating that, in this case, the hydrogenation reaction was still incomplete. For the samples milled at higher pressures, no significant differences in the X-ray diffraction patterns were observed. In such cases, the following phases were identified: nanocrystalline α-MgH_2, metastable phase β-MgH_2, minor amounts of MgO (significantly lower than 5 wt. %) and Ni.

3.4. *LiBH4/MgCl2*

Finally, the synthesis of metal-borohydrides was carried out through the ball milling of a $LiBH_4$ + $MgCl_2$ mixture under 1 bar argon (Ar). In this case, $MgCl_2$ acted as catalyst [14]. The reaction between $LiBH_4$ and $MgCl_2$ at different milling times (15 min, 2 h and 15 h, respectively) showed the disappearance of the features corresponding to the $LiBH_4$ and $MgCl_2$ phases after more than 2 h and the formation of a new phase (Li-Mg-Cl), with a lattice parameter similar to that of LiCl [15]. This indicated the successful reaction between $LiBH_4$ and $MgCl_2$ during the milling process. The Raman spectrum corresponding to the sample synthesized from $LiBH_4$ + $MgCl_3$ and milled during 12 h was compared to the Raman spectrum of pure $LiBH_4$ [14]. The results confirmed the partial formation of the $Mg(BH4)_2$ phase according to Nakamori et al. [16] and Matsunaga et al. [17].

4. Conclusions

Our results show the usefulness of the high-energy ball milling method for the synthesis and decomposition of light metal and complex hydrides. Moreover, we have shown the advantages of in-situ monitoring the reaction pressure and vial temperature variations taking place during the milling process in order to obtain immediate and valuable information about the efficiency of the different catalysts.

Acknowledgements

This work was partially supported by the Helmholtz Initiative FuncHy, the Novel Efficient Solid Storage for Hydrogen (NESSHY) EU-integrated project, and the Marie-Curie Research Training network COSY (EU-RTN).

References

1. L. Zaluski, A. Zaluska, and J. O. Strom-Olsen, *J. Alloys Comp.* **290**, 71 (1999)
2. A. Zaluska, L. Zaluski, and J. O. Strom-Olsen, *J. Alloys Comp.* **289**, 197 (1999)
3. J. L. Bobet, E. Akiba, and B. Darriet, *Int. J. Hydrogen Energy* **26**, 493 (2001)
4. J. L. Bobet and B. Darriet, *Metastable, Mechanically Alloyed and Nanocrystalline Materials*, Ismanam-2000 **360-3**, 609 (2001)

5. O. Gutfleisch, N. Schlorke-de Boer, N. Ismail, M. Herrich, A. Walton, J. Speight, I. R. Harris, A. S. Pratt, and A. Züttel, *J. Alloys Comp.* **356**, 598 (2003)
6. O. Gutfleisch, S. Dal Toe, M. Herrich, A. Handstein, and A. Pratt, *J. Alloys Comp.* **404**, 413 (2005)
7. J. Huot, S. Boily, E. Akiba, and R. Schulz, *J. Alloys Comp.* **280**, 306 (1998).
8. J. Huot, G. Liang, and R. Schulz, *App. Phys. A-Materials Science & Processing* **72**, 187 (2001)
9. J. Huot, M. L. Tremblay, and R. Schulz, *J. Alloys Comp.* **356**, 603 (2003)
10. A. Zaluska, L. Zaluski, and J. O. Strom-Olsen, *J. Alloys Comp.* **288**, 217 (1999)
11. F. C. Gennari, F. J. Castro, and J. J. A. Gamboa, *J. Alloys Comp.* **339**, 261 (2002)
12. S. Doppiu, L. Schultz, and O. Gutfleisch, *J. Alloys Comp.* **427**, 204 (2007)
13. C. Rongeat, I. Llamas-Jansa, and O. Gutfleisch, p. in preparation (2007)
14. I. Llamas-Jansa, C. Rongeat, S. Doppiu, O. Gutfleisch, and L. Schultz, *Int. J. Mat.* p. submitted (2008)
15. M. Au, A. Jurgensen, and K. Zeigler, *J. Phys. Chem. B* **110**, 26482 (2006)
16. Y. Nakamori, K. Miwa, A. Ninomiya, H. W. Li, N. Ohba, S. I. Towata, A. Zuttel, and S. I.Orimo, *Phys. Rev. B* **74**, 045126 (2006)
17. T. Matsunaga, F. Buchter, P. Mauron, M. Bielman, Y. Nakamori, S. Orimo, N. Ohba, K. Miwa, S. Towata, and A. Züttel, *J. Alloys Comp.* p. in press (2007)

DEVELOPMENT OF METAL HYDRIDES FOR HIGH PRESSURE MH TANK

T. MATSUNAGA*,**, T.SHINOZAWA*, K.WASHIO*, D.MORI*, M.ISHIKIKIYAMA*

**Higashifuji technical center, Toyota Motor Corporation, 1200 Mishuku, Susono, Shizuoka, 410-1193 Japan **Physics Department, University of Fribourg, Pérolles, 1700 Fribourg, Switzerland*

High-pressure metal hydride (MH) tank is a possible hydrogen storage system for fuel cell vehicles. The merit of the high-pressure MH tank system is improved by the use of a metal hydride with high dissociation pressure. In this study, TiCrV and TiCrVMo alloys with BCC structure have been developed for the system. The developed TiCrVMo alloy shows 2.5mass% of effective hydrogen capacity in the pressure range between 0.1MPa and 33MPa at 298K. In TiCrV, the dissociation pressure of the alloy increases with the decrease of the lattice size. This trend is consistent with a general trend often observed for other metal hydrides. However, for TiCrVMo alloy, the dissociation pressure is sensitive not only to the lattice size but also to the content of Mo. As a result, it turned out that Mo has a special effect to increase the dissociation pressure of the hydride. Combined with the developed TiCrVMo alloy, hydrogen charging/discharging properties as a high pressure MH tank was also investigated. The whole tank system has a potential to store 5kg of hydrogen within 95L and 225kg, which means $0.053kgH_2/L$ and $0.022kgH_2/kg$ as a total system, respectively.

1. Introduction

1.1. *High pressure metal hydride (MH) tank*

Metal hydride is one of the most promising materials for hydrogen storage of a fuel cell vehicle because of its high gravimetric density [1]. Recently, high-pressure metal hydride (MH) tank has been reported as a possible hydrogen storage system for fuel cell vehicles [2] [3] [4]. Figure 1 shows a schematic view of a high-pressure MH tank. In many cases, as hydrogen storage materials are in powder forms, the packing densities of the materials are limited. Therefore, more than 50% of the inner volume of the tank remains empty even when the tank is filled with the maximum amount of the alloy. At that time, by filling this empty space with high-pressure hydrogen gas, volumetric hydrogen storage density as a whole tank system can be improved considerably. Mori et al. has reported that 35MPa of high pressure compressed gas tank combined with heat exchanger and metal hydride (TiCrMn: effective H_2 capacity:1.9mass% [5]) can store more than twice the amount of hydrogen as compared to a normal 35MPa compressed gas tank at the same volume [3]. However, due to the heavy weight of the metal hydride, high-pressure metal hydride tank is so heavy that further improvement of gravimetric hydrogen storage capacity is expected.

The merit of a high pressure MH tank system is improved by the use of a metal hydride with high dissociation pressure. It is important for a hydrogen storage tank of a fuel cell vehicle to supply hydrogen even at low temperature. Using a metal hydride with high dissociation pressure, hydrogen can be easily supplied even at low temperature. Moreover, there is another merit for using a metal hydride with high dissociation pressure from the viewpoint of heat exchange. By increasing the dissociation pressure of a metal hydride, the reaction heat during hydrogen desorption (ΔH) is decreased, which makes heat exchange easier in charging and discharging of hydrogen [3] [4].

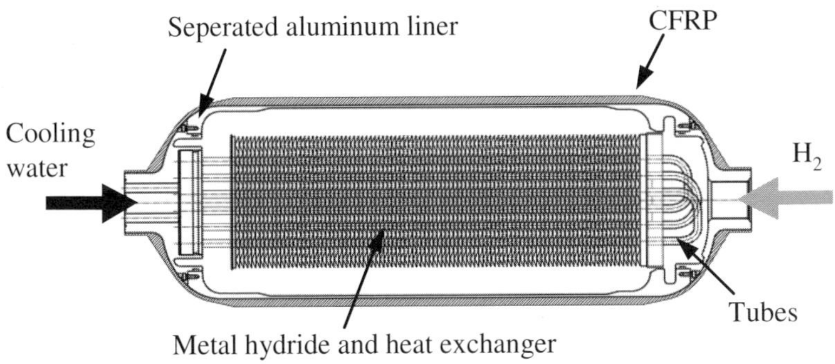

Fig. 1 A schematic view of a high-pressure MH tank

1.2. *TiCrV alloy with bcc phase*

TiCrV alloy with BCC structure has been studied for years as a promising hydrogen storage material [6] [7]. It has two plateau pressures, where only the upper plateau pressure area can be used for reversible hydrogen storage in normal condition. Recently, Arashima et al. has reported the alloy with 2.7mass% of reversible hydrogen capacity [7]. However, as most of the previous works on these alloys has been aimed for low pressure (e.g. less than 1MPa) MH tank system, dissociation pressures of these materials are not enough to be applied for above-mentioned high pressure MH tank system. In this study,

TiCrV and TiCrVMo alloys with BCC structure are developed in order to apply for high-pressure MH tank system.

2. Experimental

2.1. *Material development*

Several compositions of TiCrV and TiCrVMo alloys were prepared by arc melting from pure Ti, Cr, V and Mo elements. Subsequently, the alloys were kept at 1473K for 2 hours in Argon gas. Structures and lattice constants of the synthesized alloys were examined by powder X-ray diffraction analysis. Hydrogen storage properties of the materials were investigated using a specially designed Sievelts type apparatus (maximum pressure: 33MPa). 10g of the samples were used for each measurement. Effective hydrogen capacity was defined as the reversible hydrogen capacity in the pressure range of 0.1-33MPa at 298K. Dissociation pressure of each hydride was decided as the pressure at the center of the plateau of hydrogen desorption.

2.2. *Tank system test*

Figure 2 shows a schematic view of the test tank used for this study. Inner volume of the tank is 13 liter. The metal hydride and the heat exchanger are integrated into the tank. The heat exchanger has a fin and a tube structure with a smaller chamber for the packed bed of the metal hydride, which is the same size as the full size tank. In the experiments, 9.2kg of the metal hydrides were filled into the tank. The tubes were connected to the on-board cooling system and carried out internal transportation of coolant. The whole system was placed inside an explosion-proof test chamber with a barrier structure. Charge and discharge of hydrogen were performed using a high-pressure hydrogen filling device at a maximum flow rate of 12,500 NL/min. and a maximum pressure of 25MPa. The temperature of the heating medium was controlled between 233K and 368K using a temperature regulator.

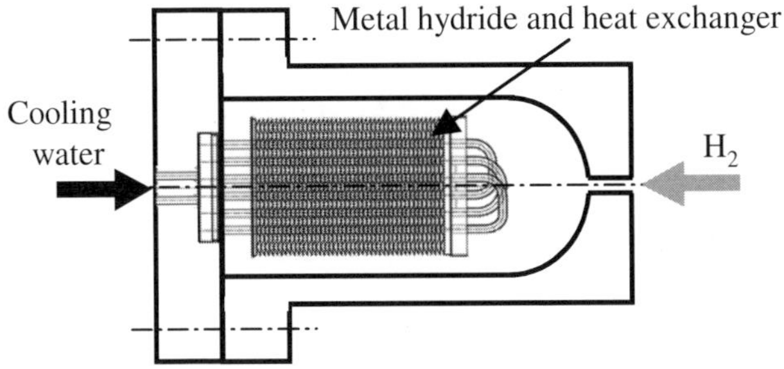

Fig. 2 A schematic view of the test tank

3. Results and discussions

3.1. *Material development*

3.1.1. *TiCrV alloy*

Six compositions of TiCrV ternary alloys were synthesized as is shown in Table 1. After the heat treatment, all of the alloys were found to be BCC single phase by X-ray diffraction analysis. The lattice constants of them are shown in Table 1. PC isotherms at 298K of the alloys are shown in Fig.2. Dissociation pressure of the alloy increases with the decrease of the lattice size. This trend is consistent with a general one often observed for other metal hydrides. However, the effective hydrogen capacity decreases with the increase of the lattice constant. As a result, in TiCrV ternary alloys, both high effective hydrogen capacity and high dissociation pressure are not satisfied simultaneously.

Table 1. Compositions and lattice constants of the synthesized TiCrV ternary alloys.

Sample No.	Composition [mol%] Ti	Cr	V	Lattice contant [Angstrom]
1	15	40	45	2.995
2	17	48	35	2.997
3	17	43	40	3.001
4	20	45	35	3.006
5	25	50	25	3.010
6	25	40	35	3.026

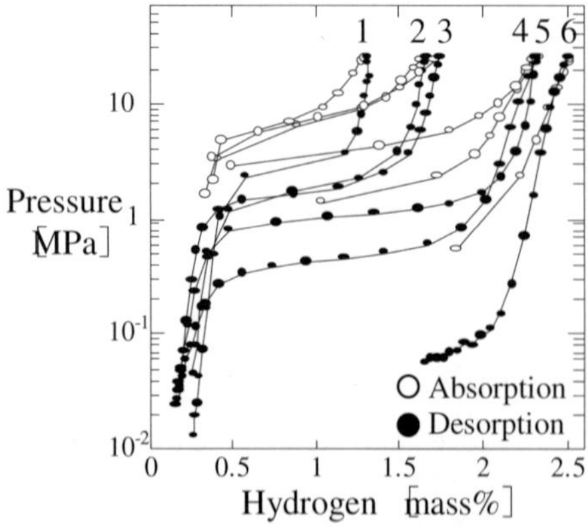

Fig. 3 PC isotherms of TiCrV alloys at 298K

3.1.2. *TiCrVMo alloy*

Figure 4 shows a correlation of dissociation pressures and lattice constants in TiCrV and TiCrMo alloys. Both of TiCrV and TiCrMo are solid solutions consisting of BCC phase. However, in TiCrMo, the correlation of dissociation pressure and lattice constant has a different trend from TiCrV [8]. This implies that Mo has an effect to increase the dissociation pressure of the metal hydride.

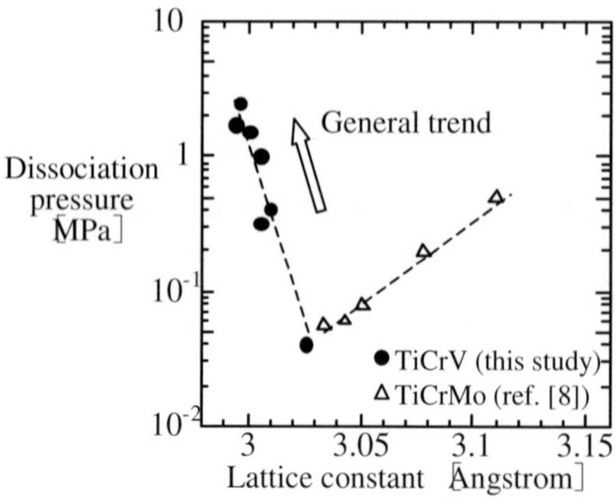

Fig. 4 Correlation of lattice constants and dissociation pressures in TiCrV and TiCrMo

In order to confirm this assumption, the alloys which 5mol% of Mo was substituted for V from the TiCrV alloys were synthesized. PC isotherms of $Ti_{25}Cr_{50}V_{25}$ and $Ti_{25}Cr_{50}V_{20}Mo_5$ at 298K are shown in Fig.5. By substituting 5mol% of Mo for V, the dissociation pressure increased. Note that at that time effective hydrogen capacity did not decrease.

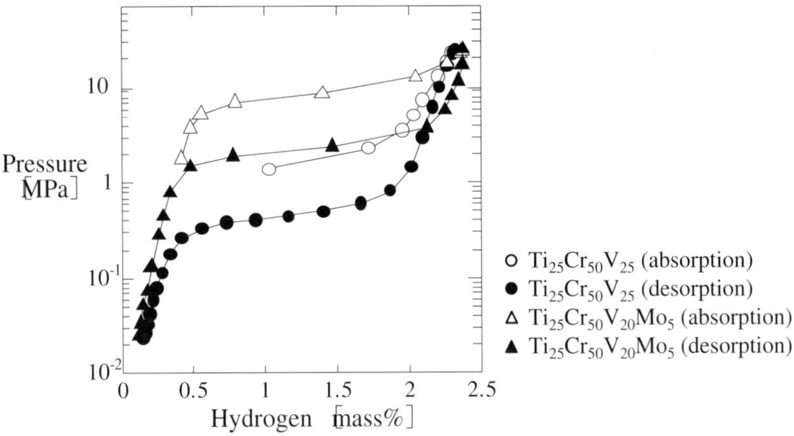

Fig. 5 PC isotherms of TiCrV and TiCrVMo

To examine the reason for this result, the lattice constants of the Mo substituted alloys were investigated. Figure 6 shows a correlation of dissociation pressures and lattice constants in TiCrV and TiCrVMo alloys synthesized in this study. By substituting 5mol% of Mo for V, lattice constant does not change as is shown in Table 2. This is probably because the atomic radius of Mo is similar to that of V (the atomic radius of V is 1.32A, where that of Mo is 1.36A). However, at that time, dissociation pressure increases drastically. In TiCrV ternary system, it seems that the dissociation pressures only depend on the lattice constants of the alloys. This result indicates that Mo has a special effect on increasing the dissociation pressure of the metal hydride. Therefore, using this effect, the dissociation pressure of a metal hydride can be controlled without changing its lattice constant, which will help keeping high effective hydrogen capacity while increasing dissociation pressure.

Table 2. Lattice constants change by substituting 5mol% of Mo for V.

Sample No.	Composition [mol%] Ti	Cr	V	Mo	Lattice contant [Angstrom]
1	25	50	25		3.010
2	25	50	20	5	3.012
3	25	40	35		3.026
4	25	40	30	5	3.026

3.2. *Tank test results*

Two types of metal hydrides (TiCrMn and TiCrVMo) were used for the tank test. The TiCrMn alloy has been developed in the previous work [5], whereas the TiCrVMo has been developed in this study. The properties of the metal hydrides are shown in Table 3.

Table 3. Hydrogen storage properties used for the tank test.

Composition	Structure	Effective hydrogen capacity [mass%]	Absorption pressure at 298K [MPa]	Dissociation pressure at 298K [MPa]
Ti35Cr34Mn31	AB2	1.9	5.0	4.0
Ti25Cr50V20Mo5	BCC	2.5	8.5	2.3

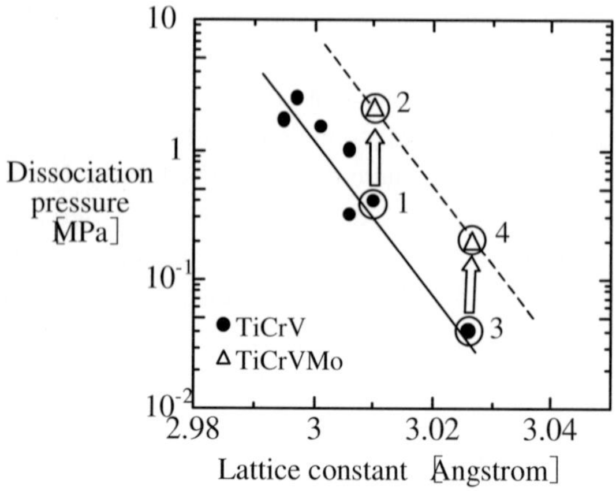

Fig. 6 Correlation of lattice constant and dissociation pressure of TiCrV and TiCrVMo

After filling the metal hydrides into the packed bed, they were activated. Subsequently, hydrogen gas was filled into the test tank up to the pressure of 25MPa at 298K. Figure 7 shows the hydrogen storage capacities of the tanks with TiCrMn or TiCrVMo. The amounts of the hydrogen stored as gas phase and as metal hydride are shown in Fig.8. The calculated storage capacity of TiCrVMo was 2.4mass%, which indicated that 96% of the alloy was activated. Figure 9 shows the rapid hydrogen filling test results. The tank with TiCrMn can store 80% of the maximum capacity in 5min., whereas the tank with TiCrVMo can store 72% in 5 min. The maximum temperatures of the alloys during filling were 365K and 348K for TiCrMn and TiCrVMo, respectively. Although the dissociation pressure of TiCrVMo is lower than that of TiCrMn as is shown in Table 3, the absorption pressure of TiCrVMo is higher than that of TiCrMn because of the big hysterisis. This difference of the absorption pressure probably results in the maximum temperature difference during filling, which influences on the filling rates in large part.

Based on the tank test results, on-board system performance with the developed TiCrVMo has been estimated as in shown in Table 4. Using TiCrVMo alloy combined with high-pressure MH tank, the whole system can store 5kg of hydrogen within 95L and 225kg, which means 0.053kgH_2/L and 0.022kgH_2/kg as a total system, respectively.

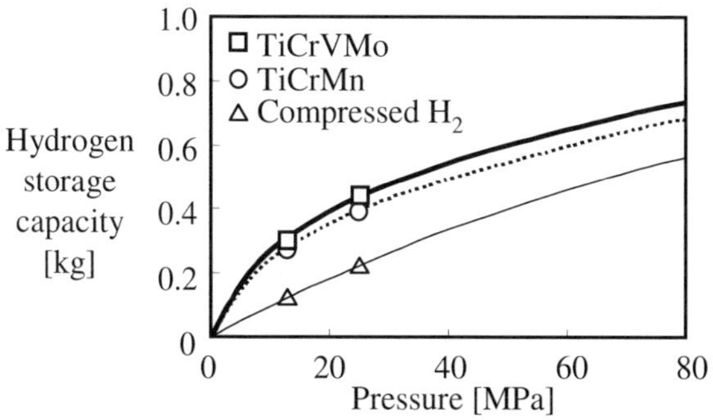

Fig. 7 Hydrogen storage capacity of the test tank(13L) with TiCrMn and TiCrVMo

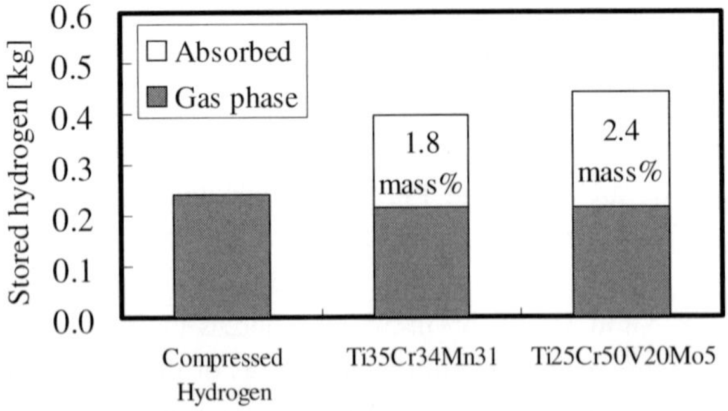

Fig. 8 The amount of stored hydrogen at 25MPa in the test tank

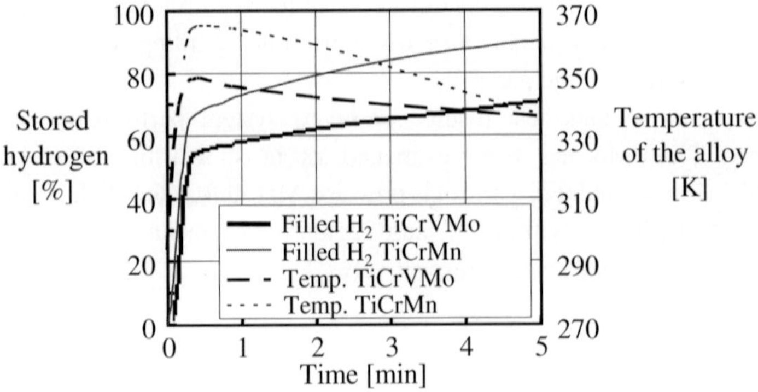

Fig. 9 Hydrogen charging speed of the tank with TiCrMn and TiCrVMo

Table 4. Estimation of on-board system performance with TiCrVMo alloy.

	High-pressure gas tank	High-pressure MH TiCrMn (1.9mass%)		High-pressure MH TiCrVMo (2.5mass%)	
Hydrogen storage capacity (35MPa)	3.0kg	7.3kg	5.0kg	9.5kg	5.0kg
Tank volume	180L	180L	125L	180L	95L
Tank weight	<100kg	420kg	290kg	420kg	225L
Hydrogen filling time	5-10min.	80% of maximum capacity in 5 min.		72% of maximum capacity in 5 min.	

4. Conclusion

TiCrV and TiCrVMo alloys with BCC structure have been developed for high-pressure MH tank system. The developed TiCrVMo alloy shows 2.5mass% of effective hydrogen capacity in the pressure range between 0.1MPa and 33MPa at 298K. In TiCrV, the dissociation pressure of the alloy increases with the decrease of the lattice size. This trend is consistent with a general trend often observed for other metal hydrides. However, for TiCrVMo alloy, the dissociation pressure is sensitive not only to the lattice size but also to the content of Mo, and it turned out that Mo has the special effect to increase the dissociation pressure of the hydride.

Combined with the developed TiCrVMo alloy, hydrogen charging/discharging properties as a high pressure MH tank was also investigated. The whole tank system can store 5kg of hydrogen within 95L and 225kg, which means 0.053kgH_2/L and 0.022kgH_2/kg as a total system, respectively.

Acknowledgments

The authors would like to thank Prof. A. Züttel for valuable discussion.

References

1. L.Schlapbach and A.Züttel, *Nature*, **414**, 353(2001)
2. N.Takeichi, H.Senoh, T.Yokota, H.Tsuruta, K.Hamada, H.T.Takeshita, H.Tanaka, T.Kiyobayashi, T.Takano and N.Kuriyama, *Int. J. Hydrogen Energ.*, **28**, 1121(2003)
3. D.Mori, N.Haraikawa, N.Kobayashi, H.Kubo, K.Toh, M.Tsuzuki, T.Shinozawa and T.Matsunaga, *Mater. Res. Soc. Symp. Proc.* Vol.884E GG6.4.1
4. D. Mori, K. Hirose, N. Haraikawa, T. Takiguchi, T. Shinozawa, T.Matsunaga, K. Toh, K. Fujita, A. Kumano and H. Kubo, JSAE 20077268/SAE 2007-01-2011, pp. 560-564.
5. Y.Kojima, Y.Kawai, S.Towata, T.Matsunaga, T.Shinozawa and M Kimbara, *J. Alloys Comp.* **419,** 256(2006).

6. E.Akiba and M.Okada, *MRS Bulletin*, **27**, 699(2002).
7. H.Arashima, F.Takahashi, T.Ebisawa, H.Itoh and T.Kabutomori, *J. Alloys Comp.* **356-357**, 405(2003).
8. A.Kamegawa, T.Tamura, H. Takamura and M. Okada, *J. Alloys Comp.*, **356-357**, 447(2003)

SYNTHESIS OF NOVEL METAL-COORDINATED FULLERENES FOR VEHICULAR HYDROGEN STORAGE

E. WHITNEY, C. ENGTRAKUL, C. J. CURTIS, Y. YAN, P. A. PARILLA, K. J. O'NEILL, L. J. SIMPSON, M. J. HEBEN, Y. ZHAO, Y. –H. KIM, S. B. ZHENG, AND A. C. DILLON[†]

National Renewable Energy Laboratory, 1617 Cole Blvd.
Golden, CO 80401, USA

Experimental wet chemical approaches have been demonstrated in the synthesis of a new chainlike $(C_{60}\text{-}Fe\text{-}C_{60}\text{-}Fe)_n$ complex. This structure has been proposed based on ^{13}C solid-state nuclear magnetic resonance, high-resolution transmission electron microscopy, energy-dispersive spectroscopy, and X-ray diffraction. Furthermore, this structure has been shown to have unique binding sites for dihydrogen molecules with the technique of temperature-programmed desorption. The new adsorption sites have binding energies that are stronger than that observed for hydrogen physisorbed on planar graphite, but significantly weaker than a chemical C-H bond. Volumetric measurements at 77 K and 2 bar show a hydrogen adsorption capacity of 0.5 wt%. Interestingly, the BET surface area is ~31 m^2/g after degassing, which is more than an order of magnitude less than expected given the measured experimental hydrogen capacity. Nitrogen and hydrogen isotherms performed at 75 K show a marked selectivity for hydrogen over nitrogen for this complex, indicating hidden surface area for hydrogen adsorption. Various Li_xC_{60} compounds have also been synthesized, inspired by theoretical predictions of an optimized $Li_{12}C_{60}$ compound with substantial hydrogen capacity. Unfortunately, the theoretical structure was not experimentally achieved, nor was the predicted hydrogen capacity reached.

1. INTRODUCTION

A hydrogen-based economy offers the pollution-free promise of using entirely renewable resources.[1] For example, hydrogen can be generated through the electrolysis of water using electricity derived from wind power, photovoltaics, or thermo-chemical processing of biomass. Once produced, hydrogen can then be used in fuel cells that convert hydrogen and oxygen back into water and produce electricity in the process. Hydrogen can also be combusted in an engine to generate mechanical energy or even burned to produce heat. Regardless of the scenario, water is produced in a virtually pollution-free cycle.[1]

However, one of the biggest challenges facing a future hydrogen economy is that of onboard vehicular hydrogen storage. Hydrogen is a nonpolarizable gas, making reversible solid state hydrogen storage a difficult challenge.

Furthermore, neither compression of H_2 to 10,000 p.s.i. or liquid hydrogen will satisfy all of the United States Department of Energy's 2015 targets for hydrogen storage of 9 wt% or 81 kg H_2/m^3.[2,3] Thus, in recent years, research has focused on novel carbon-based nanostructured materials, among others, as candidates for vehicular storage.[4,5] Carbon is promising because it is a light element and thus a step towards the goal of lightweight hydrogen storage for transportation.

Also inherent in the goal of hydrogen storage are the issues of near-room temperature operation at reasonable pressures. For an adsorption system, these challenges dictate a moderate binding energy for managing the heat load during refueling. Furthermore, the entire process must be completely reversible.[4] Although not typically appreciated, the adsorption energies for hydrogen bound to carbon surfaces are, in general, quite weak or quite strong. Non-dissociative physisorption, due purely to van der Waals interactions, involves a binding energy of only ~4 kJ/mol, whereas a C-H chemical bond is typically close to 400 kJ/mol. The desired binding energy range for reversible vehicular storage (~10-40 kJ/mol) therefore dictates that hydrogen be stabilized in an atypical fashion.

Hydrogen adsorption using carbon-based nanostructured materials has previously been explored on singled-wall nanotube (SWNT) structures,[6] with a binding energy of ~19 kJ/mol, as well as multi-wall nanotube (MWNT) structures grown with an iron catalyst (~50 kJ/mol).[5] In particular, the observation of enhanced hydrogen storage capacities in MWNT structures with small amounts of iron has fueled investigation of other potential metal-containing storage structures. While theory has predicted that isolated transition metal atoms can complex with up to six dihydrogen ligands via a Kubas interaction,[7-11] these metal atoms are predicted to form a bulk material upon removal of the hydrogen. To overcome this challenge and yet also harness the large storage potential of transition metal atoms, fullerenes have been proposed as stabilizing ligands because of their symmetric arrangement of cyclopentadiene rings, which have been shown to complex with transition metals through Dewar coordination.[12] (Individual cyclopentadiene rings coordinated to transition metals will polymerize without the presence of the fullerene matrix.)

Fullerenes, or "buckyballs," are closed structures comprised of unsaturated carbon atoms arranged in 5- and 6-membered rings, providing a number of possible bonding modes for metal coordination. The most famous fullerene, C_{60}, was discovered by Kroto et al in 1985.[13] Calculations have shown that an iron atom can form an organometallic complex with a C_{36} fullerene, sharing charge with only four carbon atoms of a bent five-membered ring. Three molecular H_2 ligands then coordinate with the iron atom with a binding energy of ~43 kJ/mol. Notably, stable transition metal-coated buckyballs (Ti, V, Nb, Ta) have been recently synthesized.[14]

In an optimized fullerene-based transition metal complex, scandium has been predicted to complex with the twelve five-membered carbon rings of C_{60}, sharing charge with all of the carbon atoms in the pentagon (η^5 coordination). For example, a $C_{60}[ScH_2(H_2)_4]_{12}$ organometallic fullerene complex (OFC) is predicted to be a minimum energy structure with ~7.0 wt% reversible hydrogen capacity.[15] Doping this OFC with boron results in a $C_{48}B[ScH(H_2)_5]_{12}$ OFC with a reversible hydrogen capacity of ~9 wt%. The complexes are arranged symmetrically on a buckyball in a minimum-energy structure, and the hydrogen is stored reversibly with a binding energy of ~30 kJ/mol, ideal for vehicular applications.

Density functional theoretical calculations predict that $Li_{12}C_{60}$, with each Li centered around a pentagon of C_{60} and binding five H_2 molecules, will have an optimized gravimetric hydrogen capacity of 13 wt% at low temperature.[16] The binding energy is calculated to be ~6.4 kJ/mol. In the dehydrogenated configuration, a Li atom stabilized on a pentagon of one fulleride interacts with a hexagon on an adjacent fulleride. The Li atom forming this linkage loses its hydrogen storage capacity. It is projected that in a stable crystal, four Li atoms per $Li_{12}C_{60}$ structure would cluster, resulting in ~9 wt% hydrogen capacity.[16]

Although stable transition metal-coated buckyballs have been synthesized, the synthesis of these abovementioned η^5 complexes is unprecedented and many hurdles must be overcome. For example, the chemistry of C_{60} is generally olefinic (i.e., η^2 coordination, in which the metal is coordinated to the fullerene through two carbon atoms contributing two electrons to the bonding).[17-22] Thus, the synthesis of the predicted fullerene-metal-H_2 complexes, where the metal is coordinated to five carbon atoms, is not expected to be easy. In fact, η^5-C_{60} coordination has only been achieved through wet chemical methods by isolating the carbon atoms of a C_{60} pentagon through five-fold addition of alkyl-groups to neighboring carbon and protonation of one of the pentagon carbons.[23] Thus, the synthesis of the predicted fullerene-metal-H_2 complexes, where the metal is coordinated to five carbon atoms, is not expected to be easy.

However, the calculations described above, together with others that have recently appeared, indicate that non-olefinic metallofullerenes[16] and metal-doped nanostructures[24] are stable. Because the synthesis of these complexes is relatively unexplored and there are no guiding precedents, it has been necessary to discover the bonding preferences of the fullerene system and to open new synthetic pathways to the desired complexes. Here we describe the characterization of an iron atom, complexed with C_{60} ligands in a chainlike structure, as well as a series of Li·C_{60} structures with different Li/C60 stoichiometries. New adsorption sites for dihydrogen molecules on carbon surfaces are clearly demonstrated.

2. EXPERIMENTAL PROCEDURES

To make the reactive fulleride compound K_6C_{60} in the Fe-C_{60} synthesis, fullerenes and a slight excess of potassium were sealed in a glass tube under vacuum and heated for approximately four days at 250 °C. Both solid-state ^{13}C NMR and Raman spectroscopy were employed to determine that the K_6C_{60} compound was in fact synthesized. The K_6C_{60} product was then reacted in an inert atmosphere with cyclopentadienyl-iron-dicarbonyl-iodide ($CpFe(CO)_2I$) in tetrahydrofuran (THF) to form the complex. The recovered product was dried in an inert atmosphere. Manipulations of air-sensitive materials were carried out in a glove box or using standard Schlenk techniques. THF was distilled just prior to use from sodium benzophenone ketyl. C_{60} was obtained from Aldrich, and $CpFe(CO)_2I$ was obtained from Strew.

For the Li·C_{60} syntheses, different stoichiometries of Li and C_{60} were dissolved in liquid ammonia. Initially, a Li compound was synthesized by dissolving Li and C_{60} in liquid ammonia at -78 °C, and the reaction did not go to completion as evidenced by visible residual Li. An Li·C_{60} compound with excess Li was then formed. If the reaction was allowed to warm to room temperature before removing the ammonia, the reaction appeared to go to completion. The recovered products were dried in inert atmosphere, stored in a glove box, and transferred to sealed ampoules for structural and H_2 capacity determination.

It was difficult to dissolve the new Fe/Li-C_{60} complexes in any organic solvent, making solid-state nuclear magnetic resonance (NMR) necessary. For these studies, a BRUKER AVANCE 200 spectrometer operating at 200 MHz was employed. Solid-state ^{13}C NMR spectroscopy under fast magic angle spinning (MAS) was required to obtain high-resolution spectra of the complexes.[25] Transmission electron microscopy (TEM) was performed on a F-20 UT Transmission Electron Microscope with dry samples on a grid. X-ray diffraction (XRD) was performed on a Scintag PTS 4-circle goniometer (Bragg-Brentano geometry) using Cu Kα radiation (0.15406 nm) generated at 45 kV and 36 mA and detected with a liquid-nitrogen-cooled solid-state germanium detector. The source slits were 4 mm and 2 mm at 290 mm goniometer radius and the detector slits were 1.0 mm and 0.5 mm at the same radius. The sample powder was mounted onto a low-x-ray-background quartz substrate using diluted Duco cement. (The sample mount is vertical so the glue is necessary; the diluted glue adds almost no background signal and is amorphous.) The scan rate was 0.12 degrees/min. (25 seconds/step) from 5 to 125 degrees two theta in 0.05 degree steps (total time = 15.3 hours).

The new hydrogen binding sites were examined with temperature-programmed desorption (TPD) spectroscopy. The sample was first dosed with 500 Torr hydrogen for ~5 minutes after pumping at a pressure of approximately 5×10^{-8} Torr overnight. The sample was then cooled to liquid nitrogen temperatures and systematically degassed to temperatures up to 250 °C. For the TPD technique, samples weighing between 1-10 mg are placed in a packet formed from 25 μm thick platinum foil and mounted at the bottom of a liquid nitrogen cooled cryostat. The packet is resistively heated with a programmable power supply, and the sample temperature is measured with a thin thermocouple spot-welded to the platinum packet. A mass spectrometer measures desorbing species and insures that only hydrogen is observed during desorption. The TPD instrument is calibrated by thermally decomposing known amounts of TiH_2. The amount of evolved hydrogen is linear with the weight of decomposed TiH_2. The TPD system is also calibrated by H_2 desorption from Pd that is charged *in situ* to a literature predicted capacity. Finally a calibrated H_2 flow is employed as a further check of the calibration standards. All of the methods have been confirmed with an in-house volumetric technique within ±3 %. A control TPD experiment, using C_{60}, was also done as a comparison. Additionally, using the same dosing and cooling techniques, the new compounds were heated at a variety of different rates in order to extract a desorption activation energy E_d.[6] Total H_2 capacity measurements for the new complexes were also obtained at 77K and 2 bar with single point measurements in a volumetric apparatus.

3. RESULTS AND DISCUSSION

3.1. *Structural Characterization of Fe-C_{60}*

Figure 1 displays the ^{13}C solid-state NMR spectrum characteristic of C_{60}. The sharp peak with a chemical shift of 143.7 ppm is consistent with C_{60}, and the broad feature at approximately 110 ppm is simply due to the Teflon cap sealing the NMR rotor.

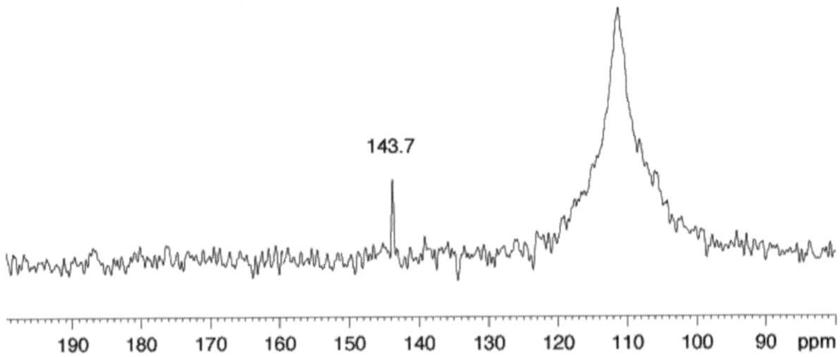

Figure 1: ${}^{13}C$ NMR spectrum of C_{60} fullerenes.

In comparison, Figure 2 displays the ${}^{13}C$ solid-state NMR spectrum of the final recovered product from synthesis reactions designed to unfold new organometallic chemistry for C_{60}. The sharp peak at 143.7 ppm may be attributed to residual unreacted C_{60}. However, it was not possible to extract unreacted C_{60}, once the compound was synthesized. The broad peak shifted to higher ppm is consistent with C_{60} coordinated with an iron atom. An inset in the figure of these two peaks is also provided for clarity.

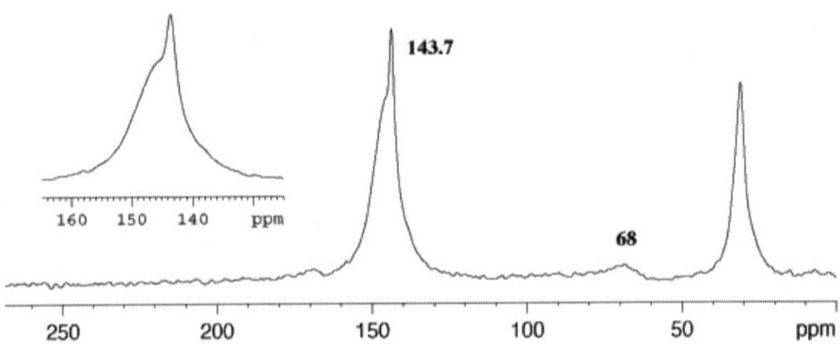

Figure 2: ${}^{13}C$ NMR spectrum of Fe-C_{60}.

In order to better elucidate the precise structure of the Fe-C_{60} compound, high-resolution transmission electron microscopy (HRTEM) and energy dispersive x-ray (EDX) spectroscopy were performed with a field emission microscope. As shown in Fig. 3, the study revealed areas in the sample, highlighted by pink circles, which were consistent with small quantities of oxidized iron in the phase Fe_2O_3. The sample was exposed to air as the complex was not found to be air-sensitive during hydrogen adsorption studies. However, it was not surprising that small amounts of residual iron were oxidized immediately. More importantly, regions of C_{60} molecules, circled in yellow, containing ~1-1.5 at% iron were also observed. The fact that such low levels of iron were stable against oxidation suggests that the iron is complexed to the C_{60} molecules and is consistent with the formation of C_{60}-Fe-C_{60}-Fe-C_{60}-chain structures of a yet-undetermined length. From the HRTEM image in Fig. 3, some ordering in the C_{60} chain-like structures may be detected. Also, no large metal clusters were observed with extensive HRTEM analyses.

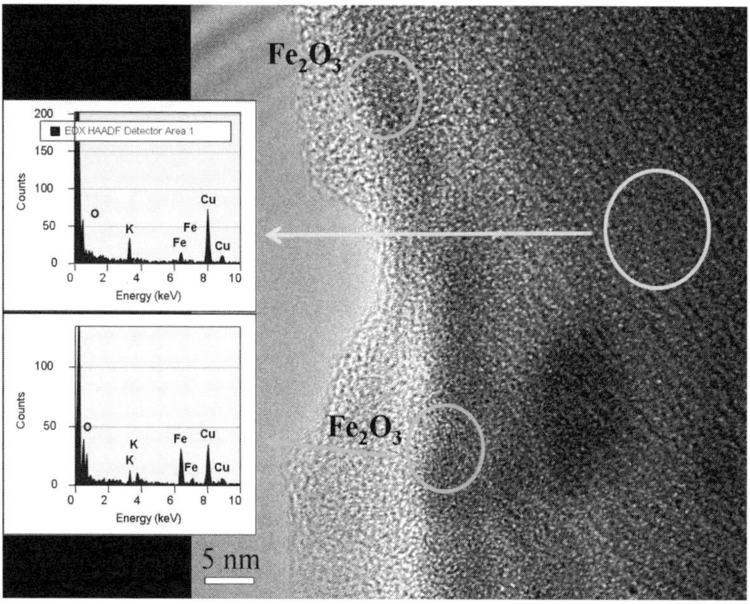

Figure 3: HRTEM of air-exposed Fe-C_{60} sample.

HRTEM analysis was repeated after washing the sample with dilute acid to remove Fe_2O_3. As shown in Figure 4, ordered chains emerge very distinctively, and an electron diffraction pattern was also obtained. Iron was still detected at ~0.5-1 at% with small spot EDX, again consistent with C_{60}-Fe-C_{60}-Fe-C_{60}-chain structures.

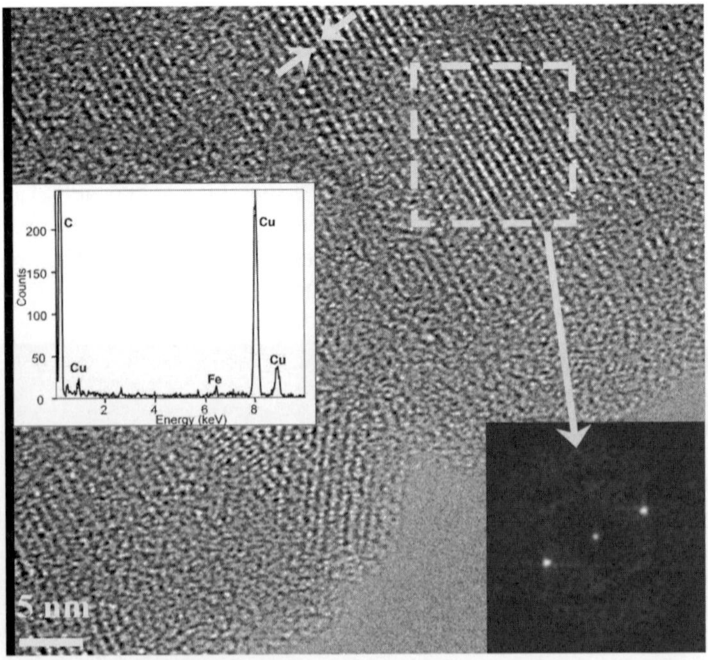

Figure 4: HRTEM of acid-washed Fe-C_{60} sample.

Since some ordering was detected in the HRTEM studies, XRD measurements were therefore performed on the powdered Fe(C_{60}) sample. Fig. 5(a) displays the XRD pattern for the Fe(C_{60}) sample. Several very broad features are observed that could be consistent with either disordered C_{60} or the disordered fulleride (used as a reactive precursor in the initial reaction). However, these broad features could also be consistent with disordered C_{60}-Fe-C_{60}-Fe-C_{60}-chain structures. Furthermore, the sharper feature occurring at low angle is consistent with a crystalline d-spacing of ~ 13.3 Å. This d-spacing is not consistent with the FCC packing of either C_{60} or the fulleride and suggests that a new packing of C_{60} is observed and may be attributed to Fe(C_{60}) chain-like structures. The broad features in the XRD pattern of Fig. 5(a) are also similar to features previously reported for carbon single wall nanotubes (SWNTs), as shown in Fig. 5(b). In the case of SWNTs, features are detected with XRD because the nanotubes pack into crystalline bundles. The features are broad, however, because the bundles are highly disorderd; i.e. there is slippage along the axis of the tubes such that the graphitic structures are not perfectly aligned.

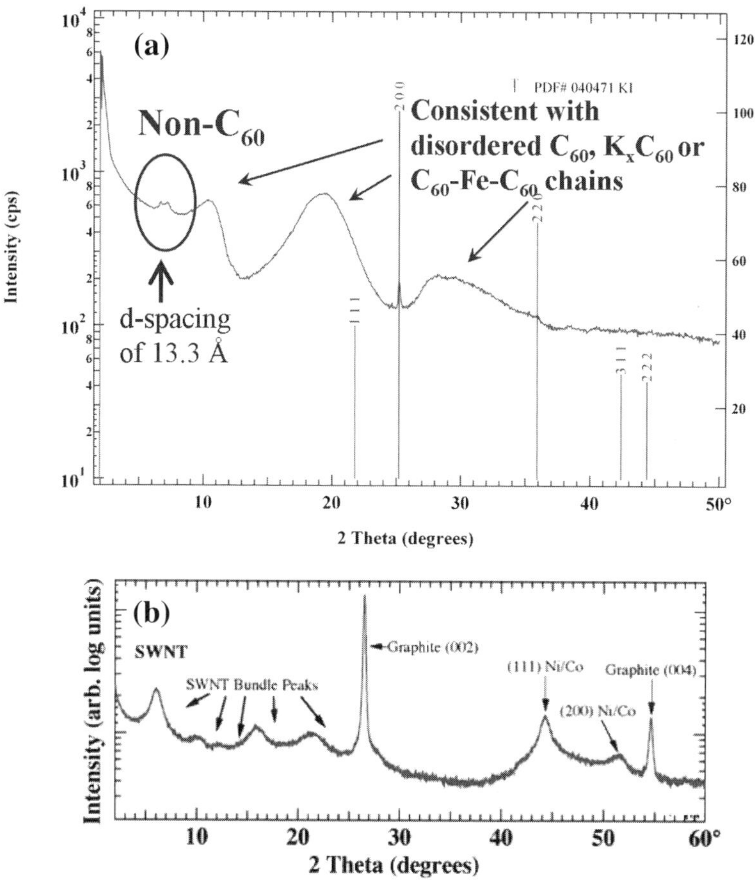

Figure 5: XRD spectra of (a) Fe-C_{60} and (b) SWNTs.

Collectively this data suggests that the C_{60}-Fe-C_{60}-Fe-C_{60}-chain structures pack in loosely ordered bundles similar to SWNTs. It is possible that the 13.3 Å d-spacing represents the interstitial spacing between the chains. Fig. 6 displays a cartoon representation of Fe-C_{60} chains. (Note that based on the XRD in the actual materials, the degree of alignment is not expected to be this high) It has been proposed that SWNTs that are atomically doped with metals are still a promising hydrogen storage material. The FeC_{60} chain structures are very similar to SWNTs doped with atomic metal. The focus of future work will be the production of these materials at higher yield so that they may be more readily purified, as well as methods to increase their alignment. An optimized

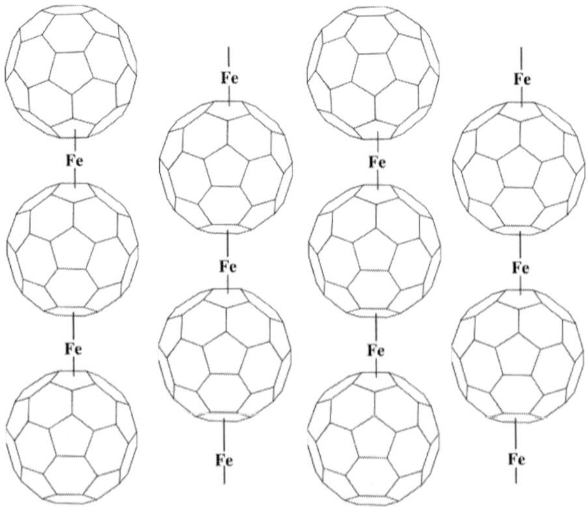

Figure 6: Cartoon representation of Fe-C_{60} chains.

porous framework for hydrogen storage at ~77 K with a moderate over-pressure may then be realized.

3.2. *Structural Characterization of Li_xC_{60}*

Figure 7 displays the ^{13}C solid-state NMR spectra of four new organometallic Li_xC_{60} compounds. The number preceding the $Li{\cdot}C_{60}$ is the number of Li atoms that was reacted with each C_{60}. In the case of the $20Li{\cdot}C_{60}$, excess Li was observed so it is not possible to determine the number of Li atoms coordinated with each C_{60}. For the $16Li{\cdot}C_{60}$, the reaction appeared to go to completion, and it is assumed that 16 Li atoms are coordinated with each C_{60} molecule. In the spectrum of the 20 $Li{\cdot}C_{60}$ compound, the feature at ~40 ppm indicates the presence of sp^3-hybridized carbon, suggesting polymerization of the fullerene molecules and the possible formation of a porous framework. This feature is almost absent in the spectrum of the $16Li{\cdot}C_{60}$ compound. In the experimental $12Li{\cdot}C_{60}$ compound, evidence for sp^3-hybridized carbon is clearly observed, indicating polymerization. It should thus be noted that the theoretical structure has not yet been made.

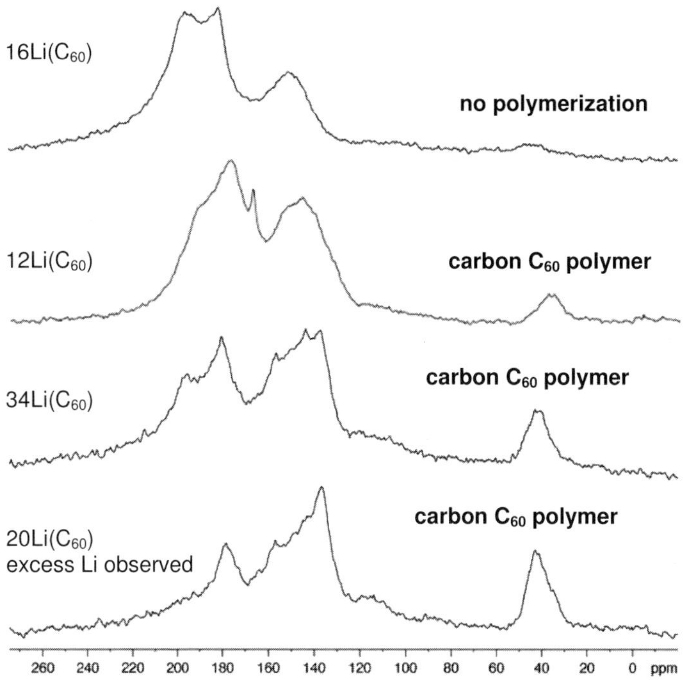

Figure 7: A variety of Li·C_{60} compounds with unique ^{13}C NMR signatures were made.

3.3. *New Hydrogen Adsorption Sites: Fe-C_{60}*

Figure 8 displays temperature-programmed desorption spectra where an Fe-C_{60} sample was exposed to hydrogen at room temperature after pumping at a pressure of approximately $5x10^{-8}$ Torr overnight and then after sequentially degassing the sample to 100, 200 and 250 °C. The lower temperature peak is centered at approximately 100 °C. This is slightly above the peak desorption temperature that is generally observed for hydrogen adsorbed on carbon surfaces. (In fact, it is generally the case that the true desorption peak for physisorbed hydrogen is not obtained due to the inability to cool the sample below -140 °C while exposing the sample to H_2 at 500 Torr.) Perhaps more interesting, however, is the appearance of the peak centered at approximately -50 °C as the sample is degassed to 250°C. (The sample was not degassed above this temperature as organometallic complexes are known to decompose at temperatures $\geq$ 300 °C.) The appearance of this new peak shows that hydrogen is stabilized at a temperature significantly above that expected for physisorption.

However, the temperature is significantly low enough that the probability of the formation of C-H bonds is essentially zero.

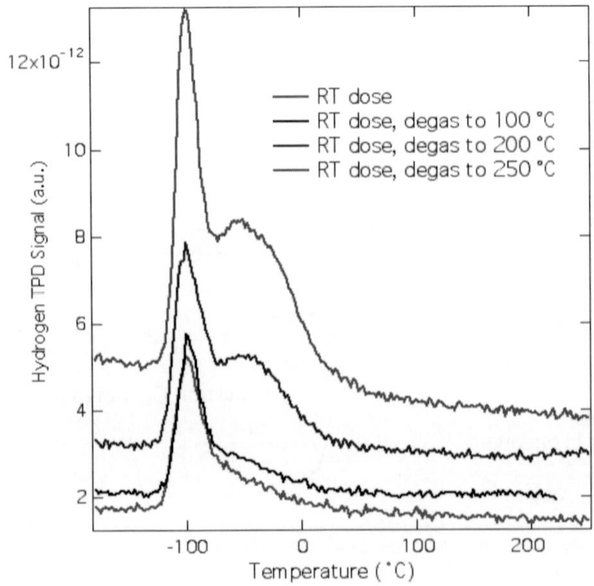

Figure 8: TPD spectra of Fe-C_{60} following a room temperature exposure to hydrogen, (500 Torr, 5 min.) after pumping at a pressure of ~ $5x10^{-8}$ Torr overnight and then after sequentially degassing the sample to 100, 200 and 250 °C.

The appearance of this new peak has also been compared with H_2 adsorption on C_{60} alone, as shown in Figure 9. Again, temperature-programmed desorption reveals H_2 adsorption sites following exposure to hydrogen at 500 torr for five minutes and then cooling approximately to -180 °C. Both samples were degassed to 250 °C prior to H_2 exposure. As the figure shows, the pure fullerene exhibits almost no hydrogen adsorption under these conditions. Thus, new adsorption sites for dihydrogen molecules have been revealed on C_{60}-Fe-C_{60}.

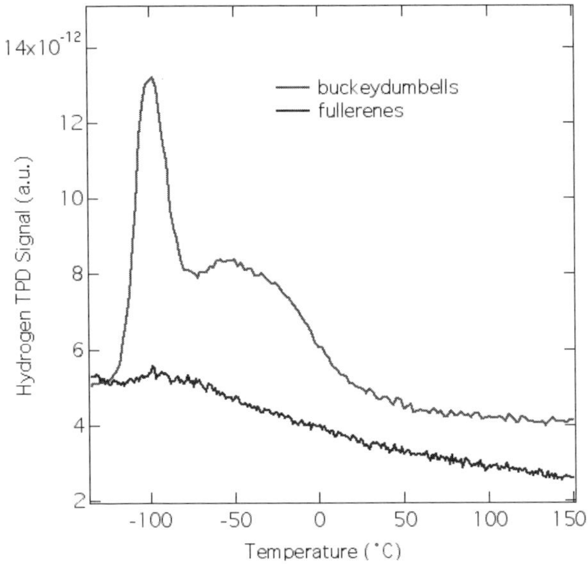

Figure 9: Comparison of H_2 adsorption on C_{60} and C_{60}-Fe-C_{60}. Sample was dosed at 500 torr for 5 minutes before cooling to -180 °C. Both samples were degassed to 250 °C prior to exposure.

For the unique adsorption site occurring at a higher temperature in the new complex, the exact binding energy (or desorption activation energy, E_d) can be determined by measuring the desorption peak temperature at different heating rates. Figure 10 shows a subsequent plot, described by $\ln T_m = E_d/RT_m$. Each point is derived from a different heating rate, which shifts the peak temperature of desorption. The slope of the line indicates an enhanced binding energy of ~6.2 kJ/mol, near the desired binding energy range for reversible onboard vehicular hydrogen storage. Since the binding energy was found to be only slightly enhanced over that observed for physisorption, non-dissociative adsorption is assumed and E_d is equivalent to the binding energy.[26] Complimenting the temperature programmed desorption studies, volumetric analyses conducted at 77K and 2 bar showed that the Fe-C_{60} compound had a hydrogen adsorption capacity of ~0.5 wt %. Furthermore, essentially zero hydrogen uptake was observed on pure fullerenes under the same conditions.

BET surface areas were calculated for the Fe-C_{60} complex, as synthesized and after degassing. For the as-synthesized compound, the surface area was measured twice, yielding results of 8.7 and 9.8 m^2/g.

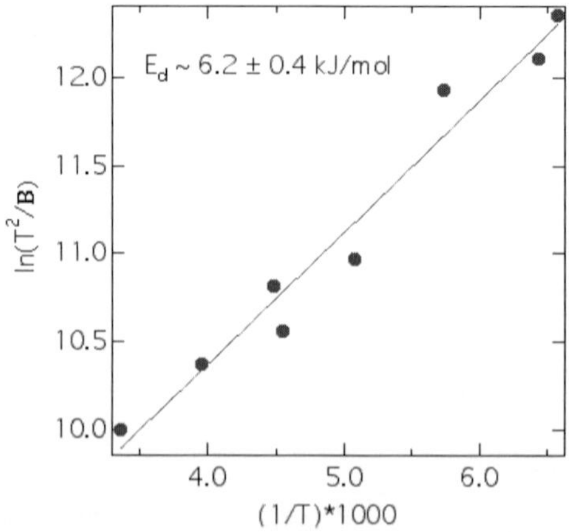

Figure 10: A plot of desorption activation energy indicated an enhanced binding energy of ~6.2 kJ/mol for the desorption peak centered at -50 °C, as shown in Fig. 7.

The gravimetric hydrogen capacity was 0.004 wt% at room temperature and 0.26 wt% at 77 K. After degassing the sample to 285 °C, the surface area was 31.1 m^2/g, with gravimetric hydrogen capacities of 0.004 wt% at room temperature and 0.5 wt% at 77 K. These results possibly suggest a mechanism other than simple physisorption and also violate Chahine's rule.[27,28] If there is not a unique mechanism, then there are surface sites which are not accessible to the N_2 used in these surface area measurements.

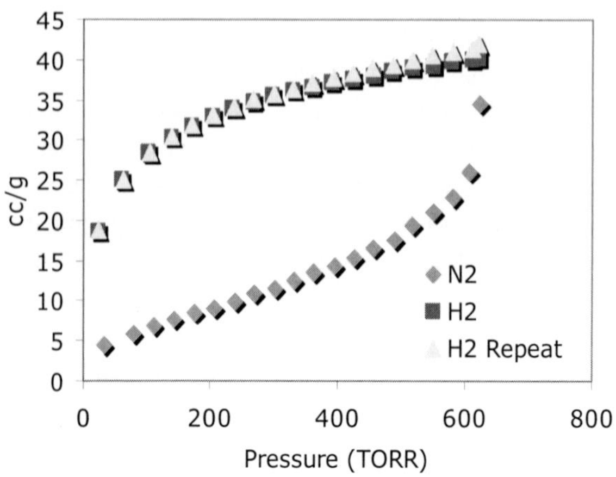

Figure 11: Fe-C_{60} H_2 and N_2 isotherms at 75 K.

To further understand these results, N_2 and H_2 isotherms were run on the Fe-C_{60} complex at 75 K. As displayed in Figure 11, the H_2 isotherm reveals a dramatically faster uptake than that of N_2, suggesting that pore size may play a role in the hydrogen capacity of this compound. To further illuminate the debate between surface area effects and bonding effects, a CO_2 isotherm at 0 K was also performed (see Figure 12) and yielded a surface area of ~170 m^2/g. The rationale for using CO_2 is its higher kinetic energy and diffusion coefficient. Finally, in addition to the previously mentioned H_2 isotherm at 75 K, another H_2 isotherm at 80 K was conducted in order to estimate an enthalpy of adsorption of ~5 kJ/mol.

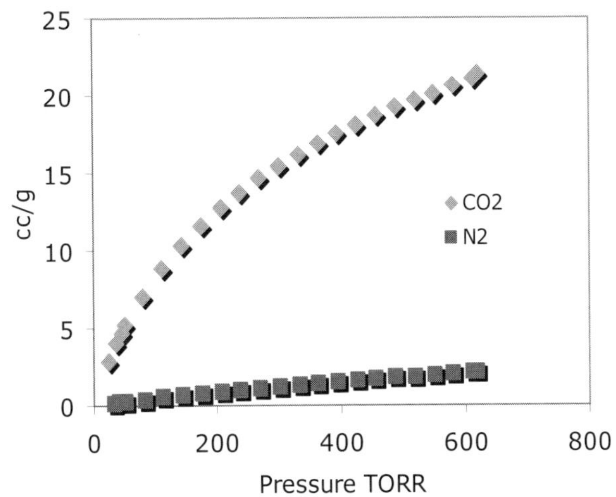

Figure 12: F-C_{60} CO_2 and N_2 isotherms at 0 °C.

3.4. *New Hydrogen Adsorption Sites: Li_xC_{60}*

All of the various Li·C_{60} compounds have a reversible hydrogen capacity of 0.2 wt% at 77K and 2 bar. The 20Li·C_{60} and 12Li·C_{60} compounds exhibit a capacity of 0.5 wt% and 0.8 wt%, respectively, without an over pressure. As shown in Figure 13, the hydrogen desorption temperature for the hydrogen stabilized without an overpressure in the 12Li·C_{60} compound is similar to the desorption temperature found for commercial LiH. Slight shifts in peak desorption temperature are observed due to differences in structure. Furthermore, variable heating rate temperature-programmed desorption analysis of the 12Li·C_{60} compound reveals reversible hydrogen adsorption with a

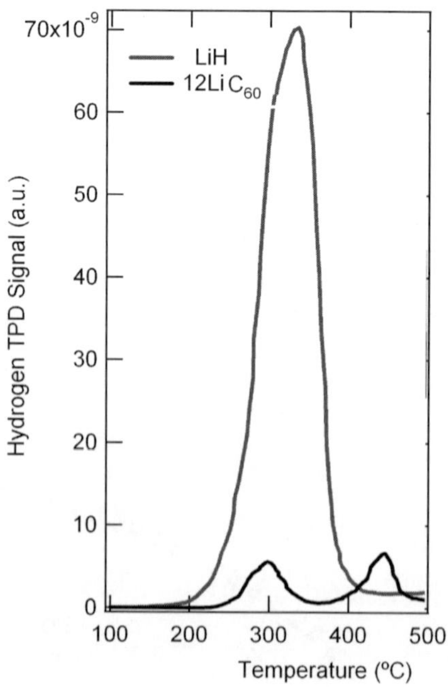

Figure 13: H_2 desorption comparison of LiH and 12Li·C_{60}.

binding energy of ~6 kJ/mol. This binding energy is in good agreement with Sun et al.[16] However, the capacity is only 0.2 wt% at 77 K and 2 bar.

4. CONCLUSIONS

The formation of a new organometallic Fe-C_{60} structure has been demonstrated to have unique hydrogen adsorption sites. This new complex has been characterized with solid state NMR spectroscopy, XRD, HRTEM, and temperature-programmed hydrogen desorption. Analysis of the iron-fullerene complex indicates the formation of C_{60}-Fe-C_{60}-Fe-C_{60} chain structures of an undetermined length, with a reversible hydrogen capacity of ~0.5 wt% at 77 K and a hydrogen overpressure of 2 bar. Interestingly, the BET surface area is ~31 m^2/g after degassing, which is more than an order of magnitude less than expected given the measured experimental hydrogen capacity. Nitrogen and hydrogen isotherms performed at 75 K show a marked selectivity for hydrogen over nitrogen for this complex, indicating hidden surface area for hydrogen adsorption. Various Li_xC_{60} compounds have also been synthesized, inspired by theoretical predictions of an optimized $Li_{12}C_{60}$ compound with substantial hydrogen capacity. The theoretical structure was not experimentally achieved, nor was the predicted hydrogen capacity reached. These results suggest that synthesis of organometallic fullerene complexes should be further explored for vehicular hydrogen storage applications.

References

1. H.-H. Rogner, Int. J. Hydrogen Energy **23,** 833 (1998).
2. *http://www.eere.energy.gov/hydrogenandfuelcells/mypp/*.
3. *http://www.sc.doe.gov/bes/hydrogen.pdf*.
4. A. C. Dillon and M. J. Heben, Appl. Phys. A **72,** 133-142 (2001).
5. A. C. Dillon, J. L. Blackburn, P. A. Parilla, Y. Zhao, Y.-H. Kim, S. B. Zhang, A. H. Mahan, J. L. Alleman, K. M. Jones, K. E. H. Gilbert, and M. J. Hebern, in *Discovering the Mechanism of H_2 Adsorption on Aromatic Carbon Nanostructures to Develop Adsorbents for Vehicular Applications*, Boston, Massachusetts, 2004 (Materials Research Society), p. 117-124.
6. A. C. Dillon, K. M. Jones, T. A. Bekkedahl, C. H. Kiang, D. S. Bethune, and M. J. Heben, Nature **386,** 377-379 (1997).
7. G. J. Kubas, R. R. Ryan, B. I. Swanson, P. J. Vergamini, and H. J. Wasserman, J. Am. Chem. Soc. **106,** 451-452 (1984).
8. G. J. Kubas, J. Organometall. Chem. **635,** 37-68 (2001).
9. T. Le-Husebo and C. M. Jensen, Inorg. Chem. **32,** 3797-3798 (1993).
10. J. Niu, B. K. Rao, and P. Jena, Phys. Rev. Lett. **68,** 2277-2280 (1992).
11. F. Maseras and A. Lledos, Chem. Rev. **100,** 601-636 (2000).
12. D. Michael and P. Mingos, J. Organometall. Chem. **635,** 1 (2001).
13. H. W. Kroto, J. R. Heath, S. C. O'Brien, R. F. Curl, and R. E. Smalley, Nature **318,** 162-163 (1985).
14. F. Tast, N. Malinowski, S. Frank, M. Heinebrodt, I. M. L. Billas, and T. P. Martin, Phys. Rev. Lett. **77,** 3529-3532 (1996).
15. Y. Zhao, Y.-H. Kim, A. C. Dillon, M. J. Heben, and S. B. Zhang, Phys. Rev. Lett. **94,** 155504 (2005).
16. Q. Sun, Q. Wang, P. Jena, and Y. Kawazoe, J. Am. Chem. Soc. **127,** 14582-14583 (2005).
17. F. J. Brady, D. J. Cardin, and M. Domin, J. Organometall. Chem. **491,** 169-172 (1995).
18. P. J. Fagan, J. C. Calabrese, and B. Malone, Acc. Chem. Res. **25,** 134-142 (1992).
19. H.-F. Hsu, Y. Du, T. E. Albrecht-Schmitt, S. R. Wlson, and J. R. Shapley, Organometallics **17,** 1756-1761 (1998).

20. M. M. Olmstead, L. Hao, and A. L. Balch, J. Organometall. Chem. **578,** 85-90 (1998).
21. L.-C. Song, G.-A. Yu, F.-H. Su, and Q.-M. Hu, Organometallics **23,** 4192-4198 (2004).
22. D. M. Thompson, M. Bengough, and M. C. Baird, Organometallics **21,** 4762-4770 (2002).
23. M. Sawamura, M. Toganoh, Y. Kuninobu, S. Kato, and E. Nakamura, Chem. Lett. **29,** 270 (2000).
24. T. Yildirim and S. Ciraci, Phys. Rev. Lett. **94,** 175501 (2005).
25. C. Engtrakul, M. R. Davis, T. Gennett, A. C. Dillon, K. M. Jones, and M. J. Heben, J. Am. Chem. Soc. **127,** 17548-17555 (2005).
26. R. J. Madix, in *Chemistry and Physics of Solid Surfaces*, edited by R. Vanselov (CRC, Boca Raton, 1979), p. 63-72.
27. E. Poirier, R. Chahine, P. Benard, D. Cossement, L. l. Lafi, E., T. K. Bose, and S. Besilets, Appl. Phys. A **78,** 961-967 (2004).
28. E. Poirier, R. Chahine, and T. K. Bose, Int. J. Hydrogen Energy **26,** 831-835 (2001).

TRENDS IN THE PROPERTIES OF SELECTED METAL-ORGANIC FRAMEWORK STRUCTURES: A THEORETICAL STUDY

AGNIESZKA KUC[1], JAN-OLE JOSWIG[2], ANDREY ENYASHIN[3], AND GOTTHARD SEIFERT[4]

Physical Chemistry, Technical University Dresden, Bergstr. 66b, 01062 Dresden, Germany

A density-functional based tight-binding method was employed for systematic studies of the structural, mechanical and electronic properties of a series of metal-organic framework (MOF) materials. We have considered cubic arrays of $Zn_4O(CO_2)_6$ (connector) units connected by different types of organic secondary building blocks (linkers). We show that these materials are stable having bulk moduli in the range of 0.5 – 24 GPa, which increase with decreasing linker size. All MOFs are semiconductors or insulators with band gaps of 1.0 – 5.5 eV. These are mainly determined by the band gaps of the linkers. First results on a Cu-based MOF are presented as well.

1. Introduction

Metal-organic frameworks (MOFs)[1-5] are a new class of porous materials with tailored properties. They are built by self-assembling of well defined building blocks[6], so-called connectors and linkers [Fig. 1]. Copolymerization of a wide range of organic molecules (linkers) with polynuclear complexes (connectors) results in the formation of coordination polymers with uniform and monodisperse pore sizes in the nanometer region. The inorganic connecting units and organic linker molecules are designed to form a rigid and stable 3D network. They both act as secondary building blocks in the extended solid. The type of connection of the MOF building blocks is as important as the molecular units themselves.[1, 7, 8]

---

[1] E-mail: Agnieszka.Kuc@chemie.tu-dresden.de
[2] Corresponding author. E-mail: Jan-Ole.Joswig@chemie.tu-dresden.de
[3] E-mail: Andrey.Enyashin@chemie.tu-dresden.de
[4] E-mail: Gotthard.Seifert@chemie.tu-dresden.de

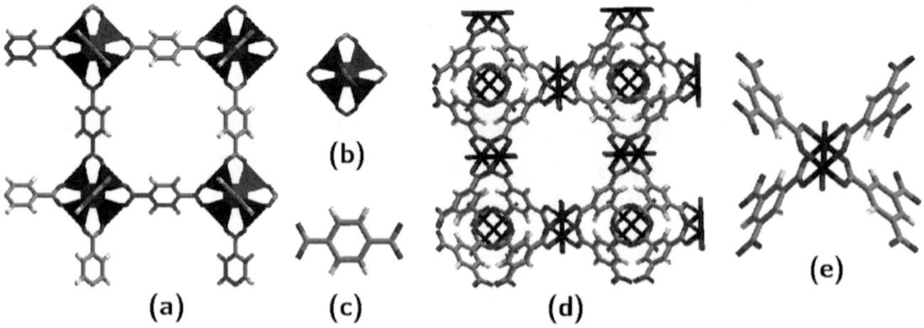

Figure 1: Examples of two topologically distinct isoreticular MOF structures. (a) A cubic fragment of IRMOF-1. Each corner of the cubes is built up of a connector (b) and linked by a linker (c) (here 1,4-benzenedicarboxylate). The linkers form the edges of the cube. The connector can be described as four distorted tetrahedra $Zn(O1)_3O$ connected by a central O2. (d) A cubic fragment of Cu-BTC ($Cu_3(BTC)_2(H2O)_3$; BTC: 1,3,5-benzene-tricarboxylate): The structure is built up of the so-called paddle-wheels (e). Each metal atom completes its pseudo-octahedral coordination sphere, $Cu(COO)_2$, with an aqua ligand.

For a given topology of a connector an isoreticular series of MOFs (IRMOFs) can be obtained. Such a series containing the $Zn_4O(CO_2)_6$ moiety in the connector have been successfully synthesized.[1] The synthesis of MOF materials is simple and based on mixing together a solution of the acid form of the linker with a simple metal salt in the desired stoichiometry.[8] The systematic variation of the pore size can be performed by using different organic molecules.

Metal-organic frameworks with large surface areas, controlled pore sizes, and easy functionalization of the organic part have a high potential in a variety of practical applications: molecular sorption, catalysis, gas separation, molecular sensors, etc.[7, 9-13] MOFs exhibit unique framework properties such as interpenetration, dynamical crystal-to-crystal transformations, chirality, and the lowest known densities for crystalline materials.[7]

The properties of MOFs can be modified by the following three approaches: (1) exchange of the linker, e.g. a benzene ring in IRMOF-1, by other organic molecules; (2) substitution of the connector by changing the metal atom in it; (3) exchange of both the linker and the connector.

Theoretical investigations of very few MOF structures have already been reported in the literature.[14-17] The exchange of the transition metal (Zn by Cd, Be, Mg, Ca) in IRMOF-1 has been studied theoretically.[15] Semiempirical, Hartree-Fock and density-functional calculations within periodic boundary conditions have been performed for several IRMOFs[14, 16, 17] focusing mainly on the geometries and the electronic properties.

In the present work we show the results of a systematic study and give an overview over the properties of a wide range of isoreticular metal-organic frameworks. First results of calculations with an exchanged metal atom (Cu) in the connector are presented as well ($Cu_3(BTC)_2(H2O)_3$; BTC: 1,3,5-benzene-tricarboxylate; in the following will be used Cu-BTC). For a more detailed study the reader is referred to Ref. 18.

The structures considered here [see e.g. Fig. 1 (a)] consist of an octahedral array of dicarboxylate organic bridges [Fig. 1 (c)] connected to a transition-metal complex [Fig. 1 (b)] of tetrahedral moieties $(OZn_4)^{6+}$. In this way the highly porous MOF materials are obtained with a well defined pore size distribution. We have studied well-known MOF structures and also hypothetical systems that may exhibit interesting properties. The exchanged linker molecules consist either of polycyclic hydrocarbons (PAHs) or carbon cages [Fig. 2]. As a reference system we have chosen the hypothetical IRMOF-M0 structure with no organic linker. This structure is built up of the inorganic clusters connected by carbon atoms C1 from the carboxylate groups [see Fig. 3 for atoms numbering].

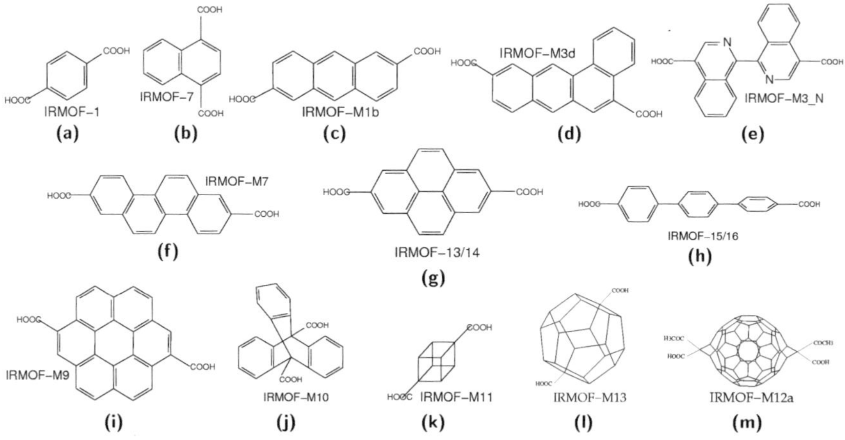

Figure 2: Some examples of organic linkers considered in the present work.[18] For clarity hydrogen atoms are not shown. Notation used in the text: IRMOF-x, where x is corresponding to the name given in the literature for experimentally known MOFs or IRMOF-Mx denoting the marks of the proposed linkers for hypothetical MOFs studied here (a ,b ,c ,etc.-indicate different isomers for a given linker.).

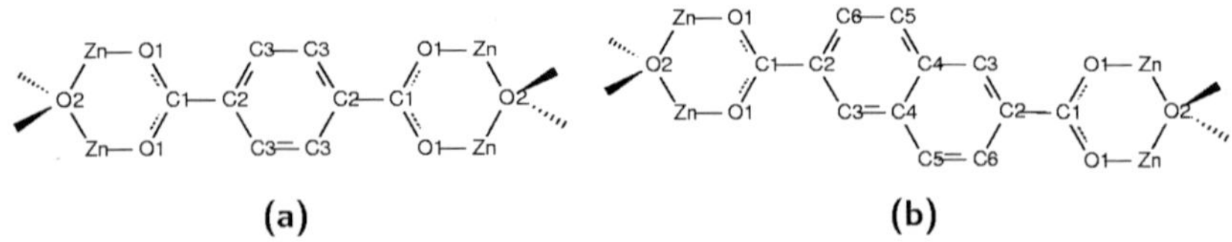

Figure 3: The zinc-oxo carboxylato ring connected with the organic ring in the 'in-plane' orientation in IRMOF-1 (a) and IRMOF-8 (b) and the atomic labels used in the text. The hydrogen atoms are not shown for clarity.

2. Computational details

We have employed a density-functional tight-binding (DFTB) scheme[19, 20] for the calculation of the total energy and several properties of a wide range of MOF crystal structures. Periodic boundary conditions (PBC) have been used to represent the infinite framework of the solid state. Lattice parameters and MOF structures have been fully optimized. The number of ***k*** points was determined by reaching convergence with respect to the total energy.

We have considered MOFs with face-centered cubic (FCC), simple cubic (SC) or body-centered cubic (BCC; catenated) crystal structures. The electronic properties have been analyzed for the FCC unit cells only. The bulk moduli ***B*** were calculated from the elastic constants, which have been obtained by calculating the total energy change after applying a suitable strain to the system. Furthermore, we have employed molecular dynamics (MD) simulations to check the thermal stability of the MOFs.

As mentioned above, a MOF network consists of connectors and linkers [see Fig. 1 (b) and (c)]. Some structural information on the studied molecules will be given in the following. The numbering convention of the atomic positions is given in Fig. 3.

The connector is a complex consisting of a fourfold coordinated central oxygen atom O2, surrounded by four fourfold coordinated zinc atoms, resulting in the $(Zn_4(O1)_{12}O2)^{6+}$ moieties. This arrangement creates six inorganic (zinc-oxo carboxylato) and six organic, e.g. benzene, rings per corner [Fig. 1 (b)]. We have chosen also a connector with a different transition metal atom [Cu-BTC; see Fig. 1 (d)]. This connector is built up from $Cu_2(COO)_4$ moieties, so-called paddle-wheels [Fig. 1 (e)]. Each metal atom completes its pseudooctahedral spheres with an axial aqua ligand.

The organic linkers, that have been studied, can be divided into four groups: (1) symmetric linkers, in which the atoms O2–C1–C1–O2 are in the same line coinciding the linkage axis [Fig. 3 (a), (g)]; (2) asymmetric linkers (relative to the linkage axis), with the O2–C1–C1–O2 atoms in the same line [Fig. 3 (b)]; (3) symmetric linkers, in which the O2–C1–C1–O2 atoms do not lie in the same line [Fig. 3 (c), (f)]; (4) asymmetric linkers, in which the O2–C1–C1–O2 atoms are not in the same line [Fig. 3 (d)].

The electronic properties of the extended crystal structures and the free building blocks have been compared. As free building blocks the linkers were represented in their dicarboxylate forms and the connector was considered as a part of the ZnO (wurtzite) crystal saturated with H atoms.

3. Results and discussion

The calculated equilibrium distances in the MOFs show that the bond lengths in the zinc-oxo carboxylato ring are almost unchanged for different organic linkers compared to the reference system IRMOF-M0. Most distances agree very well with corresponding experimental data within errors of 2-3%. Only the Zn–O distances are slightly larger than in the experiment, but this overestimation is uniform for all systems. Moreover, the calculated equilibrium lattice constants of all studied compounds are slightly larger than those, of which experimental data was available. The error is in the range 2-4%. For more details the reader is referred to the Ref. 18.

The O1–C1 distances in all MOFs correspond to values between those for the typical single (1.42 Å) and double (1.22 Å) O–C bond, while the C1–C2 bond lengths are close to the typical single C–C bond in sp^2 hybridization (~1.46 Å) independent on the linker type. We find, however, that the orientation of organic and inorganic rings (cf. Fig. 3) changes significantly for different types of linkers, going from an 'in-plane' orientation to a perpendicular one with twisted forms in between. If the connector is not distorted then the O1–C1–C2–C3 torsion angle is 'in-plane' and has one fixed O1–Zn–O1 angle (~108°). This is typical for group 1 of the linkers. For linkers from the group 3 the zinc-oxo carboxylato ring can be slightly distorted. In this case the O1–Zn–O1 angles vary in a small range (108°–113°) and both rings can be either 'in-plane' or orthogonal to each other. A wide range of values for the O1–Zn–O1 angle (108°–125°) and the twisted O1–C1–C2–C3 torsion angle is typical for a large distortion of the inorganic part (group 2 and 4). Distortions of the connector caused by some linkers may, thus, lead to a symmetry lowering.

In order to compare the energetic stability of the different MOFs, we estimated a formation energy ΔE as the difference in total energies of the products and reactants, according to the following reaction:

$$Zn_4O(OH)_6 + 3\ R(COOH)_2 \rightarrow Zn_4OR_3\ (COO)_6 + 6\ H_2O$$

The calculated formation energies are all negative suggesting that energetically a formation of these structures – even the hypothetical ones – is favorable. The most stable structure, concerning ΔE, is IRMOF-18 that contains the benzene-based linker functionalized by four methyl groups. These are arranged in a conformation, in which the steric repulsion between both rings is minimized suggesting that the functionalization of the linker is one of the factors that can stabilize the system. We also found that the BCC structures are more stable than the corresponding FCC structures only for long linkers, due to steric hindrance effects. Moreover, it can be noticed that different isomers of the same linker have similar ΔE. Generally, the formation energy depends strongly on the size and the shape of the organic linker.

We have investigated also the mechanical properties of IRMOFs. The structures are held together by strong Zn–O–C bonds. The linkage between the Zn_4O group and the organic moieties results in rather soft materials with relatively small bulk moduli (0.5–24 GPa) compared with cubic diamond (theory[21]: 441–457 GPa, exp.[22]: 443 GPa) and the wurtzite structure of zinc oxide (theory[23]: 160 GPa, exp.[24]: 183 GPa). The Cu-BTC has also a small bulk modulus (~18 GPa). However, it is twice that of IRMOF-1 (~9 GPa) that contains a similar linker. The calculated values of ***B*** indicate that MOFs are easily compressible systems. Since the inorganic basic system ZnO has a much larger bulk modulus, the lowering of ***B*** in the IRMOFs is caused by the introduction of the organic linker molecules. The length of the linker has a significant influence on the bulk modulus: IRMOFs with hypothetical cage-like linkers and IRMOF-M0 show the largest bulk moduli.

The organic linkers can rotate freely in the solid MOF at ambient conditions. The energy barrier of a linker rotation along the connection axis is $\Delta E_{rot} \approx 0.35$ eV (IRMOF-1). The rotation can, therefore, be thermally activated, which we also observed in MD simulations. These we have performed within NVT ensembles at 300 K and 1200 K. The nearly free rotation of the linker was already observed at 300 K. These simulations indicate also good thermal stability of the MOFs even at 1200 K.

In order to study the electronic properties of the metal-organic frameworks we have analyzed their Mulliken charges as well as the densities of states in the periodic crystals and the free building blocks. The calculated atomic Mulliken charges of different MOF systems show an almost unchanged charge distribution compared to the smallest in the series IRMOF-1. Moreover, the solid structures keep nearly the same atomic charges as the free linkers and free connector. This we also have found for the Cu-BTC framework.

The Zn atoms are positively charged (+0.92). This number is in between that for the free building block (+1.09) and the bulk ZnO (wurtzite; +0.82). On the other hand, the oxygen atoms O1 and O2 are both negatively charged with –0.70 ... –0.77 and –0.88 ... –0.97, respectively. In the free connector the charges are –0.92 (O1) and –1.26 (O2). Thus, in the MOF crystal the oxygen charges are decreased. Here, in the bulk ZnO the O2 charges (–0.82) are slightly smaller than that in the MOFs. In the dicarboxylate form of the free linker the charges of the carboxyl and the hydroxyl oxygen are –1.02 and –0.35, respectively, and in the MOF crystal the oxygen charges (O1) are therefore an average of both values. Finally, the carbon atoms in the free linker and the MOF crystal have in general the same charge distribution. The C1 atoms are positively charged (+0.92), while other C atoms have charges close to zero. In the case of the Cu-BTC crystal we have found that the Cu atoms are positively charged (+1.50), while the O1 atoms have charges similar to those in the IRMOFs. The C1 atoms have charges of +0.70, while for other C atoms they are close to zero.

Additionally the partial and total densities of states (PDOS, DOS) for the free linkers and the free connector in comparison with IRMOF crystals have been studied. The results show that the contributions of the carbon atoms of a free linker and the MOF system to the DOS do not significantly differ. On the other hand, the PDOS of the oxygen atoms change in both cases, since in a MOF we have only one type of O1 atoms, whereas in the free linker there are the two different oxygen atoms of the carboxyl group. Considering the free connector, the zinc PDOS is very similar to that in the MOF systems or bulk ZnO.

To analyze the differences in the densities of states for IRMOFs with different linkers IRMOF-M0 was chosen as a reference system. The IRMOF-M0 PDOS (for different states of a given atom) shows that the valence band is composed of Zn 3*d*, O 2*p* and C 2*p* states. The unoccupied band is determined by *s* and *p* orbitals of Zn, and *p* orbitals of O and C. Similar electronic properties have been found for IRMOF-1 being in agreement with the earlier work of Fuentes-Cabrera *et al.*[15] Fig. 4 (left) shows the PDOS of exemplary structures with different sizes of the linker. The results are compared with the PDOS of the reference system and IRMOF-M11 (cage-like linker). In Fig. 4 (right) the PDOS of the Cu-BTC structure is shown.

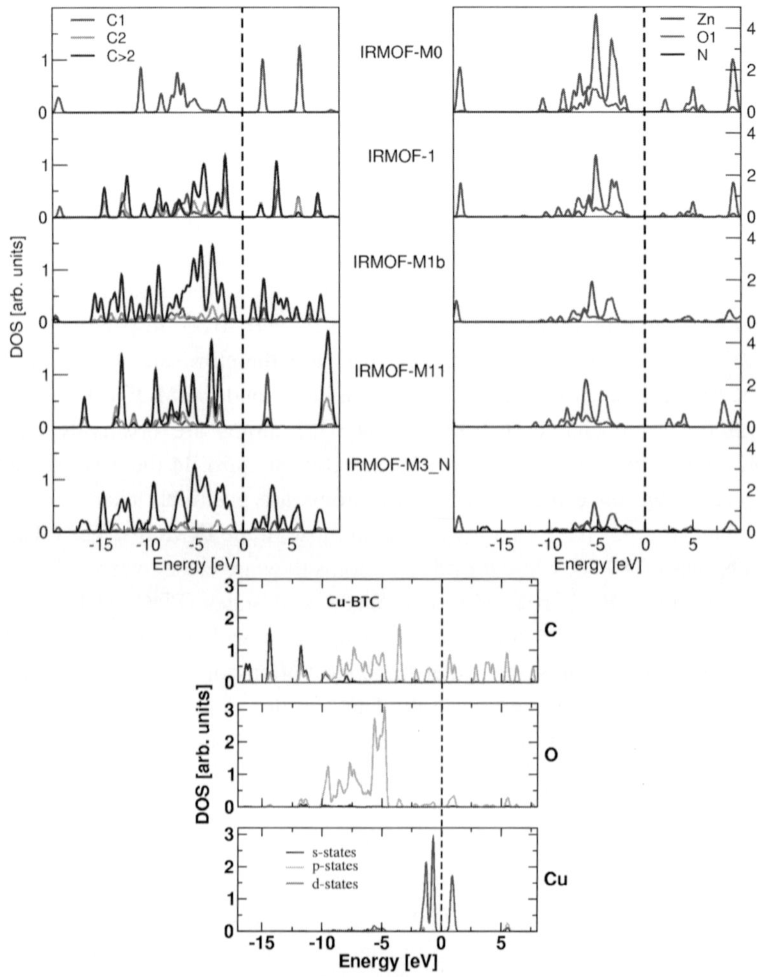

Figure 4: Left: The partial density of states of the C, Zn, O1 and N atoms in some exemplarily chosen IRMOFs. For clarity the atoms of carbon higher than 2 (crystallographic position) are presented together. Right: The PDOS of C, Cu and O1 atoms in the Cu-BTC MOF. The Fermi level is indicated by a vertical dashed line and the DOS is given in arbitrary units.

The overall electronic properties of the IRMOFs are basically characterized by $(Zn1)_4(O2)$ clusters that bring a character of a wide band-gap semiconductor (ZnO) to the system [Fig. 5]. On the other hand, the organic (especially PAH-type) linkers reduce the gap size [Fig. 6]. The values of the MOF band gaps strongly depend on the size of the linkers. Smaller band gaps are observed for linkers with larger number of conjugated carbon atoms. The largest numbers belong to those structures that do not contain linkers with conjugated sp^2 hybridized carbon atoms, e.g. IRMOF-M0, -M11 or -M13. The MOF band gaps have a range of 1.3–5.5 eV, i.e. they are in between insulating and semiconducting materials. The Cu-BTC band gap of 1.22 eV is smaller than the band gaps of Zn-based MOFs.

The results of the partial densities of states show that the band gap is dominated by the π states of sp^2 carbon atoms within the linkers. It can be noticed in Fig. 4 that with increasing number of sp^2 carbon atoms in the linker the band gap decreases. This is due to the larger contribution of π states, which means the top of the valence band is dominated by the π states of the C3, C4, etc. atoms. A similar tendency is observed for unoccupied bands, where a big influence of π states of C1 is observed for structures with a small number of sp^2 C atoms (e.g. IRMOF-M0 and IRMOF-M11).

4. Conclusions

In the present work we have investigated a set of isoreticular metal-organic frameworks with a cubic topology. The inorganic part is based on the tetrahedral moieties $(OZn_4)^{6+}$. A structure with a different arrangement of the connector (Cu-BTC) was studied as well and presented for the first time. The connectors were linked by a wide range of different organic molecules. The resulting MOF structures have been analyzed with respect to their stability, geometry and electronic properties.

All MOF systems have been found to be energetically stable. The negative energies of formation depend strongly on the size and shape of the different linkers. Symmetric linkers seem to have at most only a small influence on the geometry of the inorganic MOF part, independent of their size, whereas asymmetric linkers cause remarkable distortions of the connector. This is a possibility for MOFs existing in a non-cubic lattice. Moreover, we observed in MD simulations that a thermal activation of the free rotation of the linker is possible already at 300 K. The bulk moduli showed that MOFs are easily compressible systems.

The charge distribution within the MOFs is kept unchanged compared to the building blocks, except for the charges of the linking oxygen atoms, since these are chemically different in the free linkers and not in the periodic MOF structure. The band gaps of different systems vary over a wide range resulting in either semiconducting or insulating MOFs. Moreover, the size of the band gap is dominated by C sp^2 states of the linkers. Increasing the number of sp^2 carbon atoms in the linker decreases the band gap. However, there is a risk that systems loose their stiffness (smaller bulk moduli), when longer linkers are introduced.

Acknowledgments

The authors acknowledge financial support by Stiftung Energieforschung BW and thank Dr. Thomas Heine and Dr. Augusto F. Oliveira for fruitful discussions.

References

1. Yaghi, O. M.; O'Keeffe, M.; Ockwig, N. W.; Chae, H. K.; Eddaoudi, M.; Kim, J., *Nature* **2003,** 423, 705--714.
2. Chui, S. S. Y.; Lo, S. M. F.; Charmant, J. P. H.; Orpen, A. G.; Williams, I. D., *Science* **1999,** 283, 1148--1150.
3. Rosi, N. L.; Eddaoudi, M.; Kim, J.; O'Keeffe, M.; Yaghi, O. M., *Cryst. Eng. Comm.* **2002,** 401--404.
4. Rowsell, J. L. C.; Yaghi, O. M., *Microp. Mesop. Mater.* **2004,** 73, 3--14.
5. Schlichte, K.; Kratzke, T.; Kaskel, S., *Microp. Mesop. Mater.* **2004,** 73, 81--88.
6. O'Keeffe, M.; Eddaoudi, M.; Li, H. L.; Reineke, T.; Yaghi, O. M., *J. Solid State Chem.* **2000,** 152, 3--20.
7. Eddaoudi, M.; Kim, J.; Rosi, N. L.; Vodak, D.; Wachter, J.; O'Keeffe, M.; Yaghi, O. M., *Science* **2002,** 295, 469--472.
8. Eddaoudi, M.; Moler, D. B.; Li, H. L.; Chen, B. L.; Reineke, T. M.; O'Keeffe, M.; Yaghi, O. M., *Acc. Chem. Res.* **2001,** 34, 319--330.
9. Eddaoudi, M.; Li, H. L.; Yaghi, O. M., *J. Am. Chem. Soc.* **2000,** 122, 1391--1397.
10. Halder, G. J.; Kepert, C. J.; Moubaraki, B.; Murray, K. S.; Cashion, J. D., *Science* **2002,** 298, 1762--1765.
11. Krawiec, P.; Kramer, M.; Sabo, M.; Kunschke, R.; Frode, H.; Kaskel, S., *Adv. Eng. Mater.* **2006,** 8, 293--296.
12. Rosi, N. L.; Eckert, J.; Eddaoudi, M.; Vodak, D. T.; Kim, J.; O'Keeffe, M.; Yaghi, O. M., *Science* **2003,** 300, 1127--1129.
13. Seo, J. S.; Whang, D.; Lee, H.; Jun, S. I.; Oh, J.; Jeon, Y. J.; Kim, K., *Nature* **2000,** 404, 982--986.
14. Braga, C. F.; Longo, R. L., *J. Mol. Struct.-THEOCHEM* **2005,** 716, 33--38.
15. Fuentes-Cabrera, M.; Nicholson, D. M.; Sumpter, B. G.; Widom, M., *J. Chem. Phys.* **2005,** 123, 124713.
16. Kim, D.; Lee, T. B.; Choi, S. B.; Yoon, J. H.; Kim, J.; Choi, S. H., *Chem. Phys. Lett.* **2006,** 420, 256--260.

17. Mattesini, M.; Soler, J. M.; Yndurain, F., *Phys. Rev. B* **2006,** 73, 094111.
18. Kuc, A.; Enyashin, A.; Seifert, G., *J. Phys. Chem. B* **2007,** 11, 8179--8186.
19. Porezag, D.; Frauenheim, T.; Köhler, T.; Seifert, G.; Kaschner, R., *Phys. Rev. B* **1995,** 51, 12947.
20. Seifert, G.; Porezag, D.; Frauenheim, T., *Int. J. Quantum Chem.* **1996,** 58, 185--192.
21. Mailhiot, C.; McMahan, A. K., *Phys. Rev. B* **1991,** 44, 11578--11591.
22. McSkimin, H. J.; Andreatch, P., Jr.; Glynn, P., *J. App. Phys.* **1972,** 43, 985.
23. Ahuja, R.; Fast, L.; Eriksson, O.; Wills, J. M.; Johansson, B., *J. App. Phys.* **1998,** 83, 8065--8067.
24. Karzel, H.; Potzel, W.; Kofferlein, M.; Schiessl, W.; Steiner, M.; Hiller, U.; Kalvius, G. M.; Mitchell, D. W.; Das, T. P.; Blaha, P.; Schwarz, K.; Pasternak, M. P., *Phys. Rev. B* **1996,** 53, 11425--11438

EXPERIMENTAL TECHNIQUES TO MEASURE OF THE EQUILIBRIUM PLATEAU PRESSURES OF METAL HYDRIDES

ANDREAS BORGSCHULTE, SHUNSUKE KATO, MICHAEL BIELMANN, ANDREAS ZÜTTEL

EMPA, Materials Science and Technology, Laboratory 138, Hydrogen & Energy, Überlandstrasse 129, CH-8600 Dübendorf, Switzerland

Various experimental techniques to measure the equilibrium pressures of metal hydrides are reviewed, that is gravimetric, volumetric pcT measurements, hydrogenography and resistance measurements on thin films. The agreement of the data for the archetypical example MgH_2 is very good. Differences are explained by kinetic effects of the sorption process.

1. Introduction

The standard approach for the search of new hydrogen-storage materials is to synthesize bulk samples and to use gravimetric [1,2] or volumetric [1,3] techniques to follow their hydrogenation reaction and to record pressure–concentration isotherms (pcT). The equilibrium pressure of the metal-to-hydride transition is determined from the plateau of the pressure composition isotherms. The enthalpy of hydride formation is extracted from the temperature dependence of the equilibrium pressure, by means of the Van't Hoff relation:

$$\ln\left(\frac{p_{eq}}{p_0}\right) = \frac{\Delta H}{RT} - \frac{\Delta S}{R}, \qquad (1)$$

where ΔH is the enthalpy of formation in J/(mol H_2), ΔS is the entropy of formation in J/(K mol H_2) at standard pressure, R the gas constant, T the absolute temperature, p_0 = 1.013 bar the standard pressure, and p_{eq} the H_2 equilibrium plateau pressure of the p–c isotherm.

The technically challenging problem is the measurement of such equilibrium isotherms. Using gravimetric techniques, the mass of the sample is a measure of the exchanged hydrogen. 'Equilibrium' is practically defined, i.e. when the sample mass does not change anymore after applying a particular hydrogen pressure to the sample. Due to the slow kinetics of most systems, such measurements can take days to weeks. Accordingly, most frequently used techniques are scanning methods, in which one thermodynamic parameter (either pressure or temperature) is linearly varied and the response of the sample is recorded. As these measurements are by definition dynamic measurements, the equilibrium values have to be extrapolated. On the other hand, the scanning method allows the use of other physical properties than the direct measurement of the hydrogen content in the sample, e.g. the pressure change in the sample container accompanied with the amount of hydrogen applied (volumetric methods), the heat flux into the sample by high-pressure DSC, [4], or any other measurable change connected to hydrogen in the sample. The fact that hydrogen absorption in a metal leads to large optical changes is the basis of a new combinatorial method 'hydrogenography'[5]. Hydrogenography provides a high-throughput method to measure optically pcTs and determine the enthalpy of hydride formation. Similarly, the electric conductivity can be used to probe the electronic changes accompanied with hydrogen absorption [6]. In this paper, we will compare the various methods and discuss their particular applicability.

2. Methods

In what follows is a short description of the physical principles of the measurement of thermodynamic properties of metal hydrides. The interested reader is referred to literature references to obtain more information.

2.1. *Gravimetry*

For pressure composition isotherms one measures the concentration of hydrogen inside the sample by measuring its weight as a function of the hydrogen gas pressure around the sample. [1,2] The procedure is as follows. First, the reaction chamber is evacuated. Then hydrogen gas is added to reach a pre-set pressure and maintain it until the concentration inside the sample has reached its equilibrium value. The pressure is increased and the whole procedure is repeated at constant temperature. All the isotherms have approximately the same shape but in the case of stoichiometric hydrides there exists a set of isotherms with a well defined plateau. These plateaus indicate the presence of two coexisting phases: hydrogen dissolved in the metal (α-phase) and a concentrated hydride phase (α'-phase). When the logarithm of plateau pressures are plotted as function of the inversed temperature – the so-called van 't Hoff plot, a straight line is found. The slope of this line reveals the heat of formation, the intercept the entropy of formation (equation 1). The experimental challenge of the gravimetric method is to find a way to subtract the varying buyoncy contribution from the measured weight.

2.2. *Volumetry*

There exist two volumetry methods to measure pcT curves [1,3]. The simplest one is the so-called Sieverts method. The hydrogen pressure in the sample container is abruptly changed. The pressure in the immediately closed container is monitored. The difference between the initial and equilibrated pressure is measure of the amount of absorbed hydrogen. For dynamic pcT-measurements, hydrogen is introduced with a constant flow into the evacuated sample container, while monitoring pressure and time. The integrated flow is used for the determination of the absorbed hydrogen content. While the Sieverts method has the advantage of a simple setup, the uncertainty of the measured amount of absorbed hydrogen is larger than when using the dynamic method.

2.3. *Differential Scanning Calorimetry*

In isobaric DSC measurements the temperature of the sample is continuously increased (or decreased) and the corresponding heat flow is measured. When achieving the miscibility gap in the metal-hydrogen phase diagram, the hydrogen content will change rapidly, so that the integrated DSC signal is an estimate of the enthalpy change, if quasi-equilibrium conditions can be assumed. Due to the often sluggish sorption kinetics, this leads frequently to wrong results [4]. Therefore, simple equilibrium conditions cannot be assumed and thus the measured heat exchange does not resemble ΔH. To circumvent this problem, van‘t Hoff plots are determined with DSC by detecting the onsets of the temperature, at which the de- and hydriding reaction start [4,9]. The equilibrium value can be assumed to lie in between the two values. Repeating the measurements at various applied pressure reveals a van’t Hoff plot.

2.4. *Hydrogenography*

While in DSC measurements the exchanged heat is linked to hydrogen sorption, hydrogenography is based on the fact that most complex metal–hydrogen systems undergo a metal–insulator transition upon hydrogen exposure [5]. In good approximation, the logarithm of the transmission is proportional to the hydrogen concentration in the metal [12]. By following the optical changes during hydrogenation, one is thus able to measure the amount of absorbed hydrogen and analyze similar to the pcT-measurements in 2.1. The mapping of the hydrides formed in a large compositional gradient sample enables the investigation of a full metal-hydrogen ternary phase diagram with one sample.

2.5. *Resistance Measurements*

The metal-insulator transition induced by hydrogen causes also changes of the electric resistance of the sample. The physical relation is rather complicated as it depends on the electronic changes as well as hydride growth mechanism and morphology. Therefore, in most cases, the change in resistance behavior is used to determine the onset of hydride formation/ decomposition. Repeating the measurements at various applied pressures/temperatures reveals a van't Hoff plot, similar to DSC-measurements (2.3).

3. Comparison on MgH_2

Figure 1 shows the equilibrium plateau pressures of MgH_2 measured by various techniques. Despite the principally different measurement methods and sample preparation (bulk, thin films, additives, surface conditions), the reconstruction of a well-defined van't Hoff plot over 7 orders of magnitude is possible. Differences are visible though.

To highlight them, Fig. 2 shows an enlargement of Fig. 1. It is worth mentioning that the most significant differences occur within measurements using the same method (Sieverts: Stampfer et al [8] and Oelerich et al. [10]), while the DSC measurements of Rongeat et al [4] and Sieverts measurements by Stampfer et al [8] are in very good agreement. The reason for this is that the uncertainty of the measurements depends in first order only on the relative measurement time, which is defined by the absolute measurement time divided by the time the sample needs to reach equilibrium. In DSC measurements the absolute measurement time is determined by heating/cooling rate.

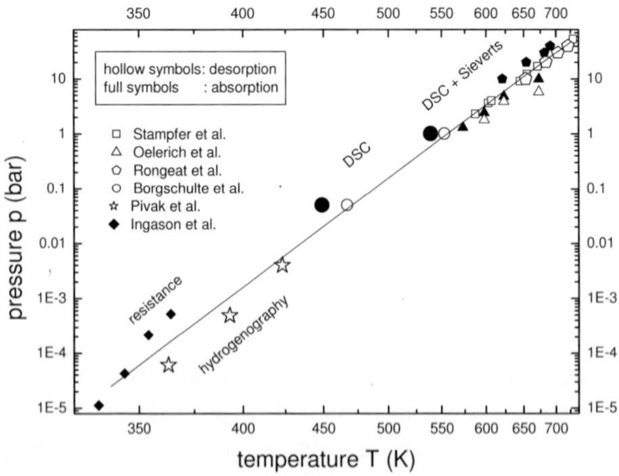

Figure 1. Van 't Hoff plots for absorption and desorption of MgH_2 measured by various techniques. References are Stampfer et al [8], Oelerich et al [10], Rongeat et al [4], Borgschulte et al [9], Pinak et al [7], Ingason et al. [6].

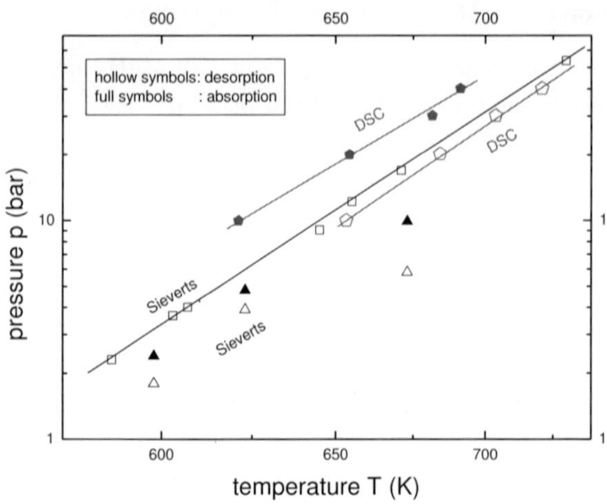

Figure 2. Enlargement of Fig. 1, highlighting the scattering of the data.

This is demonstrated in Fig. 3 by DSC measurements of MgH_2. A fast cycle leads to a kinetically destabilized MgH_2 [9]. By annealing the sample in hydrogen, the equilibrium state is reached. Accordingly, the onset temperatures defining the equilibrium plateau pressures differ significantly.

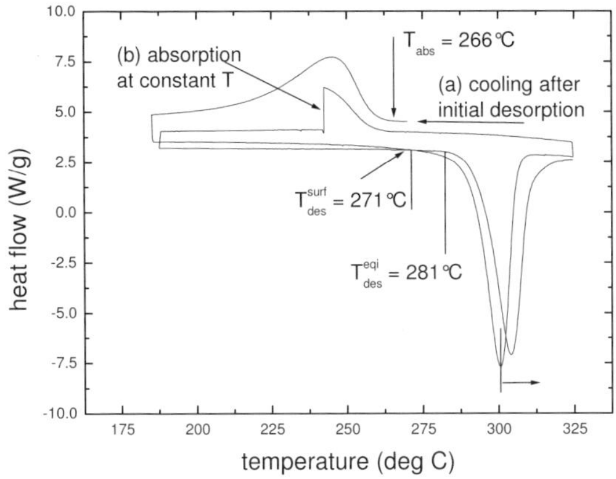

Figure 3. After the initial desorption, curve (a) is loaded during cooling with 5 K/min, while the absorption in (b) is supported by keeping the sample at constant T = 160°C for 100 min. The subsequent desorption is performed with a heating rate of 5 K/min for both curves. Hydrogen pressure is 1 bar. For better visibility, the absorption curves have been shifted. From Ref. [9].

Similar effects have been found in pressure composition isotherms of $LaNi_5H_x$ measured volumetrically [11]. Figure 4 shows variations of the absorption pressure with time to execute an entire measurement loop. As clearly visible, even after a measurement time of one week per cycle, changes in the plateau pressures are observable.

Apart from the measurement time, the kinetics of the material itself varies from experiment to experiment. Thus the relative measurement time as defined above and accordingly the accuracy of the data can vary significantly even when using the same apparatus and measurement time. Observed enthalpy changes might therefore be an artifact due to different kinetics induced by e.g. additives. Thin films need a catalytic cap layer (used for hydrogenography and resistance measurements in Fig. 1), while the DSC-measurements used Ni or Nb_2O_5 doped MgH_2.

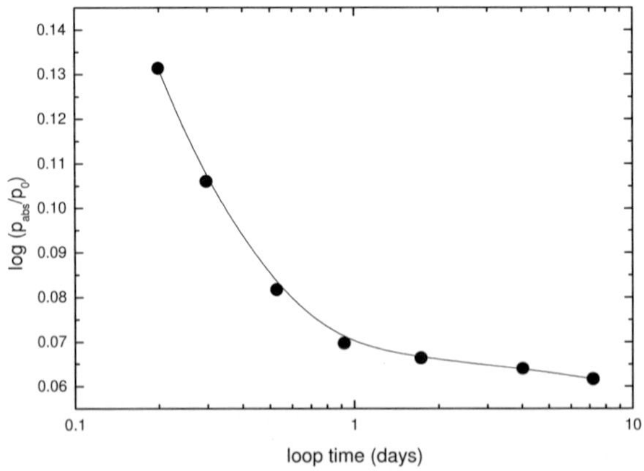

Figure 4. Variations of the absorption pressure of $LaNi_5H_5$ with time to execute an entire measurement loop. Measurement temperature is 20°C, measurement method is Sivert's method. From Ref. [11].

4. Summary

The paper compares various experimental techniques to measure the equilibrium pressure of metal hydrides. Gravimetric, volumetric pcT measurements, hydrogenography and resistance measurements on thin films are discussed. The agreement of the data for the archetypical example MgH2 is very good. Differences are explained by kinetic effects of the sorption process.

Acknowledgments

This work was financially supported by the European Commission (contract numbers MRTN-CT-2006-032474 (Hydrogen) and MRTN-CT-2006-035366 (COSY)).

References

1. E. Poirier, R. Chahine, A. Tessier, and T. K. Bose, *Gravimetric and volumetric approaches adapted for hydrogen sorption measurements with in situ conditioning on small sorbent samples.* Rev. Sci. Instrum. 76, 055101 (2005).
2. Andreas Peter Vestbø, Jens Oluf Jensen, Niels J. Bjerrum, *Development of a high-pressure microbalance for hydrogen storage materials*, J. All. Compds. 446–447 (2007) 703–706

3. T.P. Blach and E.MacA. Gray, *Sieverts apparatus and methodology for accurate determination of hydrogen uptake by light-atom hosts*, J. All. Compds. 446-447 (2007) 692-697.
4. Rongeat, C. et al, *Determination of the Heat of Hydride Formation/ Decomposition by High-Pressure Differential Scanning Calorimetry (HP-DSC)*, J. Phys. Chem. B.; (Article); 2007; ASAP Article
5. R. Gremaud et al., *Hydrogenography: An Optical Combinatorial Method To Find New Light-Weight Hydrogen-Storage Materials,* Adv. Mater. 2007, 19, 2813–2817.
6. A.S. Ingason and S. Olafsson, *Thermodynamics of hydrogen uptake in Mg films studied by resistance measurements*, J. All. Compds. 404–406 (2005) 469–472.
7. Pival et al, to be submitted.
8. J.F. Stampfer, C.E. Holley, J.F. Suttle, *The magnesium-hydrogen system,* J. Am. Chem. Soc. 82 (1960) 3504.
9. A. Borgschulte, U. Bösenberg, G. Barkhordarian, M. Dornheim and R. Bormann, *Enhanced hydrogen sorption kinetics of magnesium by destabilized* $MgH_{2-\delta}$, Catal. Today 120 (2007) 262-269.
10. W. Oelerich, *Sorption properties of nanocrystalline metal hydrides for the storage of hydrogen*, thesis, TU Hamburg-Harburg (2000).
11. C. E. Buckley, E. Mac A. Gray and E. H. Kisi, *Stability of the hydrogen absorption and desorption plateaux in* $LaNi_5$*-H Part 1: Hysteresis dynamics and location of the equilibrium isotherm,* J. All. Compds. 215, (1994) 195-199.
12. A. Borgschulte, R. J. Westerwaal, J. H. Rector, B. Dam, and R. Griessen, *Hydrogen sorption mechanism of oxidized nickel clusters*, Appl. Phys. Lett. 85, 4884 (2004).

CHARACTERIZATION OF COMPLEX METAL HYDRIDES BY HIGH-RESOLUTION SOLID STATE NMR SPECTROSCOPY

ROBERT C. BOWMAN, JR. AND JOSEPH W. REITER

Jet Propulsion Laboratory, Mail Stop 79-24, California Institute of Technology, Pasadena, CA 91109-8099, USA

SON-JONG HWANG AND CHUL KIM

Division of Chemistry and Chemical Engineering, California Institute of Technology, Pasadena, CA 91125, USA

HOURIA KABBOUR

Division of Engineering and Applied Science, California Institute of Technology, Pasadena, CA 91125, USA

Solid state nuclear magnetic resonance (NMR) studies provide very detailed information on the complicated processes involving the formation of hydride phases and their transformations including reversibility and the roles of catalysts. Examples from recent investigations on complex metal hydrides illustrate how NMR can address and often resolve diverse issues on phase formation and decomposition processes. First, the behavior of Sc-doping on the phase compositions and reactivity of the sodium tetra- and hexa-alanates has been assessed from the ^{45}Sc, ^{27}Al, and ^{23}Na spectra following mechanical milling and hydrogen absorption/desorption treatments. Second, the formation and subsequent hydrogen desorption from several borohydrides was investigated using ^{11}B MAS and CPMAS spectra that included clear evidence for the formation of highly stable intermediate "B_nH_n" species, mainly as $M_xB_{12}H_{12}$ phases, in their decomposition products that severely impact their ability to reform the initial borohydride phases. Finally, NMR was used phase identification in the Li-Mg-Al-N-H system that could not be confirmed by x-ray diffraction or other methods.

1. Introduction

Complex metal hydrides (i.e., alanates, borohydrides, and amides/imides) of the light elements (i.e., Li, Na, and Mg) appear to have the best chance of meeting the daunting hydrogen storage targets [1] issued by the US Department of Energy (DOE) for mobile applications. Although there have been extensive efforts by the international research community to develop complex hydrides with improved properties and to understand their reactions as well as catalyst effects [2], all the known complex hydrides have serious shortfalls with respect to meeting the required supply pressures and reaction kinetics at the operating temperatures for PEM fuel cells or even exhibiting suitable reversibility during hydrogen desorption/absorption cycling [1, 2]. Greater insights into the phase transformation processes and catalysis mechanisms are needed.

Solid-state nuclear magnetic resonance (NMR) has been extensively used to assess structural properties, electronic parameters and diffusion behavior of the hydride phases of numerous metals and alloys using mostly transient NMR techniques or low-resolution spectroscopy [3]. The NMR relaxation times are extremely useful to assess various diffusion processes over very wide ranges of hydrogen mobility in crystalline and amorphous phases [3]. In addition, several borohydrides [4-6] and alanates [7-11] have also been characterized by these conventional solid-state NMR methods over the years where most attention was on rotation dynamics of the BH_4^-, AlH_4^-, and AlH_6^{-3} anions; detection of order-disorder phase transitions; or thermal decomposition. There has been little indication of fast long-range diffusion behavior in any complex hydride studied by NMR to date [4-11].

The utilization of advanced solid-state NMR techniques [12] such as Magic Angle Spinning (MAS), Cross-Polarization (CP) MAS, and multi-quantum (MQ) MAS NMR, in addition to traditional measurements of nuclear relaxation times and lineshapes, permits hydride phases of light elements to be investigated more thoroughly and efficiently as reviewed recently by Bowman and Hwang [13]. While ^{27}Al MAS-NMR spectra were obtained and analyzed on a $LiAlH_4$ sample in 1990 [14], these methods have been applied to other complex hydrides only within the past few years [13, 15-18]. In this paper, the potential of these MAS techniques for phase assessments will be illustrated on various complex hydrides being evaluated by the Caltech-JPL team as part of the DOE Metal Hydride Center of Excellence.

2. Experimental Details

Multinuclear solid state MAS NMR spectra were acquired using a Bruker Avance 500 MHz spectrometer with a wide bore 11.7 T magnet and employing a boron-free Bruker 4 mm CPMAS probe. The NMR shifts were reported in parts per million (ppm) with respect to "zeros" set to standard [12] external references: tetramethylsilane (TMS) for ^{1}H, 1.0 molar LiCl aqueous solution for ^{6}Li and ^{7}Li, BF_3-Et_2O for ^{11}B, 1.0 molar NaCl aqueous solution for ^{23}Na, 1.0 molar $Al(NO_3)_3$ aqueous solution for ^{27}Al, and a 1.0 molar $Sc(NO_3)_3$-HNO_3 aqueous solution for ^{45}Sc nuclei. The powder samples were packed into 4 mm ZrO_2 rotors and were sealed with a tight fitting kel-F cap. Sample spinning was performed using dry nitrogen gas. For quantitative analyses, the MAS NMR spectra were obtained at sample spinning rates of 12-14 kHz and using short single pulses (i.e., 0.3-0.5 μs < $\pi/6(I+1/2)$, I = spin quantum number [15]) with the application of strong ^{1}H decoupling pulses. The ^{11}B CPMAS experiments were performed at various contact periods under the Hartmann-Hahn matching condition for ^{1}H (I=1/2) and ^{11}B (I=3/2) nuclear spins at a sample spinning rate of 12-13 kHz. All MAS and CPMAS spectra reported in this paper were measured at room temperature.

3. MAS-NMR Studies of Sc-Doped $NaAlH_4$

The discovery that Ti and certain other metals (e.g., Sc) greatly enhanced the reversible hydrogen absorption rates of $NaAlH_4$ stimulated the intense interest of complex hydrides as high performance storage materials [2]. In spite of extensive theoretical and experimental studies, the actual catalytic mechanisms in Ti-doped alanates still remain unresolved and very controversial [2]. Since the ^{45}Sc nucleus possesses excellent NMR sensitivity with large chemical/Knight shifts [12] as well as being a very effective catalyst with $NaAlH_4$ [19, 20], MAS-NMR studies have been performed on Sc-doped $NaAlH_4$ samples when prepared by ball milling and after two hydrogen desorption/absorption cycles. The ^{23}Na, ^{27}Al, and ^{45}Sc spectra are shown in Fig. 1 where different phases are identified. As previously seen in $NaAlH_4$ ball milled with $TiCl_3$ [2]; Al metal, Na_3AlH_6, and NaCl phases were formed during similar processing with $ScCl_3$ according to the following reaction equation:

$$NaAlH_4 + 0.04ScCl_3 = 0.88\ NaAlH_4 + 0.12\ NaCl + 0.12\ Al + 0.04\ \text{“Sc”} \quad (1)$$

Generic metallic Sc phases are denoted by “Sc”. The Na_3AlH_6 is present from enhancement of desorption by Sc-doping [19, 20]. While the fate of $ScCl_3$ after ball milling and hydrogen reactions was invisible via x-ray diffraction, which is similar to behavior of Ti-doped alanates, the peaks in the ^{45}Sc MAS-NMR spectra revealed various changes including formation of amorphous/nanophase $ScCl_3$, $Al_{1-x}Sc_x$ alloys, ScH_x, $ScAl_3$ intermetallic, as well of one or more currently unidentified phases. Additional MAS-NMR studies are in progress on samples with different amounts of $ScCl_3$ as well as performing more cycling experiments to understand better the role of the Sc additive on the reaction paths for the sodium alanate phases. These results will be reported elsewhere.

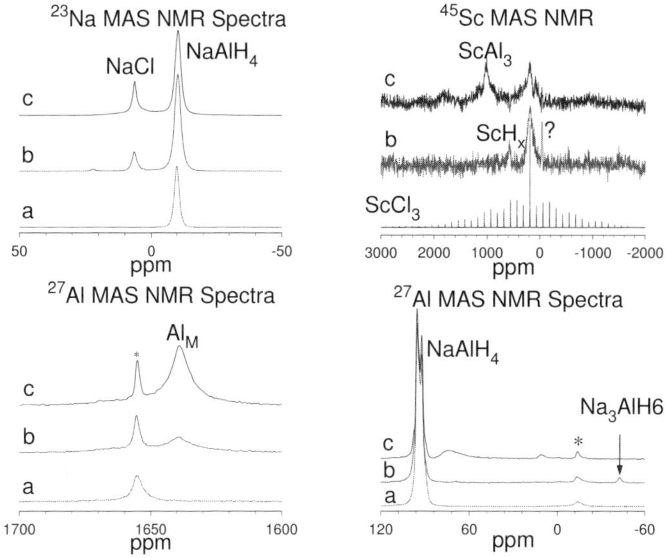

Figure 1. Multinuclear MAS NMR spectra for a) $NaAlH_4$, b) ball milled with 4 mol% $ScCl_3$, c) after two H_2 desorb-absorb cycles compared to pure $NaAlH_4$ and $ScCl_3$. Two regions for ^{27}Al spectra are for peaks from Al metal near 1640 ppm and $NaAlH_4$ and Na_3AlH_6 phases at 92 ppm and -42 ppm, respectively. Spinning side bands (rate=14.5 kHz) are marked with * while those of $ScCl_3$ were left unmarked.

4. MAS-NMR Studies of Metal Borohydrides Systems

Since light-element borohydrides $M(BH_4)_n$ have very high theoretical hydrogen storage capacities, their synthesis and properties are being extensively investigated [2]. The generic maximum release of hydrogen from $M(BH_4)_n$ during thermal desorption is commonly stated as either

$$M(BH_4)_n \rightarrow M + nB + (n/2)H_2 \tag{2}$$

$$M(BH_4)_n \rightarrow MH_x + nB + [(n-x)/2]H_2 \tag{3}$$

depending on whether a stable binary hydride (i.e., MH_x) or metal (M) is the final product and B is presumed to elemental boron, which is usually described as being in an amorphous state since it is not detected by x-ray diffraction (XRD) [2, 21, 22, 23]. The performance of some borohydrides can be improved by forming ternary borohydrides [24]. However, the reversible generation of nearly all borohydrides following desorption remains very challenging [2]. Only the $LiBH_4/MgH_2$ mixtures exhibit relatively easy reformation of the $LiBH_4$ phase below 700 K and its production is incomplete if the desorption pressures are below ~3 bar [25]. Several research groups have found that $LiBH_4$ usually decomposes in two or more stages rather than the single transition suggested by reaction (3). Hypothetical phases "$LiBH_3$", $LiBH_2$", and "LiBH" have been suggested [2] but often with little or no experimental verification due to the absence of clearly discernable x-ray or neutron diffraction peaks from the heated samples. Hence, these intermediate phases were presumed to be amorphous or nanocrystalline. By combining first principles calculations of phase stability with Raman spectroscopy measurements, Orimo, et al., [21] concluded that decomposition of $LiBH_4$ occurs via formation of one or more polyborane phases (i.e., $Li_2B_{12}H_{12}$ and perhaps other "B_nH_n" compounds) prior to yielding B and LiH as the final decomposition products. However, they were unable to support these claims by any diffraction or other measurements.

Direct confirmation for the formation of $B_{12}H_{12}$ species during the desorption of $LiBH_4$ (as purchased from Aldrich) is provided by ^{11}B MAS and CPMAS spectra in Fig. 2 in conjunction with similar NMR measurements on the reference compound $K_2B_{12}H_{12}$ (provided by Dr. S. S. Jalisatgi, U. Missouri-Columbia). It is noted that only a small amount of elemental boron is seen after the 500 °C desorption. Detailed assessments of the $B_{12}H_{12}^{-2}$ containing phases produced when various borohydrides are heated will be reported elsewhere.

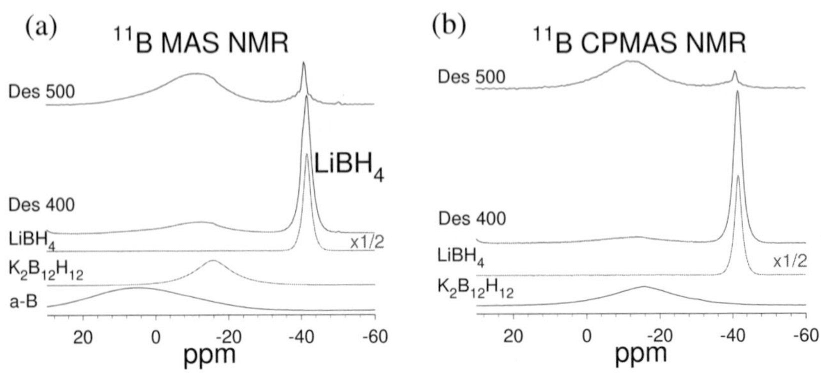

Figure 2. (a) ^{11}B MAS spectra (b) $^{11}B\{^1H\}$ CPMAS NMR spectra (contact time=0.1 ms) at spinning rate of 12-13 kHz. The ^{11}B peak for $LiBH_4$ lies at -41.5 ppm, a-B at ~5 ppm, $Li_2B_{12}H_{12}$ at -12.3 ppm, and $K_2B_{12}H_{12}$ at -15.9 ppm.

Nakamori, et al., [22-25] reported the formation of both $M(BH_4)_n$ and mixed metal $Li_mM(BH_4)_{n+m}$ phases by ball milling different ratios of $LiBH_4$ with MCl_n salts with an aim of adjusting cation electronegativity for lowering the H_2 desorption temperatures. They could not identify the products by XRD, which they assumed was caused by lack of long range order in the synthesized materials. They made phase assignments mainly from the presence and locations of BH_4^- vibration peaks in their Raman spectra. They reported formation of a $Sc(BH_4)_3$ phase with a relatively low desorption temperature [22, 23].

We have ball milled anhydrous $ScCl_3$ with three different ratios of $LiBH_4$ (i.e., 1:3, 1:4, and 1:6) and characterized the resulting products by multinuclear solid state NMR. The MAS and CPMAS spectra for ^1H, ^6Li, ^{11}B, and ^{45}Sc are shown in Figure 3. From our analyses of these spectra, we can unambiguously conclude that $LiSc(BH_4)_4$ is the primary product and not $Sc(BH_4)_3$ as was claimed previously [22, 23]. In addition to the expected LiCl and $LiBH_4$ phases for all three samples, the ^6Li and ^{45}Sc spectra indicate another phase, indicated by peaks marked by "?" in Fig. 3c and 3e, at the 1:3 ratio. This phase does not seem to contain any hydrogen since associated peaks were not observed during CPMAS experiments. It is appears the unidentified component only contains a combination of Li-Sc-Cl such as the known ternary salt Li_3ScCl_6. Further details of the phase compositions as well as the hydrogen desorption behavior of the Li-Sc-B-H systems will be presented elsewhere.

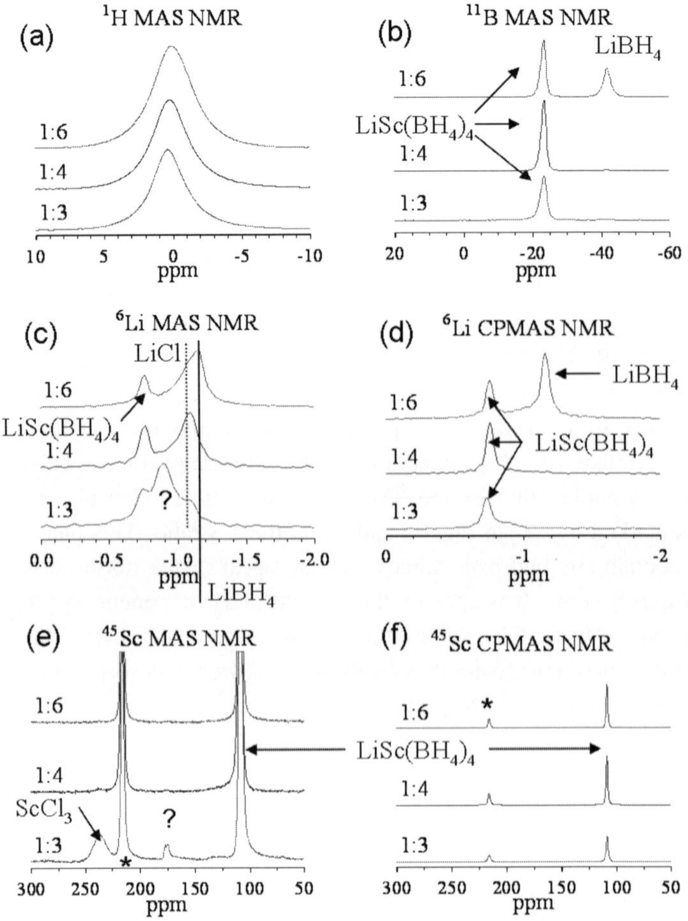

Figure 3. MAS and CPMAS spectra for ball milled mixtures of $ScCl_3/LiBH_4$ in the ratios 1:3, 1:4, and 1:6. Peak assignments were obtained from combined analyses of spectra for all nuclei. In parts (e) and (f) the asterisk (*) denote spinning side bands of the ^{45}Sc peak for $LiSc(BH_4)_4$ at 109 ppm.

5. MAS-NMR Studies of Metal Amides

Within the past year, a few papers have been published describing MAS-NMR experiments to evaluate transitions between the amide/imide phases [18] and also for amide-alanate mixtures [16, 17, 26]. Nearly all of these studies have used either the Li isotopes or ^{27}Al when the alanates were involved. As shown in Figure 4, the ^{6}Li MAS and CPMAS spectra are especially useful to discriminate between imide and amide phases as well as alanates and LiH due to tremendous improvement in resolution [27]. The counterpart ^{7}Li isotope is often used [16, 17] because of superior sensitivity and easy accessibility, but its lower resolution attributed from higher quadrupole and dipole coupling often limits their value. Since ^{6}Li NMR suffers greatly from having much longer spin-lattice relaxation times (T_1), checks for the presence of Li containing phases are more efficiently achieved by ^{7}Li MAS. Subsequently, ^{6}Li CPMAS is performed in order to save instrument time and to get insights about the detailed structure.

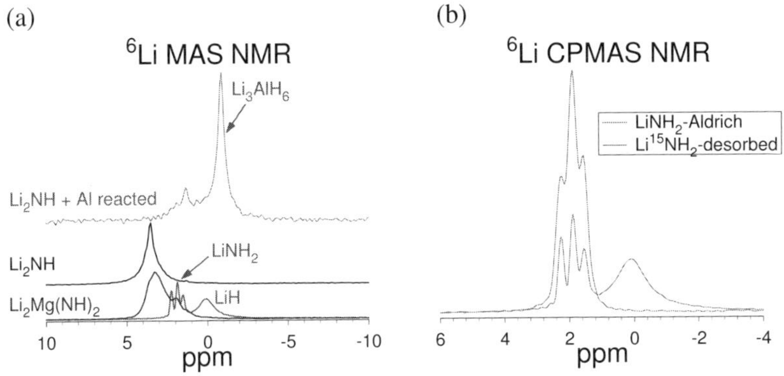

Figure 4. (a) ^{6}Li MAS spectra for amide, imide, and LiH phases that include a sample made from a mixture of Li_2NH and Al metal reacted at U. Utah with hydrogen gas to form $LiNH_2$ and Li_3AlH_6 phases [26]. (b) $^{6}Li\{^{1}H\}$ CPMAS spectrum of LiH and $LiNH_2$, where the amide was enriched with the ^{15}N isotope by reacting $^{15}NH_3$ gas with LiH. ^{6}Li CPMAS NMR spectrum of the natural $LiNH_2$ purchased from Aldrich is displaced together for comparison. Note the ^{6}Li triplet peaks of the $LiNH_2$ phase are consistent with the three distinct site locations in its crystal structure [28].

In principle, preparation of amide and imides enriched with the I = 1/2 isotope ^{15}N (which is only 0.37% naturally abundant) should permit more detailed assessments of phase conversions and chemical bonding of the NH_2^{-1} and NH^{-2} ions. Samples of ^{15}N enriched $LiNH_2$ and $Mg(NH_2)_2$ have been prepared at JPL and their ^{15}N spectra are shown in Figure 5. Unfortunately, the extremely long T_1 values in these samples have been a severe impediment to obtaining sufficient signal-to-noise ratios in reasonable measurement periods. Nevertheless, a single ^{15}N peak at -15 ppm shift found for the $LiNH_2$ phase is consistent with there being one distinct location in its crystal structure [28] while four ^{15}N peaks are apparent (albeit from relatively noisy signals) in the $Mg(NH_2)_2$ spectrum compared to three sites for the recently reported [28] crystal structure of $Mg(ND_2)_2$. This difference is currently being investigated.

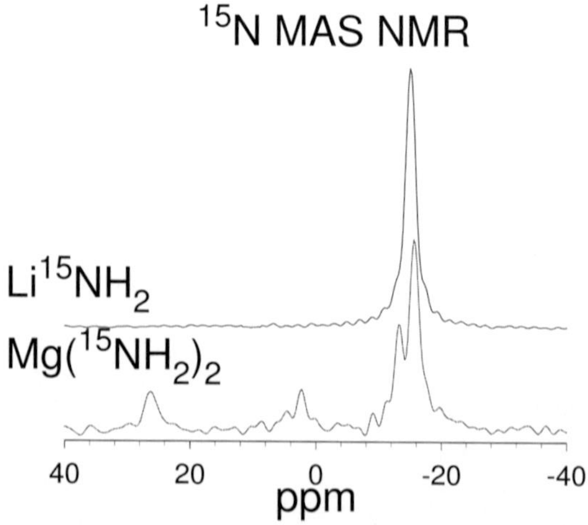

Figure 5. The ^{15}N MAS spectra for $LiNH_2$ and $Mg(NH_2)_2$ where their ^{15}N isotope contents had been enriched to ~98% by reacting $^{15}NH_3$ gas with LiH and MgH_2, respectively.

6. Summary and Conclusions

Multinuclear solid state NMR studies involving the high resolution methods based upon magic angle spinning along with other advanced techniques [12] provide powerful tools to evaluate the formation and reactions of complex metal hydrides. Since NMR involves local electron-nucleus, dipole-dipole, and quadrupolar interactions, long range crystallinity is not a mandatory requirement. Consequently, the NMR spectra are able to provide insightful characterizations of amorphous and nanophase materials that often yielded only limited structural information and phase compositions when most diffraction techniques were used. The present paper has provided brief descriptions on the nature of information now being obtained on a few current prototype complex hydrides involving alanates, borohydrides, and amides that are candidates for hydrogen storage. It is expected that many more valuable results will become available during future NMR studies.

Acknowledgments

This research was partially supported by the U. S. Department of Energy (DOE) under contract numbers DE-AI-01-06EE11105 and DEFC36-05GO15065. It was also partially performed at the Jet Propulsion Laboratory, California Institute of Technology, under a contract with the National Aeronautics and

Space Administration (NASA). The NMR facility at Caltech was supported by the National Science Foundation (NSF) under Grant Number 9724240 and partially supported by the MRSEC Program of the NSF under Award Number DMR-520565. We thank H. Brinks, B. Hauback, W. Luo, Z. Fang, and S. Jalisatgi for providing samples and reference materials. The support and contributions of C. C. Ahn are appreciated.

References

1. S. Satyapal, J. Petrovic, C. Read, G. Thomas, and G. Ordaz, Catal. Today **120**, 246 (2007).
2. S. Orimo, Y. Nakamori, J. R. Eliseo, A. Züttel, and C. M. Jensen, Chem. Rev. **107**, 4111 (2007).
3. R. G. Barnes, in *Hydrogen in Metals III: Properties and Applications*; H. Wipf. (Ed.); Springer-Verlag: Berlin; 1997; p. 93.
4. T. Tsang and T. C. Farrar, J. Chem.. Phys. **50**, 3498 (1969).
5. A. Trokiner, H. Theveneau, and P. Papon, J. Chem.. Phys. **69**, 742 (1978).
6. S. Fröberg, L. Niemelä, and E. Ylinen, Phys. Lett. **A115**, 67 (1986).
7. J. Senegas, A. M. Villepastour, and B. Bonnetot, J. Phys. Chem. Solids **42**, 1061 (1981).
8. V. P. Tarasov, S. I. Bakum, V. I. Privalov, Yu. B. Muravlev, and A. A. Samoilenko, Russ. J. Inorg. Chem. **41**, 1104 (1996).
9. V. P. Tarasov, S. I. Bakum, and S. F. Kuznetsova, Russ. J. Inorg. Chem. **42**, 694 (1997).
10. V. P. Tarasov, S. I. Bakum, and A. V. Novikov, Russ. J. Inorg. Chem. **46**, 409 (2001).
11. G. Majer, E. Stanik, L. E. Valient Banuet, F. Grinberg, O. Kircher, and M. Fichtner, J. Alloys Compds. **404-406**, 738 (2005).
12. K. J. D. MacKenzie and M. E. Smith, *Multinuclear Solid-State NMR of Inorganic Materials*; (Pergamon, Amsterdam, 2002).
13. R. C. Bowman, Jr. and S.-J. Hwang, Mater. Matters **2**(No.2), 29 (2007).
14. L. Kellberg, H. Bildsee, and H. Jakobsen, J. Chem. Soc., Chem. Commun. **1990**, 1294.
15. S.-J. Hwang, R.C. Bowman, Jr., J. Graetz, J.J. Reilly, W. Langley, C.M. Jensen, J. Alloys Compd. **446-447**, 290 (2007).
16. O. Dolotko, H. Zhang, O. Ugurlu, J. W. Wiench, M. Pruski, L. S. Chumbley, and V. Pecharsky, Act Mater. **55**, 3121 (2007).
17. Z. Xiong, G. Wu, J. Hu, Y. Liu, P. Chen, W. Luo, and J. Wang, Adv. Funct. Mater. **17**, 1137 (2007).
18. C. Lu, J. Hu, J. H. Kwak, Z. Yang, R. Ren, T. Markmaitree, and L. L. Shaw, J. Power Sources **170**, 419 (2007).

19. B. Bogdanovic, M. Felderhoff, A. Pommerin, F. Schüth, and N. Spielkamp, Adv. Mater. **18**, 1198 (2006).
20. T. Wang, J. Wang, A. D. Ebner, and J. A. Ritter, J. Alloys Compds. (2007), doi:10.1016/j.jallcom.2006.10.072.
21. S. Orimo, Y. Nakamori, N. Ohba, K. Miwa, M. Aoki, S. Towata, and A. Züttel, App. Phys. Lett. **89**, 021920 (2006).
22. Y. Nakamori, H. Li, K. Miwa, S. Towata, and S. Orimo, Mat. Trans. **47**, 1898 (2006).
23. Y. Nakamori, H.-W. Li, K. Kikuchi, M. Aoki, K. Miwa, S. Towata, and S. Orimo, J. Alloys Compds. **446-447**, 296 (2007).
24. H.-W. Li, S. Orimo, Y. Nakamori, K. Miwa, N. Ohba, S. Towata, and A. Züttel, J. Alloys Compds. **446-447**, 315 (2007).
25. F. E. Pinkerton, M. S. Meyer, G. P. Meisner, M. P. Balogh, and J. J. Vajo, J. Phys. Chem. C **111**, 12881 (2007).
26. J. Lu, Z. Z. Fang, H. Y. Sohn, R. C. Bowman, Jr., and S.-J. Hwang, J. Phys. Chem. C **111**, 16686 (2007).
27. H. Eckert, Z. Zhang, and J.H. Kennedy, MRS Symp. Proc. **135**, 259 (1989).
28. M.H. Sørby, Y. Nakamura, H.W. Brinks, T. Ichikawa, S. Hino, H. Fujii, B.C. Hauback, J. Alloys and Compds. **428**, 297 (2007).

STUDY ON THE STRUCTURE AND ELECTROCHEMICAL PROPERTIES OF NOVEL ND-MG-NI-CO HYDROGEN STORAGE ALLOYS

CHONGCHAO PAN, RONGHAI YU*

Key Laboratory of Advanced Materials of Education Ministry, Department of Materials Science & Engineering, Tsinghua University, Beijing 100084, China

$Nd_{0.75}Mg_{0.25}$ $(Ni_{0.8}Co_{0.2})$ x (x=3.8 & 4.5) hydrogen storage alloys have been prepared by using mid-frequency induction melting furnace. The structure analyses and electrochemical properties of the alloys were investigated by means of XRD, TEM and electrochemical workstation. From the results it was observed that (Nd, $Mg)_2(Ni,Co)_7$ and NdNi5 consist of superstructure (Ce_2Ni_7 type) and $CaCu_5$ structure respectively. It is found that the main phase of alloy belongs to Ce_2Ni_7 type Nd_2Ni_7 super-structure at x=3.8, Mg atoms are located only at the Laves unit of Ce2Ni7 type unit cell, while Co atoms are located only at the $CaCu_5$ unit. Furthermore, the (Nd, $Mg)_2(Ni,Co)_7$ phase was composed with $NdNi_2$(Laves type) units and $NdNi_5$($CaCu_5$ type) units and aligned along C axial direction like ABBABB, the period of the distance of the dark line pair is about 2.48nm. Electrochemical analyses showed that all the alloys had a large discharge capacity and can be easily activated. The $Nd_{0.75}Mg_{0.25}$ $(Ni_{0.8}Co_{0.2})_{3.8}$ alloy exhibits better electrochemical properties.

1. Introduction

Hydrogen storage alloys have attracted considerable attention in view of its potential as a new energy storage material [1]. Recently, hydrogen storage alloys is employed as negative electrode materials. Commercial La-Ni systems hydrogen storage alloys have better comprehensive properties, but their discharge capacity was limited by the $CaCu_5$ type structure [2], hydrogen storage capacity of PuNi3 type alloys are higher than $CaCu_5$ type alloys and the Laves phase alloys [3,4], the $PuNi_3$ type alloys have poor cycle stability, it has little improvement by element substitutions [5].

Kohno etal. reported that La_2MgNi_9, $La_5Mg_2Ni_{23}$ and La_3MgNi_{14} which were new (La, Mg)Nix=3.0-3.5 system alloys, absorbing and desorbing hydrogen at

* Corresponding author. Tel.: +86-10-62771593.

E-mail address: rhyu@tsinghua.edu.cn.

The authors would like to thanks for National Nature Science Foundation of China for financial assistance room temperature. They reported that these alloys which are replaced by Co, these alloys have high absorbing and desorbing hydrogen capacity at room temperature [6]. According to Nd-Ni phase graph, there exists the $(Nd, Mg)_2(Ni,Co)_7$ structure (Ce_2Ni_7 type)[7]. B. D. Dunlap [8] and E. Parthe [9] reported that the Ce_2Ni_7 type structure is similar with PuNi3 type structure; they are different in the number of $CaCu_5$ unit and Laves unit. It is very hard to improve the cycle stability. The Nd element has better corrosion resistance than La element and also Co element addition will increase the corrosion resistance of these alloys, and different stoichiometric proportion can control the microstructure of the alloys. So in the present work $Nd_{0.75}Mg_{0.25}(Ni_{0.8}Co_{0.2})_x$ for x=3.8 & 4.5 hydrogen storage alloys have been investigated systemically.

2. Experimental

2.1. *Alloys preparation*

$Nd_{0.75}Mg_{0.25}(Ni_{0.8}Co_{0.2})_x$ (for x=3.8 & 4.5) alloys were prepared by using mid-frequency induction melting furnace under 0.3bar pressure in high purity argon atmosphere. Then samples were cooled in the copper mould and named S1 and S2 for x=3.8 & 4.5 respectively. All metal elements are high purified with 99.9 wt. %, melting loss had been considered with Nd and Mg elements. The alloy ingots were crushed, grinded into powder and then sieved through a 150 mesh. This alloy powder was used to electrochemical tests and structure characterizations.

2.2. *Electrochemical test method*

A hydride electrode was prepared each time by cold pressing the mixture of one kind of alloy powder with Nickel powder in the weight ratio of 1:3 to form a pellet under the pressure of 25MPa. Electrochemical measurements were performed at 298K in a standard open tri-electrode electrolysis cell consisting of a working electrode, a sintered $Ni(OH)_2/NiOOH$ counter electrode and a Hg/HgO reference electrode immersing in 6M KOH electrolyte. The discharge capacities of alloys were determined by the galvanostatic method. Each electrode was charged at 60 mA/g for 7 h, rest for 5 minutes and discharged at 60 mA/g to the cut-off potential of -0.6V versus the Hg/HgO reference electrode. For investigating the high rate discharge ability, discharge capacities at different discharge current densities (150mA/g, 300mA/g, 900mA/g, 1200mA/g) were determined. Cycle stability was determined by the following method: Charged by 150mA/g for 2h and 40 minutes, and discharged at 150mA/g to the cut-off potential of -0.6V versus the Hg/HgO reference electrode.

2.3. *Structure characterization*

X-ray powder diffraction data were taken by using Rigaku D/max2500 X-rays diffractometer with CuKα radiation, scanning speed with 6°/min, and the 2θ degree in range from 20° to 90°, working voltage and current are 40kV and 200 mA respectively. Microstructures and high resolution images of alloys were examined with JEOL2011 transmission electron microscope (TEM) operated at 200 Kv. TEM foils were grinded to fine powder and dispersed into alcohol in ultrasonic equipment, then collected by micro-grid.

3. Results and Discussion

3.1. *Electrochemical properties*

The activation and the high rate capacity of the alloy electrodes is shown in Fig.1, both the different stoichiometric proportion alloys were fully activated within two cycles. The good activation under investigation would make them be

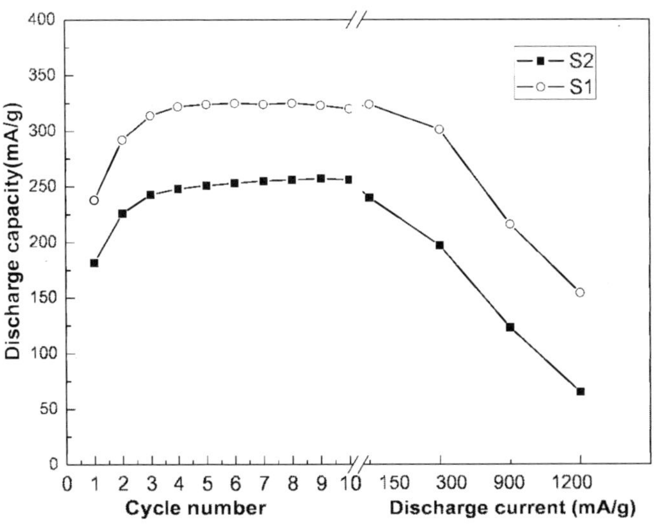

Figure 1. Activation and rate properties of $Nd_{0.75}Mg_{0.25}(Ni_{0.8}Co_{0.2})_x$ (x=3.8, 4.5) alloys

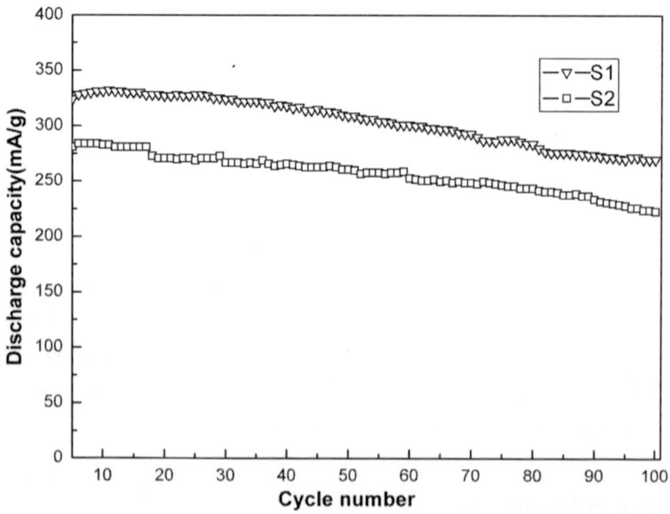

Figure 2. Cycle stability properties of Nd0.75Mg0.25（Ni0.8Co0.2）x (x=3.8，4.5) alloys

practical. Maximum discharge capacity of S1 and S2 are about 324 mA/g and 246 mA/g respectively. From the right side of the Fig.1, the high rate property of S1 is better than S2. The cyclic trends of discharge capacities of $Nd_{0.75}Mg_{0.25}$ $(Ni_{0.8}Co_{0.2})_x$ (x = 3.8 & 4.5) alloys are shown in Fig.2 which shows S1 and S2 have good cycle stabilities.

The capacity retention ratios of S1 and S2 at the 100th cycle are 83.8% and 79.3% respectively. According to the electrochemical test results, it was observed that different stoichiometric ratios affect on the maximum discharge capacity, high rate property and cycle stability.

3.2. *Phase structure*

The XRD results of S1 and S2 are shown in Fig.3. It is found that all the alloys are composed of $(Nd, Mg)_2(Ni,Co)_7$ phase (Ce_2Ni_7-type structure), $NdNi_5$ phase ($CaCu_5$-type structure) and other phases (small diffraction peaks). In addition, it can also be seen that with increasing x from 3.8 to 4.5, the main phase is different.

The main phase of S1 is $(Nd, Mg)_2(Ni,Co)_7$ phase. According to the intensity of diffraction peak of two alloys, the space group is P63/mmc and crystal plane is (1 1 24), the abundance of the $NdNi_5$ phase was less than S2

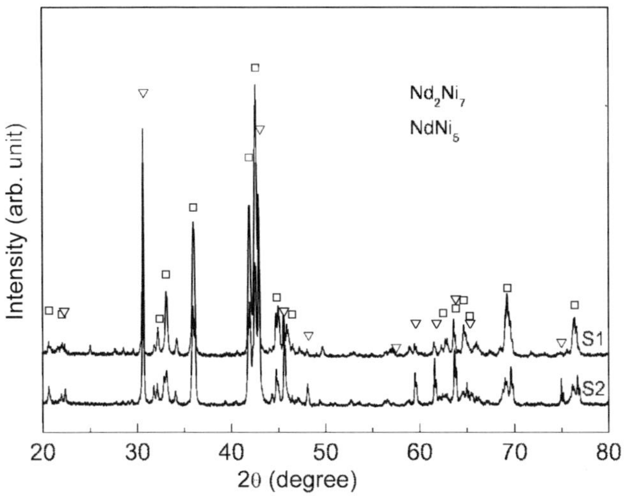

Figure 3. XRD patterns of the $Nd_{0.75}Mg_{0.25}(Ni_{0.8}Co_{0.2})_x$(x=3.8, 4.5) alloys

according to the the intensity of diffraction peak of $CaCu_5$-type structure. Two phase have different absorbing and desorbing hydrogen capacities. The hydrogen storage capacity of $(Nd,Mg)_2(Ni,Co)_7$ phase was higher than $NdNi_5$ phase because of the different number of effective interstitial positions in the cell. So S1 has better electrochemical properties than S2.

3.3. *Microstructure*

From the XRD pattern of these alloys, the phases were defined, but it was difficult to estimate the structure with high precision. The crystal structure of $(Nd, Mg)_2(Ni,Co)_7$ phase of the S1 was analyzed in detail by using TEM. The bright field image of the S1 at low magnification ratio shows that there is no grain boundary as shown in Fig.4. The electron beam diffraction image for S1 crystal grain is shown in Fig.4. It is verified that the line of very dense reflection is in the C axial direction. Which shows that periodicity is long in the C axial direction. Furthermore, it is observed that strength of 12th and 24th reflection is strong from (0 0 0).

The crystal structure of $(Nd, Mg)_2(Ni,Co)_7$ phase was composed with $NdNi_2$ (Laves type) and $NdNi_5$ ($CaCu_5$ type) units which were aligned along C axial direction like ABBABB order, the HR-TEM image is shown in Fig.5. The period of the distance of the dark line pair is about 2.48nm, which value matches the C axial length of this alloy.

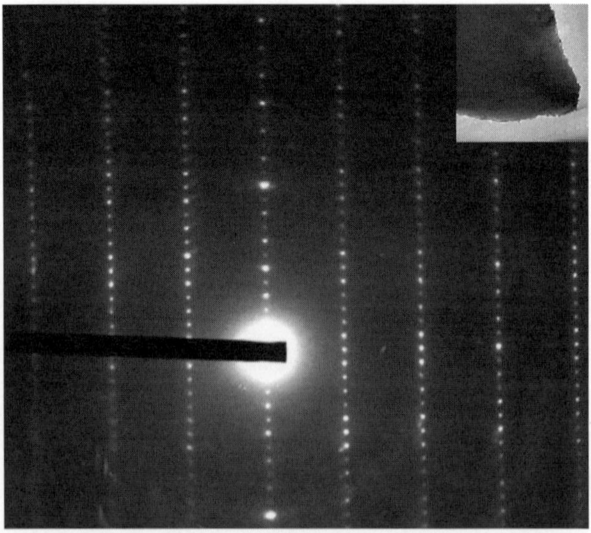

Figure 4. Selected electronic diffraction patterns of superstructure of $Nd_{0.75}Mg_{0.25}(Ni_{0.8}Co_{0.2})_{3.8}$ alloys

Figure 5. High resolution TEM of $Nd_{0.75}Mg_{0.25}(Ni_{0.8}Co_{0.2})_{3.8}$ alloys

There are clear lattice image about the stacking order in the left bottom and top corner of Fig.5, which is identical with the electron diffraction pattern. The value of distance between the periods matches the XRD results. The structure of this alloy is thought to be as shown in Fig.6, The stacking sequence of double layers were aligned along C axial direction like ABBABB order, Mg atom substitutes the position of Nd atom of $NdNi_2$ units, so the ABBABB stacking order arose. So the special structure was called super-structure, the existing super-structure may have some special advantage. According the electrochemical results, the S1 has higher hydrogen storage capacity than S2 because there are more super-structure existence in S1.

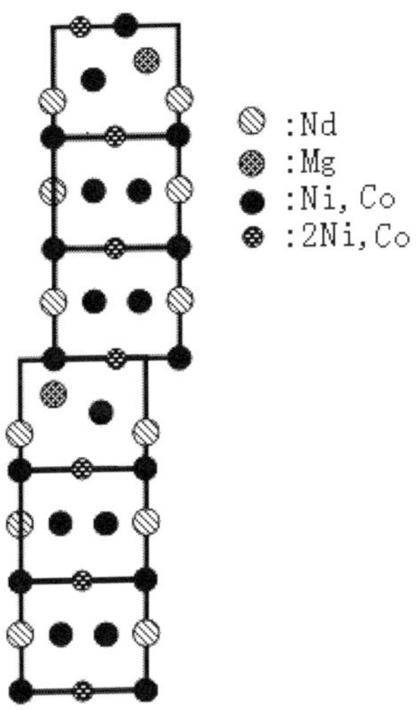

Figure 6. Super-structure model of (Nd, $Mg)_2(Ni,Co)_7$ phase

4. Conclusion

Two Nd-Mg-Ni-Co alloys were prepared and investigated their crystal structure by using XRD and TEM techniques. Both of alloys consist of (Nd, $Mg)_2(Ni,Co)_7$ phase(Ce_2Ni_7-type structure) and the $NdNi_5$ phase ($CaCu_5$-type structure), but the content of phase in different alloys are different because of their different stoichiometric proportion. Furthermore, super-structure is existed according to the microstructure and electron transmission pattern of S1. The (Nd, $Mg)_2(Ni,Co)_7$ phase was composed with $NdNi_2$(Laves type) units and $NdNi_5$($CaCu_5$ type) units and aligned along C axial direction like ABBABB, the period of the distance of the dark line pair is about 2.48nm. The S1 sample has better comprehensive electrochemical properties owing to the higher content of super-structure than S2 sample.

References

[1] Y.Q. Lei, Q. Wan, D.S. Shi, New Powder Materials, Tianjing: Publication of Tianjing University press, 2000: 52–55.
[2] T. Sakai, H. Miyamura, N. Kuriyama, etal. Journal of the Electrochemical Society, 137(1990): 795–799.
[3] J. Chen, N. Kuriyama, N.T. Takashita, Electrochemical and Solid State Lett, 3(2000) 249-252.
[4] C.H. Peng, M.Zhu J. Alloys and Compounds, 375(2004): 324-329.
[5] B. Liao, Y.Q. Lei, L.X. Chen, etal. J. Alloys and Compounds, 376(2004):186-195.
[6] T. Kohno, H. Yoshida, F. Kawashima, etal. J. Alloys and Compounds, 311(2000):L5-L7.
[7] Y. Y. Pan, C.S. Cheng, Acta Physica Sinica,34(3):384-389.
[8] B. D. Dunlap, P. J. Viccaro, G.K. Shenoy. J. Less Common Metals, 74(1980):75-79.
[9] E. Parthe, R.Lemaire, B31(1975):1879-1889.

ANALYSIS AND MODELLING OF THE BURST PRESSURE OF HIGH PRESSURE HYGROGEN TANKS

D. CHAPELLE, F. THIEBAUD, D. PERREUX

LMARC-FEMTO-ST, Université de Franche-Comté, Besançon, France

Email: contact@mahytec.com

The present study deals with the analysis of the cylindrical part of a high pressure hydrogen storage vessel, combining an aluminium liner and an over wrapped filament wound composite. The first is a barrier against the hydrogen permeation whereas the second allows to reinforce the structure and to decrease the weight. This widely used technique still requires important time investments. Regarding at mobile application and taking into account 1 kg hydrogen per 100 km consumption of the full cell, one of the main goals is to store at least 5 kg of hydrogen in the smallest volume. Based on mechanical considerations, the model provides an exact solution for stresses and deformations on the cylindrical section of the vessel under thermo-mechanical static loading. The liner is assumed to behave as an elastic plastic material whereas the laminate is supposed to be an elastic damageable material; the Tsaï-Wu criterion is introduced to predict the failure of each layer, and finally the burst of the structure. The effect of the stacking sequence on the gap occurrence, on the residual stress magnitude and on the structure durability may be then investigated. In the present paper, after an overview of the theoretical background, a comparison between experimental investigations and results obtained with the model is presented. A preliminary discussion is attempted in order to assess the initial stress state of the structure. Further works should then lead us to predict the mechanical response of the vessel when submitted to cycle loadings, when experiments are still carried out, and before an optimisation of the composite laminate staking sequence is performed.

1. Introduction

Hydrogen energy vector turns out to be one of the main challenges of the next decades. Among the breakthroughs to fulfil, the technological developments around the hydrogen storage still require some relevant improvements. Three hydrogen storage media are currently quoted (1): a) liquefied hydrogen b) hydrogen storage materials, and finally c) compressed hydrogen gas undoubtedly the most successfully completed and effective solution.

Unfortunately, this storage medium, in the classical range of storage pressure for gas, allows small volumetric density. Based on a 0.8 to 1 kg of hydrogen for 100 km consumption, the hydrogen storage vessel of the fuel cell powered vehicle (FCV) should contain at least 5 kg of hydrogen to be a competitive solution. As an instantaneous consequence, the major disadvantage of the compressed hydrogen gas technique is the high pressure required in order to reduce volume. So, it is clear that no relevant breakthrough should be expected from an increase of internal hydrogen pressure, but attention must be paid on the gravimetric hydrogen density to be reduced.

To do so, this paper aims, on the basis of previous works (3, 4, 5), to model analytically the hydrogen storage vessel which combine a thin metal liner to prevent gas diffusion and a composite laminate made by filament winding with Carbon/Epoxy to ensure the vessel strength. The laminate is assumed to be a damageable elastic material whereas the metal liner is considered as an elastic plastic material. The plasticity is introduced by the way of the von Mises criterion, and the laminate failure happens when the Tsai-Wu criterion (6) is satisfied. When the mechanical model is written, the radial gap occurrence is investigated; the residual stresses are then assessed considering the complete manufacturing process. Finally the mechanical response of a prototype of hydrogen tank is simulated and discussed. Obviously, these results should be repeated with other experimental specimens before going on the next step that is the optimisation of the hydrogen storage vessel.

2. Mechanical analysis

2.1. Displacements, strains and balance

Attention focuses on the cylindrical section of the hydrogen vessel subjected to internal pressure with close-end effect loading (Figure 1). The inner radius R_0 and its thickness e are constant. The vessel strength is ensured by n_s layers of filament wound composite. The kth layer is characterised by its thickness ep(k) and winding angle β(k). The radial, the hoop and the axial coordinates are respectively denoted by r, θ and z. Regarding at the usual assumption, the displacement fields is expressed as:

$$u = u(r), \quad v = v(r,z), \quad w = w(z) \tag{1}$$

Where, u, v and w are radial, hoop and axial displacements.

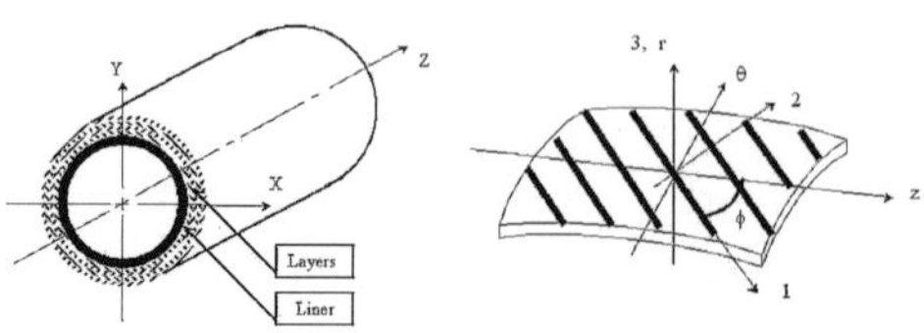

Figure 1: Cylindrical section of the vessel, coordinate systems on the laminate plane.

Assuming section remains cylindrical, the axial strains of all layers and of the liner are equal to a constant ε_0 and the shear strain $\gamma_{z\theta} = 2\varepsilon_{z\theta}$ do not depend on the z coordinate, so $\gamma_{z\theta} = r\,\gamma_0$, where γ_0 is twist per unit length.
Consequently, the strain-displacement relations can be written as:

$$\begin{cases} \varepsilon_r = \dfrac{du}{dr}, \quad \varepsilon_\theta = \dfrac{u}{r}, \quad \varepsilon_z = \varepsilon_0 \\ \varepsilon_{rz} = 0, \quad \varepsilon_{r\theta} = \dfrac{\partial v}{\partial r} - \dfrac{v}{r}, \quad \varepsilon_{z\theta} = \dfrac{1}{2} r\gamma_0 \end{cases} \tag{2}$$

and equilibrium equations are reduced to:

$$\frac{\partial \sigma_r}{\partial r} + \frac{\sigma_r - \sigma_\theta}{r} = 0 \tag{3}$$

All previous relations have to be written for both composite layers and liner.

2.2. Stress-strain relations

The liner is an elastic plastic isotropic material. Besides, the laminate behaviour is different from a layer to another and each layer behaves according to the fibre direction. The fibre is assumed to have a transverse isotropy and equivalent properties in the (2-3) plane which normal axis (1) refers to the fibre longitudinal direction, as shown in Figure 1.
General stress-strain relations can be described as:

$$\begin{aligned} \varepsilon &= \boldsymbol{S}\,\sigma + \alpha\Delta T \\ \sigma &= \boldsymbol{C}\left(\varepsilon - \alpha\Delta T\right) \end{aligned} \tag{4}$$

Where,

$$\sigma = \left(\sigma_1\ \sigma_2\ \sigma_3\ \sigma_4\ \sigma_5\ \sigma_6\right), \qquad \varepsilon = \left(\varepsilon_1\ \varepsilon_2\ \varepsilon_3\ \varepsilon_4\ \varepsilon_5\ \varepsilon_6\right) \tag{5}$$

and

$$\begin{cases} \varepsilon_1 = \varepsilon_{11} \quad \varepsilon_2 = \varepsilon_{22} \quad \varepsilon_3 = \varepsilon_{33} \quad \varepsilon_4 = 2\varepsilon_{23} \quad \varepsilon_5 = 2\varepsilon_{13} \quad \varepsilon_6 = 2\varepsilon_{12} \\ \sigma_1 = \sigma_{11} \quad \sigma_2 = \sigma_{22} \quad \sigma_3 = \sigma_{33} \quad \sigma_4 = \sigma_{23} \quad \sigma_5 = \sigma_{13} \quad \sigma_6 = \sigma_{12} \end{cases} \tag{6}$$

C and ***S*** are respectively the stiffness and the compliance tensors.

2.2.1. *Liner behaviour*

Considering the liner as an isotropic elastic plastic material, the incremental stress-strain relations can be rewritten as:

$$d\varepsilon^e + d\varepsilon^p = \left(\boldsymbol{S}_e^L + \boldsymbol{S}_p^L\right)d\sigma + \alpha\Delta T \tag{7}$$

where e and p denotes the elastic and plastic contribution, L designs the liner.

with

$$\begin{cases} S_{e11}^{L} = S_{e22}^{L} = S_{e33}^{L} = \dfrac{1}{E}, \quad S_{e23}^{L} = S_{e13}^{L} = S_{e12}^{L} = \dfrac{-\nu}{E} \\ S_{e44}^{L} = S_{e55}^{L} = S_{e66}^{L} = \dfrac{1}{G} \end{cases} \tag{8}$$

E is the Young modulus and ν the Poisson ratio of the metal.
After some calculations and depending on the hardening law, the plastic contribution $\boldsymbol{S}_p^L$ can be written.

2.2.2. *Laminate behaviour*

Assuming (1) axis is the longitudinal direction of the fibre, the compliance tensor $\boldsymbol{S}_c$ takes the same form than $\boldsymbol{S}_e^L$, whereas, taking into account the transversal isotropy, the compliance constants have the following expressions:

$$\begin{cases} S_{11}^{c} = \dfrac{1}{E_1}, \quad S_{22}^{c} = S_{33}^{c} = \dfrac{1}{E_2} \\ S_{12}^{c} = S_{13}^{c} = \dfrac{-\nu_{12}}{E_1}, \quad S_{23}^{c} = \dfrac{-\nu_{23}}{E_2} \\ S_{44}^{c} = 2\left(S_{22}^{c} - S_{23}^{c}\right) = \dfrac{2(1+\nu_{23})}{E_2}, \quad S_{55}^{c} = S_{66}^{c} = \dfrac{2(1+\nu_{12})}{E_1} \\ E_2 = E_3, \quad \nu_{12} = \nu_{13} \end{cases} \tag{9}$$

where E_i refer to Young modulus and ν_{ij} to Poisson ratios.
If T_e and T_s are respectively the transformation matrices for the strain and stress vectors the transformation matrix which allows writing the vectors in the cylindrical coordinate, we get:

$$\begin{aligned} \varepsilon' &= T_e \varepsilon \\ \sigma' &= T_s \varepsilon \\ \boldsymbol{C}' &= T_s \boldsymbol{C} T_e^{-1} \\ \alpha' &= T_e \alpha = \left(\alpha_1' \; \alpha_2' \; \alpha_3' \; \alpha_4' \; \alpha_5' \; \alpha_6'\right) \end{aligned} \tag{10}$$

The quote denotes the vector or tensor in the cylindrical coordinate system.
The damage of laminate is introduced by adding the damage contribution $\boldsymbol{H}$ tensor to the compliance tensor of composite (7, 8, 9). The only non zero component of $\boldsymbol{H}$ are H_{22} and H_{66} (H_{44} has no influence in the present analysis):

$$H_{22} = \frac{D_I S_{22}^{c}}{1 - D_I}, \qquad H_{66} = D_I \left(\frac{S_{22}^{c\,2}}{1 - D_I} \right)^{1/2} \tag{11}$$

where D_I is the damage parameter. The damage kinetics is obtained using the thermodynamics of the irreversible process framework (not presented here).

According the Tsai-Wu criterion (6), the non-failure is ensured as far as the constraint

$$F_{11}\sigma_1^2+F_{22}\sigma_2^2+F_{66}\sigma_6^2+2F_{12}\sigma_1\sigma_2+F_1\sigma_1+F_2\sigma_2-1\leq 0 \qquad (12)$$

is satisfied (F_{11}, F_{22}… are the classical Tsai-Wu parameters which are calculated from the material properties $\sigma_{LU}, \sigma'_{LU}$ …).

2.3. Problem to solve

Taking into account a progressive plasticization of the liner involves to arbitrarily considering n_l layer through the metal thickness. Consequently, the structure is assumed to be made of (n_l + n_s) layers. Moreover, to be rigorous in the present analysis the same relations should have been expressed under an increment form. The symmetric stiffness tensor of the kth layer takes the following expression in the cylindrical coordinate system:

$$\boldsymbol{C}_e^L = \begin{pmatrix} C_{11}'^{(k)} & C_{12}'^{(k)} & C_{13}'^{(k)} & C_{14}'^{(k)} & C_{15}'^{(k)} & C_{16}'^{(k)} \\ & C_{22}'^{(k)} & C_{23}^{(k)} & C_{24}'^{(k)} & C_{25}'^{(k)} & C_{26}'^{(k)} \\ & & C_{33}'^{(k)} & C_{34}'^{(k)} & C_{35}'^{(k)} & C_{36}'^{(k)} \\ & & & C_{44}'^{(k)} & C_{45}'^{(k)} & C_{46}'^{(k)} \\ & & & & C_{55}'^{(k)} & C_{56}'^{(k)} \\ & & & & & C_{66}'^{(k)} \end{pmatrix} \qquad (13)$$

Then, the coefficients $K_i^{(k)}$ are introduced:

$$K_i^{(k)} = \sum_{j=1}^{6}\left(C_{ij}'^{(k)}\,\alpha_{ij}'\right) \qquad (14)$$

According the stress-strain relations, the axial, hoop, radial and shear stresses can be expressed as:

$$\begin{aligned}
\sigma_z^{(k)} &= C_{11}'^{(k)}\,\varepsilon_z^{(k)}+C_{12}'^{(k)}\,\varepsilon_\theta^{(k)}+C_{13}'^{(k)}\,\varepsilon_r^{(k)}+C_{16}'^{(k)}\,\gamma_{z\theta}^{(k)}-K_1^{(k)}\Delta T \\
\sigma_\theta^{(k)} &= C_{21}'^{(k)}\,\varepsilon_z^{(k)}+C_{22}'^{(k)}\,\varepsilon_\theta^{(k)}+C_{23}'^{(k)}\,\varepsilon_r^{(k)}+C_{26}'^{(k)}\,\gamma_{z\theta}^{(k)}-K_2^{(k)}\Delta T \\
\sigma_r^{(k)} &= C_{31}'^{(k)}\,\varepsilon_z^{(k)}+C_{32}'^{(k)}\,\varepsilon_\theta^{(k)}+C_{33}'^{(k)}\,\varepsilon_r^{(k)}+C_{36}'^{(k)}\,\gamma_{z\theta}^{(k)}-K_3^{(k)}\Delta T \\
\sigma_{z\theta}^{(k)} &= C_{61}'^{(k)}\,\varepsilon_z^{(k)}+C_{62}'^{(k)}\,\varepsilon_\theta^{(k)}+C_{63}'^{(k)}\,\varepsilon_r^{(k)}+C_{66}'^{(k)}\,\gamma_{z\theta}^{(k)}-K_6^{(k)}\Delta T
\end{aligned} \qquad (15)$$

Equation (3) provides the differential equation to be solved along the (r) direction in order to guaranty the balance between each layer:

$$\frac{d^2u^{(k)}}{dr^2}+\frac{1}{r}\frac{du^{(k)}}{dr}-\frac{N_1^{(k)}}{r^2}u^{(k)}=\left[N_2^{(k)}\varepsilon_0+N_3^{(k)}\Delta T\right]\frac{1}{r}+N_6^{(k)}\gamma_0$$

with

$$N_1^{(k)}=\frac{C_{22}'^{(k)}}{C_{33}'^{(k)}},\quad N_2^{(k)}=\frac{C_{12}'^{(k)}-C_{13}'^{(k)}}{C_{33}'^{(k)}},\quad N_3^{(k)}=\frac{K_3^{(k)}-K_2^{(k)}}{C_{33}'^{(k)}},\quad N_6^{(k)}=\frac{C_{26}'^{(k)}-2C_{36}'^{(k)}}{C_{33}'^{(k)}} \tag{16}$$

and, where temperature is supposed to be homogeneous through the thickness. The usual solution is:

$$u_r^{(k)}=D^{(k)}r^{M_1^{(k)}}+E^{(k)}r^{-M_1^{(k)}}+r\left(M_2^{(k)}\varepsilon_0+M_3^{(k)}\Delta T\right)+M_6^{(k)}\gamma_0 r^2 \tag{17}$$

with

$$M_1^{(k)}=\sqrt{N_1^{(k)}},\quad M_2^{(k)}=\frac{N_2^{(k)}}{1-N_1^{(k)}},\quad M_3^{(k)}=\frac{N_3^{(k)}}{1-N_1^{(k)}},\quad M_2^{(k)}=\frac{N_6^{(k)}}{6-N_1^{(k)}} \tag{18}$$

and, where $E^{(k)}$ and $D^{(k)}$, $k\in[1,n_l+n_s]$, are the integration constants.

2.4. Boundary conditions

The number of unknown constants is $2(n_l+n_s)$. In the following, inner and outer radii of each layer are introduced, respectively $R_{int}^{(k)}$ and $R_{ext}^{(k)}$ with $R_{int}^{(1)}=R_0$ and $R_{ext}^{(n_l+n_s)}=R_0+e+\sum_{i=1}^{n_s}ep(i)$. Boundary conditions are then summarised as:

Displacement continuity:

$$\forall k\in[1,n_l+n_s-1],\quad u^{(k)}\left(R_{ext}^{(k)}\right)=u^{(k+1)}\left(R_{ext}^{(k)}\right) \tag{19}$$

Radial stress continuity –P_0 is the internal pressure:

$$\begin{aligned}\forall k\in[1,n_l+n_s-1],\quad &\sigma_r^{(k)}\left(R_{ext}^{(k)}\right)=\sigma_r^{(k+1)}\left(R_{ext}^{(k)}\right)\\ &\sigma_r^{(1)}(R_0)=-P_0\\ &\sigma_r^{(n_l+n_s)}\left(R_{ext}^{(n_l+n)}\right)=0\end{aligned} \tag{20}$$

Axial equilibrium and zero torsion condition:

$$\sum_{k=1}^{n_l+n_s}\int_{R_{int}^{(k)}}^{R_{ext}^{(k)}}\left(r\sigma_z^{(k)}(r)\right)dr=\frac{R_0P_0}{2},\qquad \sum_{k=1}^{n_l+n_s}\int_{R_{int}^{(k)}}^{R_{ext}^{(k)}}\left(r^2\sigma_{z\theta}^{(k)}(r)\right)dr=0 \tag{21}$$

And finally, the problem can be reduced to a linear system under the form:

$$\mathbf{A.X}=\mathbf{B}$$

with

$$\mathbf{X}=\left(D^{(1)}\quad E^{(1)}\quad D^{(2)}\quad E^{(2)}\quad \ldots\quad D^{(n_l+n_s)}\quad E^{(n_l+n_s)}\quad \varepsilon_0\quad \gamma_0\right)$$

3. Results and discussion

3.1. The prototype and the material properties

This section is devoted to the definition of the vessel dimension as long as the material properties. Typically, the dimension of the prototypes which are manufactured in our Laboratory is a 1 litre bottle of 250 mm long for a 75.3 mm inside diameter. The thickness of the aluminium liner is 1.85 mm and the thickness of each layer of the laminate is 0.27 mm. The stacking sequence of the laminate is the following: $[\pm30] + [\pm50]_4 + [90]_3$. This sequence means the liner is reinforced with 13 layers of composite: 2 layers with a 30° angle, 8 layers with a 50° angle and finally 3 layers with a circumferential winding.

Figure 2 shows an aluminium liner (old version) on which the first layer is deposed. This equipment allows controlling the displacement of the fibre distribution cell along 4 axes: x, y, z and w, that is the rotation of the mandrel.

Figure 2: Manufacturing process of a type III high pressure vessel.

Table one presents the material properties that are introduced in the model. Units for dilatation coefficients, for Young modulus and for other parameters (except Poisson ratio and α_H) are respectively 10^{-5}°C^{-1}, GPa and MPa.

Table I: Material properties used to simulate the structure behaviour.

	α_L	α_T	E_L	E_T	G	ν_{LT}	σ_{LU}	σ'_{LU}	σ_{TU}	σ'_{TU}	σ_{TLU}
C/E	-0.065	2.7	150	11	4	0.3	1500	1500	50	250	70
	α	E	ν	σ_0	σ_r	K	α_H				
Al alloy	2	72	0.25	200	250	310	0.09				

3.2. Residual stress after manufacturing process and experimental results

The manufacturing process is assumed to have the successive following effects. It is commonly said that residual stresses appears during the cooling after curing. We assume there is no stress in the structure before cooling; at this moment, the structure diameter should take into account the dilatation of the liner submitted to the 120°C curing temperature. The outer liner diameter spread from 79 mm to 79.048 mm what must be neglected. Then depending on the stacking sequence and due to the difference of the dilatation coefficients, a gap generally happens between the liner and the composite. That means no residual stress is present in the liner after curing. At the opposite, this process generates residual stresses in the laminate. Figure 3 presents the residual stress state that can be predicted with the previous analysis only for the composite laminate: (a) shows the radial displacement along the radial direction, (b), (c) and (d) respectively refers to the axial, hoop and radial-hoop shear residual stresses according the radial direction.

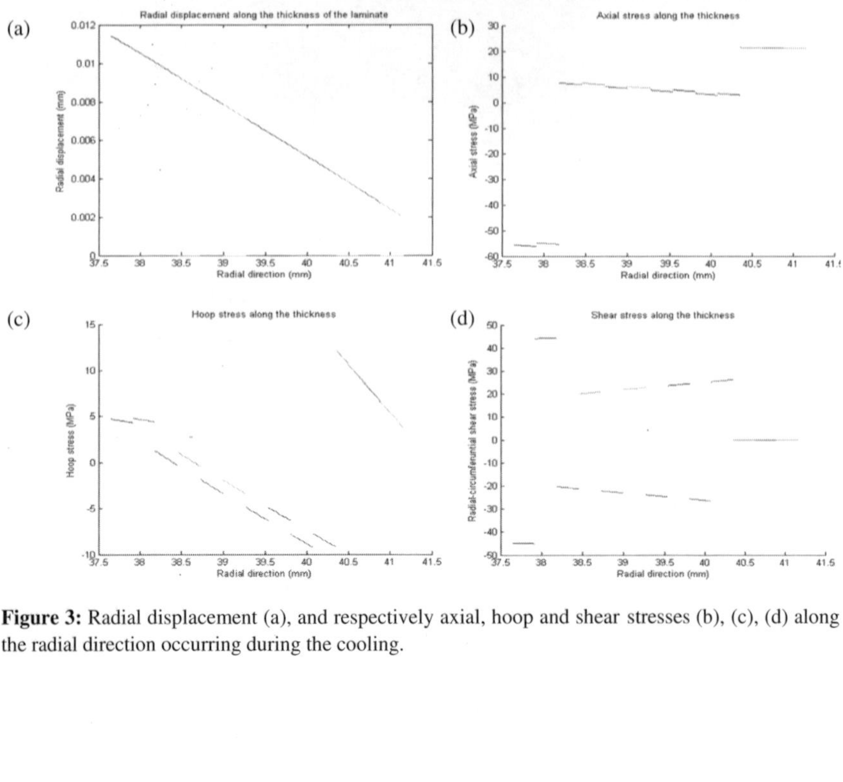

Figure 3: Radial displacement (a), and respectively axial, hoop and shear stresses (b), (c), (d) along the radial direction occurring during the cooling.

In this case, a gap occurs between the liner and the composite and it can be assessed to 0.059 mm (0.048 + 0.011) because the composite spread during the cooling phase. These phenomena and the will to ensure the structure to be in a compressive state fully justify to have the vessel maintain (30 seconds) to a pressure higher than the pressure of use during the manufacturing process. In our case, the burst pressure is one of the investigated parameters so the initial test pressure is limited to 200 bars.

Figure 4 shows the predicted hoop strain and axial strain according the increasing pressure. Due to the close-end effect, the axial deformation remains low and exhibits a backup once the liner plasticization started around 200 bars. Only a change of slope is observed for the hoop strain at the same point, even if some non linear phenomena occur.

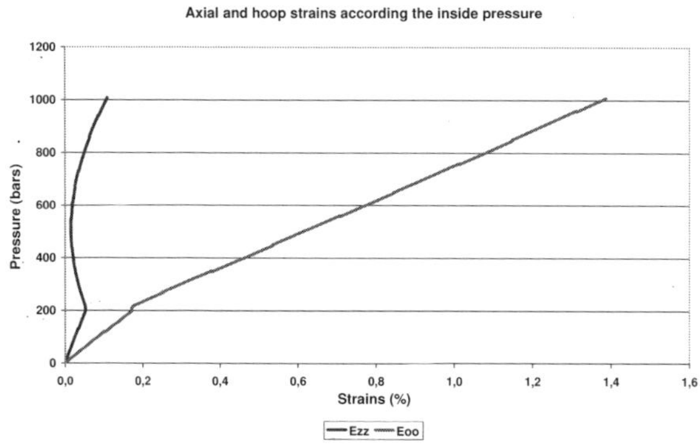

Figure 4: Deformations from simulation, Eoo for the hoop strain and Ezz for the axial strain.

In the simulation, because of the delay observed for the failure of the 30° oriented fibre, the complete failure is assumed as far as all the other layers collapsed. At this time, some relevant improvements must be brought in regard to this discrepancy between all the layers failure.

In spite of the previous comment, comparison between simulations and experiments (not presented here!) is fully convincing as far as the purpose of such a model is to allow predicting the burst pressure of a vessel when the thickness of the liner and the stacking sequence are known. It means for every burst pressure and consequently for any use pressure, one or several stacking sequences may be found to ensure the strength of the structure.

4. Conclusion

In this paper, an elastic plastic analytic modelling of the {liner+composite} structure, that also regards at the composite damage, has been presented. This analysis allows to predict the structure response both for thermal and mechanical loadings, and consequently to study the durability of the cylindrical section of an hydrogen storage vessel taking into account the thermal gap but also the residual stresses due to the manufacturing process. These results are to be now compared with some extended experimental investigations before including cycling life of the vessel and an optimization procedure.

References

1. Züttel A., Materials for hydrogen storage. Materials Today, Volume 6, Issue 9, September 2003, 24-33
2. Carter T.J., Cornish L.A., Hydrogen in metals, Engineering Failure Analysis 8 (2001) 113-121
3. Xia M., Takayanagi H., Kemmochi K., Analysis of multi-layered filament-wound composite pipes under internal pressure. Composite Structures 53 (2001) 483-491
4. Lifshitz J.M., Dayan H., Filament-wound pressure vessel with thick metal liner. Composites Structures 32 (1995) 313-323
5. Chapelle D., Perreux D., Optimal design of a Type 3 hydrogen vessel: Part I - Analytic modelling of the cylindrical section, International Journal of Hydrogen Energy 31 (2006) 627 – 638
6. Tsai S.W., Wu E.M., A general theory of strength for anisotropic materials. J. Compos. Mater. 5 (1971) 58-80
7. Perreux D., Oytana C., Continuum damage mechanics of microcracked composites. Journal of Composites Engineering Vol.3 2 (1993) 115-122
8. Perreux D., Lazuardi D., The effects of residual stress on the non-linear behaviour of composite laminates. Part I. Experimental results and residual stress assessments. Composites Science and Technology 61(2) (2001) 167-175
9. Perreux D., Lazuardi D., The effects of residual stress on the non-linear behaviour of composite laminates. Part II. Layer, laminate non-linear models and the effect of residual stress on the model parameters. Composites Science and Technology 61(2) (2001) 177-190

HYDROGEN BEHAVIOR AND COLORATION OF TUNGSTEN OXIDE FILMS PREPARED BY MAGNETRON SPUTTERING AND PULSED LASER DEPOSITION*

SHINJI NAGATA†, BUN TSUCHIYA, TATSUO SHIKAMA

Institute for Materials Research, Tohoku University
Sendai 980-8577, Japan

AICHI INOUYE, SHUNYA YAMAMOTO

Japan Atomic Energy Agency,
Takasaki 370-1292, Japan

The relation between hydrogen and gasochromic properties were investigated by measuring simultaneously hydrogen concentration depth profiles and optical absorption in tungsten oxide films prepared by RF magnetron sputtering and pulsed laser deposition. A large amount of hydrogen was contained in the amorphous WO_3 films prepared by the both methods. The excellent gasochromic properties were found in amorphous H_xWO_3 films with the maximum value of x about 0.8, while poorer coloration was observed in the films with less hydrogen. Under hydrogen exposure, hydrogen concentration increased with increasing the optical absorption in the wavelength of 600 – 1000 nm.

1. Introduction

Hydrogen fuel is considered to be a clean energy resource for the future. Because the hydrogen gas has a relatively low explosive limit in the atmosphere, development of a sensor of hydrogen gas is very important to handle hydrogen safely. Tungsten tri-oxide films covered with a thin catalyst layer is one of the candidates for hydrogen sensing devices that show a reversible coloration phenomenon under hydrogen exposure. Meanwhile, the mechanism of the gasochromic phenomenon is not fully understood. The electrochromic properties of the tungsten oxide film are of great interest from scientific and technological point of view [1]. The optical switching of the gasochromic film based on an electrochromic layer offers wide range industrial applications such as smart windows, an optically based hydrogen detector [2, 3]. There is a widely accepted model [4], in which protons and electrons are simultaneously

† Work partially supported by a Grant-in-Aid for Scientific Research (C) No. 18560789 from the Japan Society for the Promotion of Science

injected into a WO_3 film and reduceWO^{6+} toWO^{5+}, changing the optical absorption in the oxide layer. In another model [5], dissociated hydrogen is transferred into a pore or grain boundary of WO_3 and subsequently creates water and an oxygen vacancy. Despite extensive investigations on the optical and electrical properties of tungsten oxides, the role of the hydrogen on the gasochromic mechanism is still not clearly understood. So far, hydrogen transport was measured mainly by electrochemical techniques and by infrared absorption measurements [6, 7]. Besides, the gasochromic characteristics depended on the preparation methods of theWO_3 films, such as a sputtering evaporation and sol–gel coatings [8, 9].

In the present work, the hydrogen incorporation behavior in tungsten oxide films with different composition and structure prepared by RF magnetron sputtering and by pulsed laser deposition (PLD) was examined using ion beam analysis techniques. Also, the relation between the hydrogen and gasochromic property was investigated by measuring simultaneously hydrogen concentration depth profiles and optical absorption in tungsten oxide films.

2. Experimental procedure

Tungsten oxide thin films were prepared on a SiO_2 glass or glassy carbon substrates by RF magnetron sputtering with a metal W target (purity: 3N, Furuuchi Chemical Corp.), and by pulsed laser deposition (PLD) with a WO3 target (Furuuchi Chemical Corp.). Each process was performed in a deposition camber under the base pressure of about 1×10^{-5} Pa, and the substrate temperature during the deposition was kept at 300 K. For the sputtering procedure, a mixture of argon and oxygen gases was introduced into the chamber through a mass-flow controller and the Ar:O_2 gas flow ratio was adjusted to be about 4:1. An ArF laser beam of 193 nm was used for PLD deposition with a power of 150 mJ/pulse with duration of 10 Hz. Only the O_2 gas was introduced in the deposition chamber for PLD. The deposition rate of the tungsten oxide layer was typically about 0.1 nm/s. The thickness of the deposited films was in the range of 400–600 nm. The crystal structure of deposited films was examined by X-ray diffractometry using Cu-K_α radiations using gracing geometry. The composition of W and O in the film was determined by Rutherford Backscattering Spectroscopy (RBS) using 2 MeV He^{++} ions. The concentration depth profiles of hydrogen in the near surface layer were measured by the Elastic Recoil Detection Analysis (ERDA) technique. The RBS and ERDA experiments were performed in a scattering vacuum chamber, connected to a tandem accelerator. For the ERDA measurement, an analyzing beam of 2.8MeV He^{++} was incident on the specimen at an angle of 75◦ to the surface normal and the recoiled hydrogen atoms were detected at an angle of 30◦ with respect to the analyzing beam. An Al foil of 12μm thickness was placed in front of the detector to stop the forward scattered He ions. Since the probing depth of the ERDA in the present experimental condition was about 800 nm in the tungsten oxide, the whole thickness of the deposited films was analyzed. To evaluate the hydrogen concentration, a plate of titanium hydride was employed as a standard sample having a known content of hydrogen. For the measurements of gasochromic characteristics, the tungsten oxide films were coated with a thin Pd layer of about 20 nm. The coloration and bleaching processes of the $Pd/WO_3/SiO_2$ sample were examined in atmospheres of diluted hydrogen in Ar (Ar +1%H2) gas and air, respectively. The optical transmittance as a function of time was measured at a wavelength of 640 nm using a red light-emitting diode (LED), by a CCD camera equipped with a monochrometer. Details of the experimental setup were described in elsewhere [10].

3. Results and discussions

Fig. 1 shows the X-ray diffraction patterns from the deposited films prepared by magnetron RF sputtering using pure metal tungsten target with different partial pressure of the oxygen. Without introducing the oxygen gas, we observed clear peaks which can be identified as beta-tungsten []. With increasing the oxygen

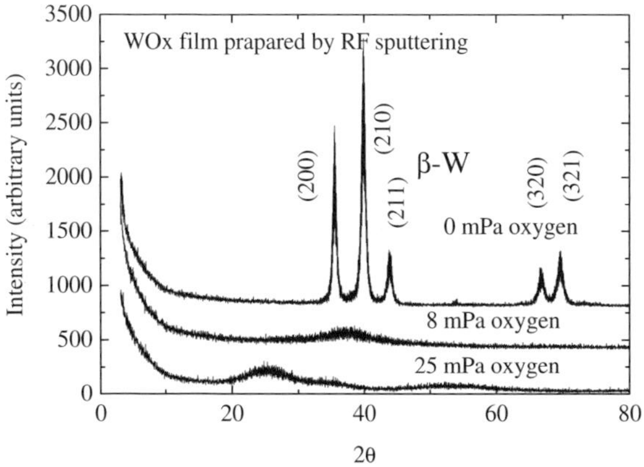

Figure 1. X-ray diffraction patterns for deposited films prepared by RF sputtering using pure W metal target with different oxygen partial pressure.

pressure, a broad peak appeared instead of the sharp reflection patterns, indicating the formation of the amorphous structure of the oxides based on the beta-tungsten. A further increase of the oxygen pressure caused a broad peak at lower angle, corresponding to the amorphous phase related with a crystal structure different from the beta-tungsten. Similar X-ray diffraction patterns and those changes by the oxygen pressure were observed for the films prepared by the PLD method as shown in Fig. 2.

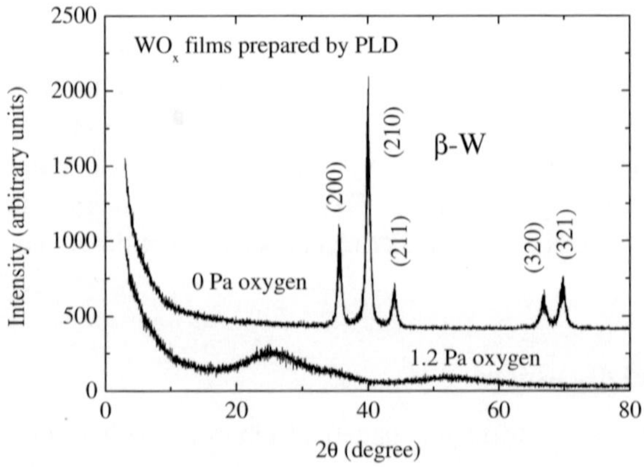

Figure 2. The X-ray diffraction patterns from the deposited films prepared by pulsed laser deposition using tungsten oxide (WO_3) target with different partial pressure of oxygen.

Although the starting target material was WO_3 plate for the laser ablation, beta-tungsten crystal structure was formed with no oxygen gas. At higher pressure of the oxygen gas, a broad peak was found at lower angle in the diffraction patterns, indicating the formation of the same amorphous structure by RF sputtering. The broad peak at around 20 degree can be assigned to (002) of monoclinic or orthorhombic structure of WO_3. The elemental composition of oxygen and tungsten in the films was determined by the RBS measurements, as shown in Fig. 3(a) and (b).

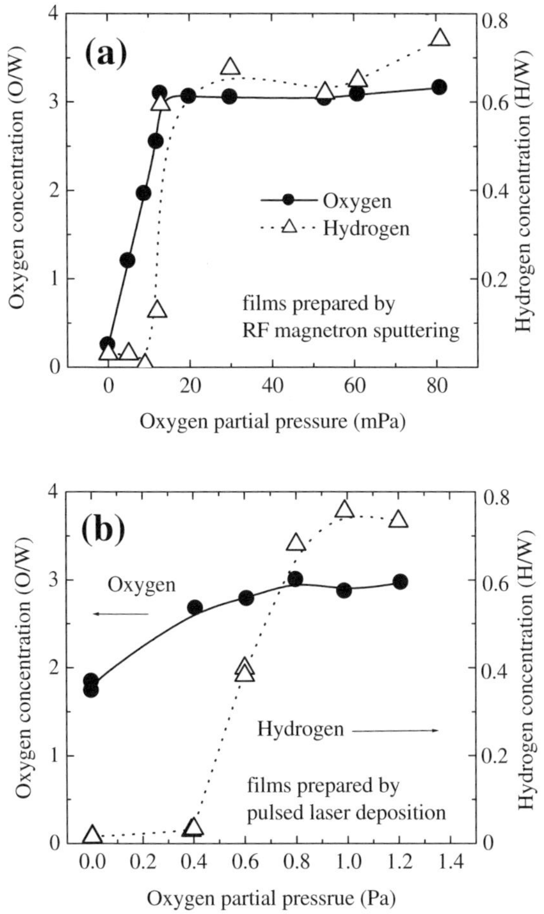

Figure 3. Concentration of oxygen and hydrogen in thin film prepared by RF sputtering and pulsed laser deposition, (a) and (b), respectively, plotted as a function of oxygen partial pressure during the deposition.

In the RF sputtering procedure, the oxygen concentration linearly increased with an increase of the oxygen partial pressure in a range of 0 to 10 mPa, and suddenly saturated to the ratio of O/W to 3.0 at higher oxygen pressure. On the contrary, the O/W ratio was already as high as 2 in the film prepared by PLD method without oxygen introduction, because of the WO_3 target for the ablation. Surprisingly, the X-ray diffraction patterns clearly showed beta-tungsten crystal structure instead of tungsten oxides, although the average composition is measured as WO_2.

The oxygen pressure dependence of the hydrogen concentration in the films was determined by detecting the recoil hydrogen atoms in the same samples, as shown in Fig. 3(a) and (b). Hydrogen concentration was very low in the sample with oxygen concentration below 3.0 O/W. When the amorphous WO_3 film was formed, the hydrogen concentration significantly raise up to 0.7 or 0.8 H/W for both preparation method. Hydrogen is uniformly distributed in the film thickness [10, 11]. The incorporation of this large amount of hydrogen in the WO_3 film may be occurred during the film preparation by the RF sputtering and PLD. Or in the atmosphere after the preparation, the hydrogen dissociated at the surface and diffuses into the films to be retained.

Fig. 4 shows changes of the transmission of 640 nm light through the film deposited on the SiO_2 substrate under Ar+1%H_2 gas exposure, plotted as a function of hydrogen originally contained in the film. In both preparation methods, the coloration occurred in the WO_3 film originally containing hydrogen about 0.7 H/W in the films.

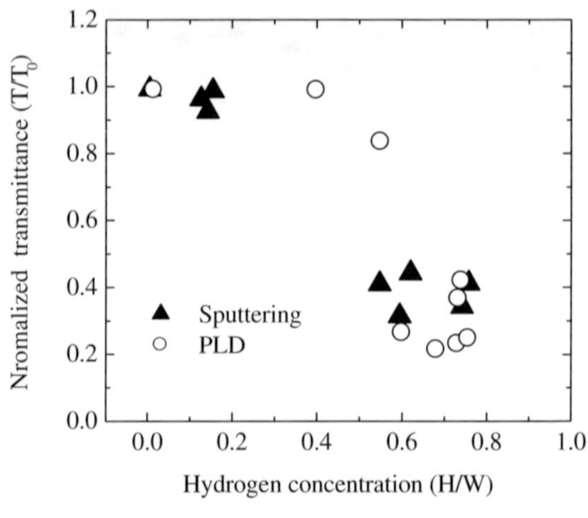

Figure 4. Changes of the transmission of 645 nm light through the film deposited on the SiO_2 substrate under Ar+1%H_2 gas exposure, plotted as a function of hydrogen originally contained in the film.

During the gasochromic experiment, the hydrogen concentration in the film increased with the coloring process. Fig. 5 shows concentration depth profiles of hydrogen in the $H_{0.7}WO_3$ films of 450 nm before and after hydrogen exposure. Long tails of the hydrogen to the deeper into the substrate are attributed to the depth resolution. The profiles include the depth resolution of about 50 nm at surface and over 100 nm at the interface between the WO_3 film and substrate. The originally retained hydrogen does not affect the optical absorption of the WO_3 films. This evidence of the coloration accompanied by the hydrogen incorporation supports the double injection model [4], in which protons and electrons are simultaneously inserted into a WO_3 film to reduceWO^{6+} toWO^{5+}.

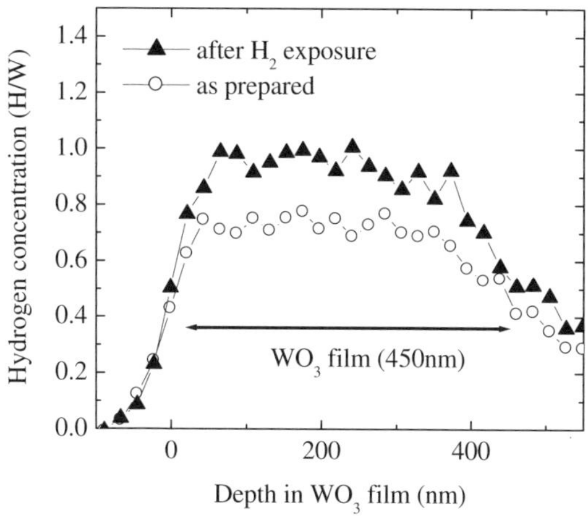

Figure 5. Concentration depth profiles of hydrogen in the WO_3 film as prepared and after exposure of Ar+1%H_2 gas.

4. Conclusions

The structure and composition of tungsten oxide films prepared by RF magnetron sputtering and by pulsed laser deposition was examined using X-ray diffraction and ion beam analysis techniques. The correlation of the hydrogen in the film with the optical absorption characteristics was investigated to clarify the gasochromic mechanism.

The same amorphous structure of WO_3 films were prepared by both methods, under the sufficient pressure of oxygen gas during the deposition. The ion beam analysis revealed that the amorphous WO_3 films contain uniformly distributed hydrogen of 0.7-0.8 H/W, which might be up-taken in the film during the deposition and/or after the exposure of air. Those H_xWO_3 film shows excellent gasochromic characteristics by exposure of Ar+1%H_2. The hydrogen insertion during the coloration was substantiated by measuring simultaneously hydrogen concentration depth profiles and optical absorption in WO_3 films. This incorporation of hydrogen supports the double injection model of the gasochromic process, in which the dissolved protons play a role for reducing W^{+6}.

Acknowledgments

This work is partially supported by a Grant-in-Aid for Scientific Research (C) No. 18560789 from the Japan Society for the Promotion of Science.

References

1. H. Shanak, H. Schmitt, J. Nowoczin, C. Ziebert, Solid State Ionics **171**, 99 (2004).
2. A. Georg, W. Graf, D. Schweiger, V. Wittwer, P. Nitz, H.R. Wilson, Sol. Energy **62**, 215 (1998).
3. K. Itoh, T. Ohgami, Appl. Phys. Lett. **60**, 938 (1992).
4. S.-H. Lee, H.M. Cheong, P. Liu, D. Smith, C.E. Tracy, A. Masarenhas, J.R Pitts, S.K. Deb, Electrochim. Acta **46**, 1995 (2001)
5. A. Georg,W. Graf, R. Neumann, V.Wittwer, Thin Solid Films **384**, 269 (2001).
6. D.-J. Kim, S.-I. Pyun, Y.-M. Choi, Soild State Ionics **109**, 81 (1998).
7. X.Q. Xu, H. Shen, X.Y. Xiong, Thin Solid Films **415**, 290 (2002).
8. M. Stolze, D. Gogova, L.-K. Thomas, Thin Solid Films **476**, 185 (2005).
9. C.O. Avellaneda, L.O.S. Bulh˜oesi, Solid State Ionics **165**, 117 (2003).
10. S. Nagata, A. Inouye, S. Yamamoto, B. Tsuchiya, K. Takano, K. Toh, T. Shikama, J. Alloys and Comp. **446-447**, 558 (2007).
11. A. Inouye, K. Takano, S. Yamamoto, M. Yoshikawa, S. Nagata, Trans. MRS-J **31**, 227 (2006).

HIGH HYDROGEN ABSORPTION IN TITANIUM ETHYLENE COMPLEXES AT ROOM TEMPERATURE

ADAM PHILLIPS AND B.S. SHIVARAM

Department of Physics, University of Virginia, Charlottesville, VA. 22901, USA.

We report observation of high weight percentage hydrogen uptake at room temperature in a titanium-ethylene complex formed through vapor deposition in a ultra high vacuum chamber. An uptake upto 14% by weight is observed in monolayer films. Substitution of hydrogen with deuterium confirms the gravimetric percentage via a doubling. Such molecular level gravimetric observations are made possible with a high resolution quartz sensor based technique. The validity of this quartz sensor based gravimetric method is verified through H_2 absorption measurements on a standard sample of samarium and through O_2 absorption measurements on vanadium nanoparticles.

Carbon nanotubes (CNTs) soon after their discovery were heralded as the ultimate hydrogen storage material. However, recent more definitive experiments have shown that CNTs absorb less than a 1% by weight of H_2[1]. Nevertheless interest in carbon based materials for hydrogen storage has persisted. They retain their attraction due to the light mass, abundance, and favorable chemistry of carbon. While carbon (albeit organic) chemistry is a rich and thoroughly investigated field of science, most hydrocarbons are not friendly towards easy and reversible release of hydrogen. They either disintegrate or make it energetically extremely expensive to get the hydrogen out. A more promising approach to preserve the hydrogen entrained in its molecular form while at the same time allowing it to desorb at around temperature was worked out by Kubas through the discovery of his organo-metallic complexes[2].

With ideas that can be traced back to this discovery, several theoretical groups have focused on using density functional theory and/or first principles quantum mechanical calculations to explore high hydrogen storage in various carbon systems doped with metal atoms. Thus, carbon nanotubes[3], buckyballs[4], metcars[5], and carbon containing polymers[6] have been theoretically studied with many of them predicted to absorb hydrogen in excess of 6 wt%. Among the various theoretical papers there is general agreement that such metal doping methods should lead to high hydrogen absorbing complexes. But there are also concerns that it might not be possible to prevent the high hydrogen absorbing isolated organo-metallic entities from clustering together[7]. Such clustering is generally expected to lead to a degradation of hydrogen absorption. In the context of the present work the recent work of Durgun and collaborators[8] is most relevant. These investigators established that transition metals (such as titanium) in reaction with ethylene should form new complexes through the Dewar interaction and have the potential to absorb as many as 5 H_2 molecules per transition metal atom through the Kubas interaction. To our knowledge there have been no experimental reports of room temperature hydrogen absorbing properties of such complexes.

In this report we present results of a study of a titanium - ethylene complex formed in an ultra high vacuum chamber and collected on a quartz sensor for mass analysis. The technique we use for mass analysis is unique. It is based on the application of a frequency modulation method to unambiguously track the resonant frequency of a quartz sensor to better than 0.1 ppm[9]. A downward shift in the resonant frequency of the sensor, either due to the initial deposition of the complex or due to the absorption of hydrogen gas, indicates a mass accumulation on its face. This shift can be measured with a 0.1 ppm resolution,

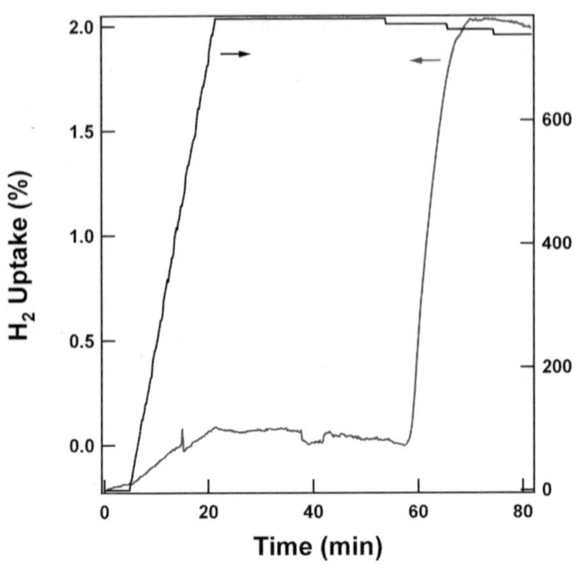

Figure 1: Shows the weight % uptake of H_2 by a 35 nanogram samarium sample measured with a quartz gravimeter.

and in our case this translates to a mass change of 4 picograms. This level is crucial in the context of hydrogen absorption measurements with a monolayer or nanogram level material.

In Figure 1 we show the percentage mass increase measured on a pure samarium sample to establish the validity of the quartz sensor method. The right vertical axis shows the hydrogen gas pressure. The "activation time" of approximately 60 minutes observed in these nanoscale samples is consistent with similar observations on bulk samples[10].

In Figure 2 we show the hydrogen uptake in a titanium-ethylene complex. This figure also illustrates the kinetics of the H_2 absorption process. The kinetics is quite rapid and comparable to that shown by the leading materials being considered for storage today. To verify that it is indeed hydrogen that we are absorbing into the complexes formed, we have performed deuterium loading experiments. With D_2 introduced into the chamber the titanium-ethylene complex exhibits a near doubling of the gravimetric percentage[11]. This is a direct confirmation of hydrogen uptake by the titanium-ethylene complex we are forming in the chamber. Furthermore, we observe that during ablation of the titanium a systematic reduction in ethylene pressure in the chamber occurs

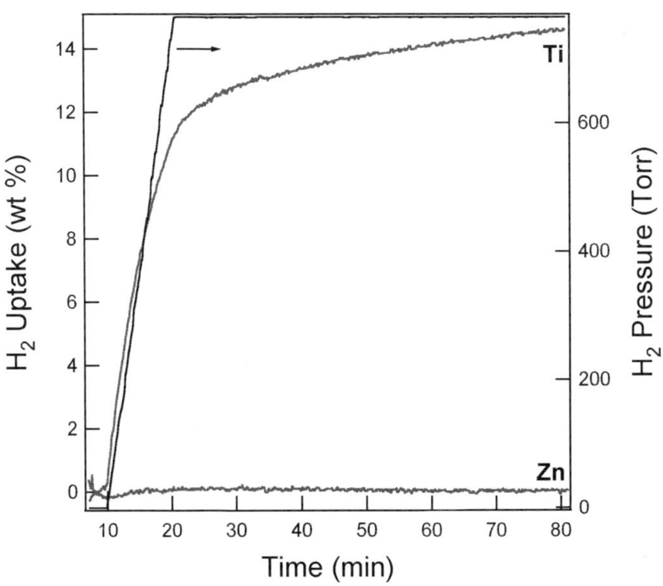

Figure 2: The weight % uptake of H_2 in an ethylene complex formed from Ti is shown along with the data from Zn.

(Fig. 3). From the known measured pressure reduction we calculate the number of ethylene molecules consumed. From the mass accumulated on the substrate we can calculate the number of Ti atoms liberated from the target. These calculated values are comparable and suggest that we are forming a complex with 1 C_2H_4 – 2 Ti atoms bonded together. As a final verification, the Ti was

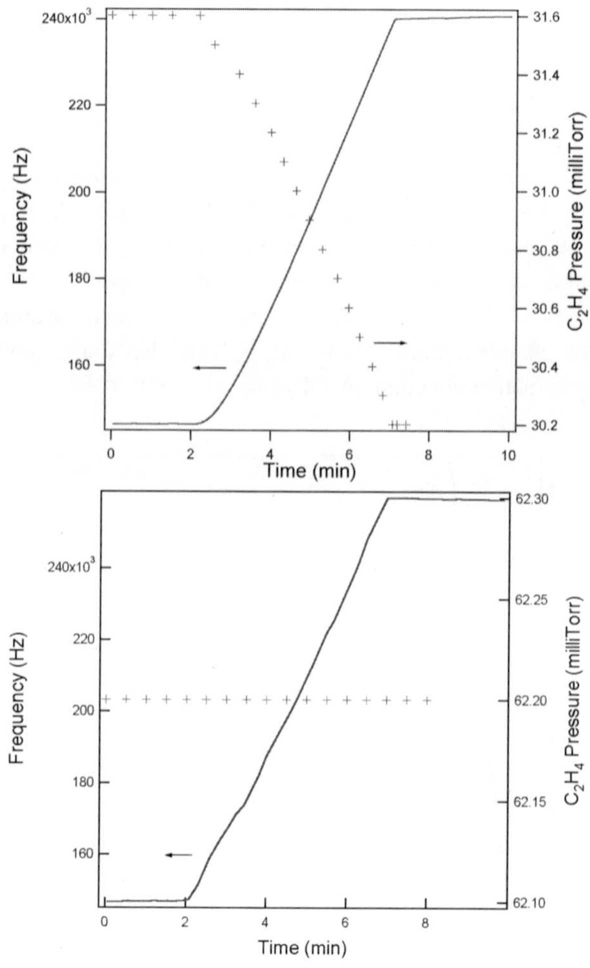

Figure 3: Shows the decrease in the ethylene pressure in the chamber on the continued deposition of titanium-ethylene complex. The bottom part of the figure shows that no such decrease occurs when Zn is vaporized in the presence of C_2H_4.

ablated under pure Ar (no ethylene) and observed to show no significant hydrogen uptake in the same time frame. In contrast to Ti is the behavior of Zn, which exhibits no decrease in ethylene pressure on ablation (lower part of Fig. 3). Its subsequent hydrogen absorption at room temperature is also negligible.

In conclusion we have observed titanium–ethylene complexes with high hydrogen absorption capacity at room temperature. These complexes also have excellent absorption kinetics. With deuterium replacing hydrogen we obtain the expected near doubling of the mass absorbed. Further experiments to characterize the materials produced in our work through spectroscopic means are clearly necessary.

Acknowledgments

This work was partially supported by NSF DMR-007456 and support was also derived through a graduate fellowship to Adam Phillips by the US-DOE administered through SURA.

References

1. S. Orimo, A. Züttel, L. Schlapback, G. Majer, T. Fukunaga, and H. Fujii, J. Alloys Compd. **356-357**, 716 (2003).
2. G.J. Kubas, J. Chem. Soc., Chem. Commun., 61 (1980).
3. T. Yildirim and S. Ciraci, Phys. Rev. Lett. **94**, 175501 (2005).
4. Y. Zhao, Y.-H. Kim, A.C. Dillon, M.J. Heben, and S.B. Zhang, Phys. Rev. Lett. **94**, 155504 (2005).
5. N. Akman, E. Durgun, T. Yildirim, and S. Ciraci, J. Phys.: Cond. Mat. **18**, 9509 (2006).
6. H. Lee, W.I. Choi, and J. Ihm, Phys. Rev. Lett. **97**, 056104 (2006).
7. Q. Sun, Q. Wang, P. Jena, and Y. Kawazoe, J. Am. Chem. Soc. **127**, 14582 (2005).
8. E. Durgun, S. Ciraci, W. Zhou, and T. Yildirim, Phys. Rev. Lett. **97**, 226102 (2006).
9. A.B. Phillips and B.S. Shivaram, Rev. Sci. Instrum., Rev. Sci. Inst., to be published.
10. K. Hirano, J. Kadano, S. Yamamoto, T. Tanabe and H. Miyake, J. Alloys and Compounds, **408-412**, 351, (2006).
11. A.B. Phillips and B.S. Shivaram, to be published.

A COMPARATIVE STUDY OF DEHYDROGENATION ENERGETICS OF B_2H_6, AL_2H_6 AND GA_2H_6 BASED ON DENSITY FUNCTIONAL THEORY*

J. LIU, J. AESCHLEMAN, L. M. RAJAN, C. CHE, Q. GE†

Department of Chemistry and Biochemistry, Southern Illinois University Carbondale, IL 62901, USA

The M_2H_6 molecules, with M=B, Al, Ga, as well as their dehydrogenation derivatives, M_2H_n (n = 0 to 5), have been studied using the B3LYP/6-311++G(2d,3p). Based on the optimized minimum energy structures at each n value, we determined reaction energies for the $M_2H_6 \rightarrow M_2H_n + \frac{6-n}{2} H_2$ reactions. These reaction energies represent the low limit of the energy cost to form molecular hydrogen from the corresponding M_2H_6. Transition state analysis for the first unimolecular step, $M_2H_6 \rightarrow M_2H_4 + H_2$, showed that the critical bond for B_2H_6 dehydrogenation is different from that for Al_2H_6 and Ga_2H_6 dehydrogenation although the structure of the transition states appeared to be the same. These transition states lead to the formation of less stable intermediate product states.

1. Introduction

Alane (AlH_3) has been proposed as an important intermediate in hydrogen cycling in Ti-doped $NaAlH_4$ although the mechanism of AlH_3 formation is not yet understood.[1, 2] The solid state form of AlH_3 can be used as propellant for solid rocket motors.[3] In gas phase, dialane has recently been characterized by photodetachment[4] and IR spectroscopy.[5] These studies showed that Al_2H_6 has a similar μ-hydrido-bridged structure to the electron-deficient diborane, B_2H_6. The binding energy of Al_2H_6, defined as splitting into two AlH_3, was measured recently using mass spectrometry to be 138±21 kJ/mol.[6] Our density functional theory (DFT) studies of hydrogen desorption from the Ti-doped $NaAlH_4$ surfaces showed that di-hydrogen bridged structures can be formed and consequently modified the hydrogen interaction in alanate.[7, 8] Di-hydrogen bridged structures could be a precursor state that resulted to the AlH_x intermediates during dehydrogenation process. Therefore, an understanding of the intrinsic molecular properties of Al_2H_6 as well as its dehydrogenation

* This work was supported by U. S. DOE (DE-FG02-05ER46231) and ACS PRF (PRF#44103-G10).
† Corresponding author, email: qge@chem.siu.edu; fax (618) 453 6408.

energetics would provide useful insights into hydrogen interactions in the alanate-based solid-state hydrogen storage materials. Herein, we focus on the lowest energy structure of Al_2H_6 as well as its dehydrogenation derivatives, Al_2H_n (n = 0 to 5), to establish the thermochemistry of the sequential dehydrogenation reactions. We also studied the dehydrogenation energetics of similar compounds, B_2H_6 and Ga_2H_6.

As a prototype of electron-deficient molecule, the structure and properties of B_2H_6 have been well characterized both experimentally and theoretically. For example, the binding energy of B_2H_6 was determined to be in the range of 150 to 247 kJ/mol experimentally and 162 to 198 kJ/mol theoretically.[6, 9] Shen and Schaefer[10] studied molecular structure and vibrational frequencies of B_2H_6, Al_2H_6, and Ga_2H_6 using the coupled cluster method (CCSD) prior to the experimental synthesis and characterization of Al_2H_6.[4, 5, 11] There are also numerous studies on the low hydrides, such as B_2H_4,[12-14] Al_2H_4,[11, 15] Ga_2H_4,[16-18] Al_2H_3 and Ga_2H_3,[19] Al_2H_2,[20, 21] and Ga_2H_2.[22-24] Most of these studies focused on the structure and vibrational frequencies of different isomers although some dealt with the isomerization between different structural minima. Herein, we studied all these M_2H_n molecules at the same level of theory and determined the global minimum at each n value. We then determined the thermochemistry for the dehydrogenation reactions of M_2H_6 on the basis of calculated total energies and vibrational frequencies. We also determined the activation barrier of the first unimolecular step: $M_2H_6 \rightarrow M_2H_4 + H_2$.

2. Computational Methods

All calculations were performed using the Gaussian 03 computational package.[25] Geometries were optimized using hybrid B3LYP exchange-correlation functionals with the split valence basis set of 6-311++G(2d,3p), commonly referred to as B3LYP/6-311++G(2d,3p). This level of theory has been shown to provide reliable description for dialane and diborane.[5, 6, 9] Harmonic vibrational frequencies were computed for the optimized geometries at the same theory level to distinguish minima from transition states and to provide zero point energy (ZPE) corrections. As such, ZPE was included in all the energies reported in the present paper. A number of initial conformations and spin states were considered for each M_2H_n (with n = 0 to 5) to find the global minimum at each n value. Basis-set superposition errors were not corrected in the calculation of reaction energies. The relaxed structures were compared with the available results from high-level *ab initio* calculations in literature and details will be given in the corresponding section. For a transition state structure, intrinsic reaction coordination (IRC) calculations were performed to connect a specific pair of minima as reactant and product states.

3. Results and Discussion

3.1. *Stable structures*

Relaxation of initial structures and electronic states led to a number of stable structures for M_2H_n at each n value. The structures of the global minimum of each M_2H_n were summarized in Table 1. The key distances were labeled in each structure. The connection between two atoms drawn in these structures does not necessarily indicate a bond being formed between the connected pair. In fact, the

Table 1. Global minimum structures of B_2H_m, Al_2H_m and Ga_2H_m, with m = 0 to 6.

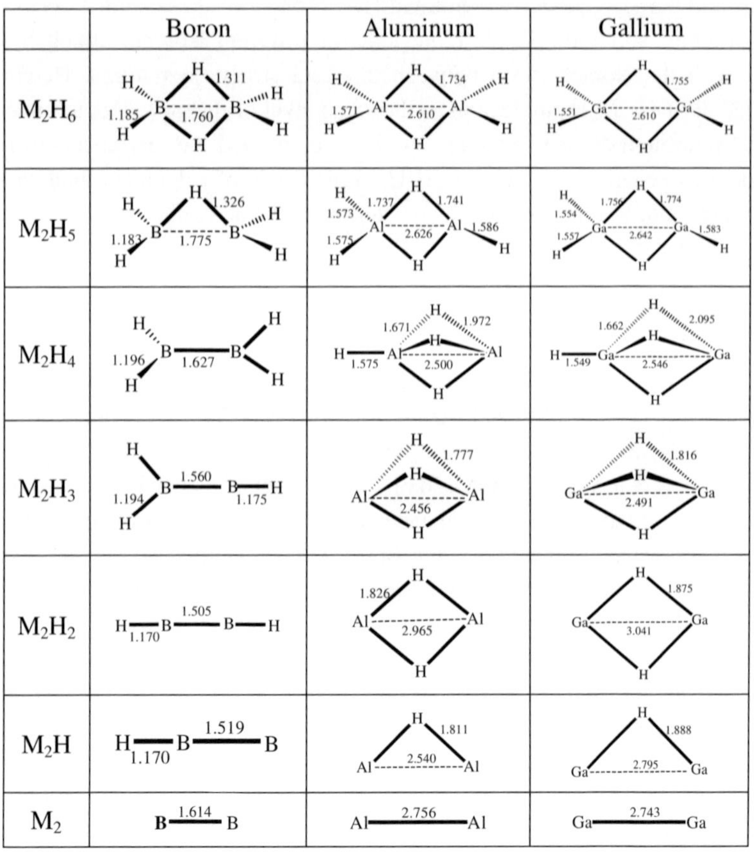

Mulliken charge and overlap population indicate that some of the pairs drawn together may be non-bonding, and in some cases, even anti-bonding (negative overlap population).

The optimized structures of M_2H_6 molecules were shown in the first row of Table 1. These structures are in good agreement with those from previous experimental and theoretical studies.[5, 6, 9, 11, 16] It has been established that all three molecules have a similar μ-hydrido-bridged structure, although the bond length and bond angles are different when M is changed from B to Al, and then to Ga. The bridging hydrogen atoms help to stabilize these electron-deficient structures. Our calculated B—B distance is 1.760 Å while both Al—Al and Ga—Ga distances are 2.610 Å. Mulliken charge analysis showed that the bridging hydrogen atoms in B_2H_6 are almost neutral whereas the corresponding atoms in Al_2H_6 and Ga_2H_6 lose their electrons and become negatively charged, indicating some ionic nature of the bonds formed between Al or Ga and the bridging H atoms. Our results and analysis are in agreement with the previous report.[6] The overlap populations indicate that the B—B bond order in B_2H_6 is about 0.5 while Al—Al bond order in Al_2H_6 is close to zero and Ga—Ga bond order is negative, i.e. antibonding. The nature of the M—H and M—M bonds in M_2H_6 determines the global minimum that the molecule adopts when hydrogen is removed, and this in turn will affect the dehydrogenation energetics of the molecules.

The structures of the hydrides with even number of hydrogen atoms, M_2H_4 and M_2H_2, have been well-studied computationally. The global minimum of B_2H_4 adopts a B—B-bonded structure although the non-planar, doubly hydrogen-bridged C_{2v} structure is less stable than the global minimum by only 13 kJ/mol at the current level of theory. Our results are in qualitative agreement with the previous report by Mohr and Lipscomb based on a MP2/6-31G** calculation who reported an energy difference of 6.4 kJ/mol.[13] On the other hand, Al_2H_4 and Ga_2H_4 favor a tri-hydrogen-bridged HM(μ-H_3)M structure.[26] In the HM(μ-H_3)M structure, the Columbic interactions between a M^+ cation and a tetrahedral MH_4^- anion were believed to be dominant. In fact, a HAl(H_3)Al species has been identified in solid hydrogen.[11] The global minimum of B_2H_2 is linear (H—B—B—H) and in a triplet state. On the other hand, Al_2H_2 and Ga_2H_2 were found to prefer a M(μ-H_2)M structure with a singlet electronic state. The ground state of all three dimers, i.e. M_2, is triplet. The results were shown in Table 1 and are in agreement with the previous reports.[20-23, 27, 28]

Previous studies on the hydrides with odd number of hydrogen atoms were mostly done for cation species.[19, 29, 30] Indeed, dehydrogenation of M_2H_6 through an elementary, unimolecular step would result in losing hydrogen atoms by pair.

On the other hand, hydrogen atoms can be striped off one at a time, starting from M_2H_6, and combine with hydrogen atoms from other M_2H_x to from a H_2 molecule. Consequently, M_2H_n with n being an odd number value becomes a possible product. The first step involves removal of one hydrogen atom from M_2H_6 to form M_2H_5. The global minimum structure of B_2H_5 is different from those of Al_2H_5 and Ga_2H_5, as shown in Table 1. To form the minimum energy B_2H_5 structure, one of the bridging hydrogen atoms in B_2H_6 has to be removed whereas one of the terminal hydrogens in Al_2H_6 or Ga_2H_6 was removed to form minimum energy Al_2H_5 or Ga_2H_5 structures. The structures of M_2H_3 and M_2H follow the same pattern: B—B-bonded structures were found in B_2H_3 and B_2H whereas the hydrogen-bridged structures are favored in Al_2H_3, Al_2H, Ga_2H_3 and Ga_2H. These molecular bonding properties may influence the cycles of hydrogen release and re-adsorption in the hydrides containing Al and B.

3.2. *Dehydrogenation energetics*

The energy required to thermally remove hydrogen from M_2H_6 to form molecular hydrogen is at the heart of this work. The dehydrogenation reactions can be represented in a general formula as:

$$M_2H_6 \rightarrow M_2H_n + \frac{6\text{-n}}{2} H_2 \tag{1}$$

The reaction energy for reaction (1) was plotted *vs* 6-n, the number of hydrogen atoms removed in Figure 1 for each M_2H_6. In order to show clearly the differences, only the reaction energies less than 400 kJ/mol were plotted in the figure. The reaction energies to form B_2H and B_2 according to reaction (1) are 584 and 834 kJ/mol, respectively, and are out of the range of the figure. In

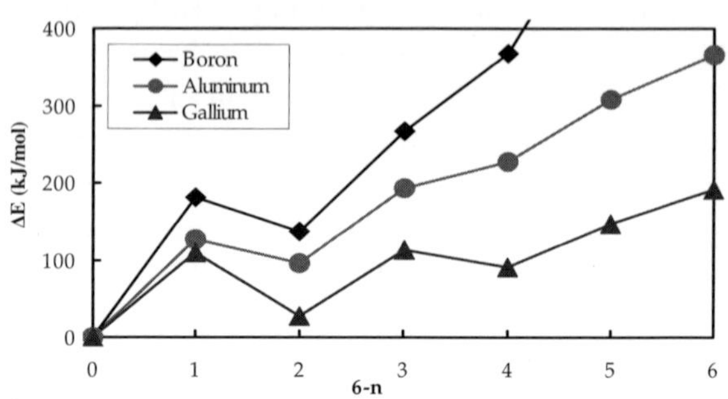

Figure 1. Overall dehydrogenation energy for reaction (1), from n=6 to n=0.

general, the order of these reaction energy curves reflects the intrinsic M—H bond strengths. The average M—H bond strengths are 340 kJ/mol, 284 kJ/mol, and less than 274 kJ/mol for boron, aluminum, and gallium, respectively[31].

Overall, the dehydrogenation process from M_2H_6 is an endothermic process. The energy cost of forming the M_2H_5 structures shown in table 1 and half H_2 molecule from M_2H_6 correlates with the intrinsic M—H bond strength, with B_2H_6 requiring the most energy and Ga_2H_6 the least. The first step energy costs for Al_2H_6 and Ga_2H_6 are 126 and 110 kJ/mol, respectively. These values were calculated according to reaction (1) with n = 5 and plotted in figure 1. In fact, removal of the first hydrogen atom from any of the M_2H_6 molecules to form M_2H_5 and a hydrogen molecule would require a bimolecular reaction step. Starting from B_2H_6, this process would require breaking four B—H bridging bonds, two from each B_2H_6 molecule. If the B—B bond energy is assumed to be the same in B_2H_6 and B_2H_5 (the bond lengths are within 0.01 Å), these bridging B—H bonds have an estimated strength of 217 kJ/mol. On the other hand, the terminal B-H bonds have an estimated strength of 456 kJ/mol. The overall balance of energy makes the removal of the bridging hydrogen atoms more favorable by 22 kJ/mol, with an overall energy cost of 180 kJ/mol.

Removing both bridging hydrogen atoms from B_2H_6 leads to the formation of a full B—B bond in B_2H_4. The formation of the B—B bond stabilizes the resulting B_2H_4 structure significantly. This stabilization makes the process of removing the second hydrogen atom an overall energetically favorable process as compared to the first one. The energy gain to form B_2H_4 from B_2H_5 is 42 kJ/mol. Subsequent dehydrogenation steps only require breaking the terminal B—H bonds and, therefore, the dehydrogenation energy increases almost linearly. The reaction energies for forming B_2H_3 and B_2H_2 are 267 and 367 kJ/mol, respectively, and further dehydrogenation resulted in reaction energies much higher than 400 kJ/mol.

Formation of Al_2H_4 and Ga_2H_4 from the corresponding M_2H_6 also resulted in some energy gains as compared with the formation of M_2H_5. In fact, the extent of stabilization for Al_2H_4, 43 kJ/mol, is similar to that of B_2H_4 but the origin of the stabilization for Al_2H_4 and Ga_2H_4 is likely different from that for B_2H_4. The stabilization for Al_2H_4 and Ga_2H_4 originates from the tri-hydrogen bridge bonds formed between two Al or Ga atoms. Moreover, the stabilization in Ga_2H_4 is more pronounced, reducing the overall reaction energy by more than 80 kJ/mol. This extra stabilization due to Ga_2H_4 formation affects the subsequent dehydrogenation energetics and makes the overall dehydrogenation energy for Ga_2H_6 to Ga_2H_2 less than the energy cost of removing one hydrogen atom from Ga_2H_6. In fact, the overall reaction energy to completely dehydrogenate Ga_2H_6 to form Ga_2 is only 192 kJ/mol. Moc studied the formation of Ga_2H_2, Ga_2H_4 and Ga_2H_6 through a step-wise hydrogenation process of Ga_2.[17, 27] We calculated the overall dehydrogenation energy based on his results and obtained a value of 189 kJ/mol at B3LYP level and 222 kJ/mol at CCSD(T) level, which provided further support to our results.

3.3. *The first unimolecular step*

The energetics of the dehydrogenation reaction shown in figure 1 was based on the assumption that the hydrogen atoms were removed sequentially and the remaining M_2H_n structure was relaxed to the global minimum. We note that the energetics obtained on the basis of these assumptions only provides the low energy limit of the dehydrogenation process. The actual dehydrogenation process will likely take place through a series of elementary steps and each of these elementary steps will overcome an activation barrier. The overall reactivity will be determined by the complexity of the potential energy landscape.

In the present study, we determined the transition states for the first unimolecular step, i.e. $M_2H_6 \rightarrow M_2H_4 + H_2$, for all three hydrides. The transitions states for the unimolecular dehydrogenation of all three M_2H_6 appear to be similar, shown schematically in Figure 2. Detailed analyses showed that the transition state of B_2H_6 dehydrogenation is different from those of Al_2H_6 and Ga_2H_6. The limiting step in B_2H_6 dehydrogenation is breaking B—H bond(s) whereas in Al_2H_6 and Ga_2H_6 the step is forming H—H bond. This difference is reflected in the structure of the transition state: the H—H distance of the desorbing H_2 is 0.785 Å for B_2H_6, whereas the corresponding distances for Al_2H_6 and Ga_2H_6 are 1.042 Å and 1.030 Å, respectively. The shortest M—H_2 distances are 1.674 Å, 1.662 Å and 1.678 Å for B_2H_6, Al_2H_6 and Ga_2H_6, respectively. IRC calculations showed that these transition states did not lead to the global minimum of M_2H_4 shown in table 1. Instead, the final states, also shown in Figure 2, are intermediate structures that can be transformed to the ground state structures. The barriers for this elementary step relative to the corresponding initial M_2H_6 in table 1 are 373 kJ/mol for B_2H_6, 222 kJ/mol for Al_2H_6 and 186 kJ/mol for Ga_2H_6, and were labeled in Figure 2. These activation barriers correlate almost linearly with the reaction energies of forming the intermediate structures but not with the overall reaction energies of forming the global minimum structures. The estimated dehydrogenation barriers for Ga_2H_6 from Moc's results[17] are 186 kJ/mol at the B3LYP level and 204 kJ/mol at the

CCSD(T) level, which are again in agreement with our results. In order to reach the global minimum structure, additional steps involving isomerization of the intermediates will have to be taken. Nevertheless, the unimolecular reaction to dehydrogenate all three hydrides needs to overcome a significantly high activation barrier. Therefore, the thermal dehydrogenation of M_2H_6 is unlikely to proceed via a unimolecular mechanism. Other reaction channels such as dissociation into monomers may precede the dehydrogenation reactions. In fact, catalysts were used in the chemical vapor deposition of B-containing a thin film with B_2H_6 as a precursor.

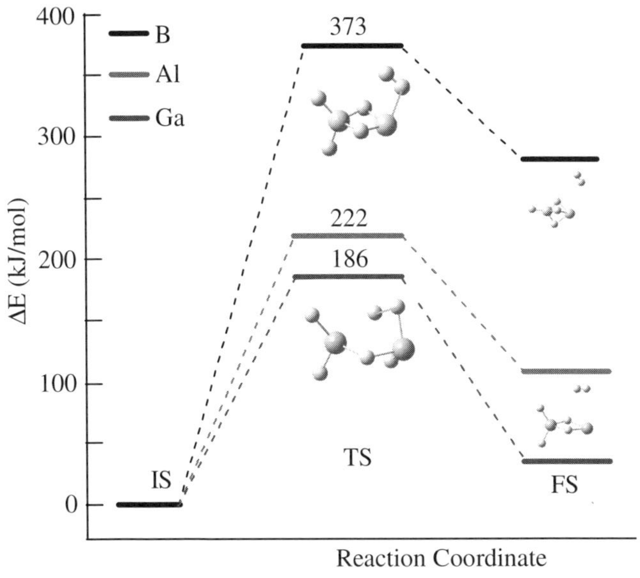

Figure 2. Potential energy profile of the first unimolecular dehydrogenation step for M_2H_6. Initial state (IS) structures were those M_6H_6 in table 1. Transition state (TS) structures for B_2H_6 (upper) and for Al_2H_6 and Ga_2H_6 (lower) were shown in the middle. The final state (FS) structures from the IRC calculations were shown on the right. Short green line segments in TS structures show the critical bonds.

4. Conclusions

We performed a DFT analysis of the global minimum structure and energetics of M_2H_6 (M = B, Al, and Ga) and the corresponding dehydrogenation products: M_2H_n, (n= 0, 5). Reaction energies for the dehydrogenation reactions were calculated based on the total energy of the global minimum structures and molecular hydrogen. These reaction energies correspond to the low limit of the energy costs to desorb molecular hydrogen from the corresponding M_2H_6. Transition states for the first unimolecular step, $M_2H_6 \rightarrow M_2H_4 + H_2$, were located for all three hydrides. The transition states appear to have a similar structure for all three hydrides but the critical bond for the dehydrogenation step of B_2H_6 is the B—H bond (breaking) whereas the critical bond for Al_2H_6 or Ga_2H_6 is H—H bond (forming). This transition state leads to the formation of an intermediate product state that is less stable than the global minimum.

References

1. S. Chaudhuri, J. Graetz, A. Ignatov, J. J. Reilly and J. T. Muckerman, *Journal of the American Chemical Society*, 2006, **128**, 11404-11415.
2. R. T. Walters and J. H. Scogin, *J. Alloy. Compd.*, 2004, **379**, 135-142.
3. T. Bazyn, H. Krier, N. Glumac, N. Shankar, X. Wang and T. L. Jackson, *J. Propul. Power*, 2007, **23**, 457-464.
4. B. K. Rao, P. Jena, S. Burkart, G. Gantefor and G. Seifert, *Phys. Rev. Lett.*, 2001, **86**, 692-695.
5. L. Andrews and X. F. Wang, *Science*, 2003, **299**, 2049-2052.
6. D. J. Goebbert, H. Hernandez, J. S. Francisco and P. G. Wenthold, *Journal of the American Chemical Society*, 2005, **127**, 11684-11689.
7. J. Liu and Q. Ge, *Chem. Commun.*, 2006, 1822-1824.
8. J. Liu and Q. Ge, *J. Phys. Chem. B*, 2006, **110**, 25863-25868.
9. N. W. Mitzel, *Angew. Chem.-Int. Edit.*, 2003, **42**, 3856-3858.
10. M. Z. Shen and H. F. Schaefer, *J. Chem. Phys.*, 1992, **96**, 2868-2876.
11. X. F. Wang, L. Andrews, S. Tam, M. E. DeRose and M. E. Fajardo, *Journal of the American Chemical Society*, 2003, **125**, 9218-9228.
12. B. Ruscic, M. Schwarz and J. Berkowitz, *J. Chem. Phys.*, 1989, **91**, 4576-4582.
13. R. R. Mohr and W. N. Lipscomb, *Inorg. Chem.*, 1986, **25**, 1053-1057.
14. L. A. Curtiss and J. A. Pople, *J. Chem. Phys.*, 1989, **90**, 4314-4319.
15. S. X. Tian, *Theor. Chem. Acc.*, 2006, **115**, 291-297.
16. X. F. Wang and L. Andrews, *J. Phys. Chem. A*, 2003, **107**, 11371-11379.
17. J. Moc, *Chem. Phys.*, 2005, **313**, 93-100.
18. K. Lammertsma and J. Leszczynski, *J. Phys. Chem.*, 1990, **94**, 5543-5548.
19. A. Guermoune and A. Jarid, *Chem. Phys.*, 2007, **333**, 1-9.
20. J. Moc, *Chem. Phys. Lett.*, 2005, **401**, 497-502.
21. T. J. Dudley and M. S. Gordon, *Mol. Phys.*, 2006, **104**, 751-762.
22. J. Moc and M. Wierzejewska, *Chem. Phys. Lett.*, 2003, **380**, 304-312.
23. H. J. Himmel, L. Manceron, A. J. Downs and P. Pullumbi, *Journal of the American Chemical Society*, 2002, **124**, 4448-4457.

24. H. J. Himmel, L. Manceron, A. J. Downs and P. Pullumbi, *Angew. Chem.-Int. Edit.*, 2002, **41**, 796-+.
25. M. J. Frisch, et al, *Gaussian 03, Revision C.02,*, (2004) Gaussian, Inc., Wallingford CT.
26. S. Aldridge and A. J. Downs, *Chem. Rev.*, 2001, **101**, 3305-3365.
27. J. Moc, *Chem. Phys. Lett.*, 2004, **395**, 38-43.
28. M. Hachey, S. P. Karna and F. Grein, *J. Phys. B-At. Mol. Opt. Phys.*, 1992, **25**, 1119-1136.
29. L. A. Curtiss and J. A. Pople, *J. Chem. Phys.*, 1989, **91**, 4809-4812.
30. L. A. Curtiss and J. A. Pople, *J. Chem. Phys.*, 1988, **89**, 4875-4879.
31. D. R. Lide, *CRC Handbook of Chemistry and Physics*, CRC Press, 2004.

COMPUTATIONAL DESIGN OF NANOMATERIALS FOR HYDROGEN STORAGE

QIANG SUN

Department of Advanced Materials and Nanotechnology, Peking University Beijing 100871, China
Department of Physics, Virginia Commonwealth University, Richmond, VA 23284

QIAN WANG AND PURU JENA

Department of Physics, Virginia Commonwealth University, Richmond, VA 23284

Based on density functional theory we have explored several nanostructures for hydrogen storage, including BN cage, metal decorated fullerenes, hybrid organic and inorganic structures, and the supramolecular assembly of Cp rings. We have shown that in nanostructures the size and shape provide additional variables to tune the bonding environment so that clustering of doped metal atoms can be avoided and adsorption energies can be improved.

1. Introduction

Hydrogen, the first element in the periodic table, is the simplest and most abundant element in the universe and exists as a gas under normal atmospheric conditions. It is odorless, colorless, and tasteless and has the potential as an alternate fuel. As an energy carrier, hydrogen can play a critical role in a new, decentralized energy infrastructure that can provide power to vehicles, homes, and industries. Hydrogen has many important advantages over other fuels. Unlike conventional petroleum-based fuels like gasoline, diesel, natural gas and coal that, when burned, contribute to greenhouse gas and other environmental pollutants, hydrogen burns clean.

Hydrogen storage is considered to be the biggest challenge in a new hydrogen economy since the storage medium must meet the requirements of high gravimetric and volumetric density, fast kinetics and favorable thermodynamics [1-7]. The current methods of storing hydrogen as compressed gas or in the liquid form does not meet the industry requirements since the energy densities are much lower than that in gasoline. Moreover, there are issues of safety and cost involved in compressing hydrogen under high pressure or liquefying it at cryogenic temperatures. Although storage of hydrogen in solid state materials offers an alternative, there are no current solid state storage materials that meet the industry requirement.

Hydrogen can be stored in solid materials either in atomic or molecular form. In metal hydrides hydrogen molecules dissociate on the metal surface and reside in interstitial positions in atomic form and can diffuse readily. In complex light metal hydrides, on the other hand, hydrogen atoms are held by strong covalent bond and their dissociation requires high temperatures. Storage of hydrogen in molecular form is characterized by weak bonding and desorption takes place at low temperatures. Recently, considerable attention has been focused on porous materials [8-14] such as clathrates, zeolites, nanocage, carbon nanotubes and fullerenes as possible materials for hydrogen storage. Early experiments on carbon nanotubes have met with some controversy and very different results for their hydrogen storing capacity have been reported. Recent experiments [15-17] have shown that maximum storage capacity in these systems is less than 1 wt%. Theoretical study has also indicated that high hydrogen content in the pure carbon nanotubes cannot be achieved through physical sorption [18]. Therefore, the main challenge for hydrogen storage is following: How to tune the structure and composition of materials to improve and balance the gravimetric density, energetics and kinetics of hydrogen storage and release? These require: (1) The binding energy of hydrogen with the substrate should be intermediate between physisorbed and chemisorbed state (0.1 ~ 1.0eV); (2) The weight percentage of stored hydrogen for the system should be greater than 6% for practical applications.

By using state-of-art simulation techniques, we have studied the interactions of hydrogen molecules with nanostructures in order to gain insight into the design of new materials for hydrogen storage. In the following we provide a summary of our results on various nanostructures composed of B-N cage, metal doped C_{60}, and hybrid structures.

2. Hydrogen storage in B-N cage

Because of the shortcomings of carbon nanotubes, recent efforts have been directed at non-carbon nano systems composed of light elements such as B and N. B-N nanostructures are an analogue of the carbon ones and offer several advantages. For example, carbon nanotubes are oxidized at 600 °C in air while B-N nanotubes are stable up to 1000 °C. In addition to their heat resistance in air and structural stability, B-N nanotubes are semiconducting with wide band gaps

(5.5 eV) which is nearly independent of tube diameter or helicity. With the advancement in synthesis techniques, many novel forms of B-N nanostructures such as nanotubes [19-26], bamboo-like wires [27], nanocages and nanocapsules [28] have been discovered. Furthermore, several authors have also studied the hydrogen uptake and reversibility issues of B-N nanostructures [26, 28-31]. It has been found experimentally that at 10 MPa, the B-N nanotubes can store as much as 2.6 wt % of hydrogen while bulk B-N powder can only store 0.2 wt %. This clearly shows that nanostructures provide added advantage in storing hydrogen. Although calculations at the semi-empirical level have been performed on the interaction of hydrogen with B-N cages [30, 31], full understanding of this system is lacking. For example: (1) Does hydrogen prefer to reside on the surface of the cage or does it enter into the cage? (2) Once hydrogen enters into the cage structure, does it remain molecular or dissociate? (3) What is the maximum number of hydrogen molecules that can be stored inside a cage before the cage breaks? (4) As more hydrogen is stored, how are the geometry and electronic structure of the cage changed? We have studied these questions by using $B_{36}N_{36}$ cage as an example [11]. The calculations were carried out using a plane-wave basis set with the projector augmented plane wave (PAW) method [32] as implemented in the Vienna *ab nitio* Simulation Package (VASP) [33]. The details of the numerical procedure are given in Ref. 11.

We first discuss the interaction of a single H_2 molecule with $B_{36}N_{36}$ cage and determine if it binds associatively or dissociatively and if it remains on the outer surface of the cage or inside? In the later case, we are interested in knowing how it enters into the cage – through the square or hexagonal face. To calculate these we have studied three different configurations by placing H_2 on the top of B or N site on the cage surface. H_2 was always seen to fly away [Fig. 1(a)] from the cage. In the second configuration, we placed H_2 initially in the hollow site of a 4-membered ring with one hydrogen atom inside the cage and the other outside the cage. After the structure optimization, we found that the 4-membered ring is broken [Fig. 1(b)], indicating that the 4-membered ring is too small for H_2 to go through.

The situation is different for the third configuration where H_2 was placed in the hollow site of a 6-membered ring. Here we found that one H_2 can go through the 6-memberd ring from the outside. In the equilibrium configuration the encapsulated H_2 resides at the center of the cage resulting in D_{2d} symmetry for the complex [Fig. 1(c)]. However, the formation energy, which is defined as the energy difference between $H_2@B_{36}N_{36}$ and the separated $B_{36}N_{36}$ and H_2 is 0.0

eV, within the accuracy of our calculation. As subsequent H_2 molecules were introduced, the formation energies continued to rise up to n=18 H_2 molecules.

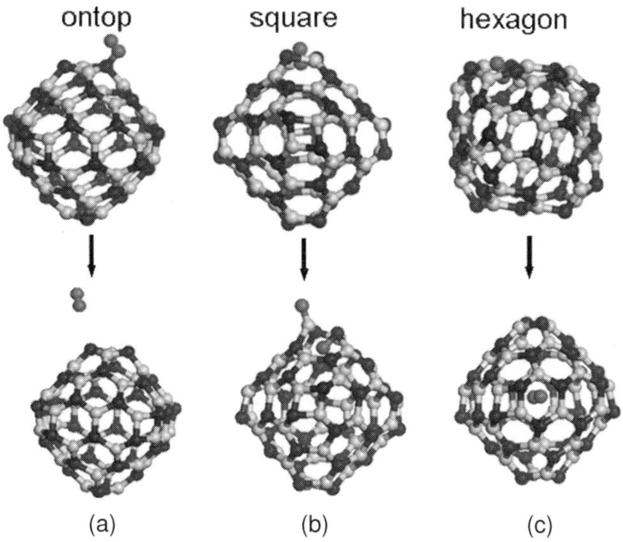

Figure 1. Starting and optimized structures of H_2 interacting with the BN cage.

When the number of H_2 molecules increases to 19, one of the bonds in the cage is broken, but all the H_2 molecules still remain inside the cage. With further increase in the number of H_2 molecules, more BN bonds are broken and finally when the number of H_2 reached 36, the B-N cage is totally broken. This process is depicted in Fig. 2(d). The large energy cost in storing hydrogen inside the $B_{36}N_{36}$ cage raises an important question: Are these materials suitable for practical applications? In particular, do these materials have thermal stability? To address this question, we have carried out molecular dynamics simulation by using Nose algorithm [34] at finite temperatures. First, we studied the thermal stability of $(H_2)_{18}@B_{36}N_{36}$ structure at room temperature (T=300K) using molecular dynamics simulation with 0.4 fs time steps. After 0.4 ps simulation, we found that four H_2 molecules escaped out of the $B_{36}N_{36}$ cage. We then reduced the number of H_2 molecules from 18 to 13, and repeated the calculations. However, after 1.3 ps of simulation three H_2 molecules were found to escape from the BN cage. This indicates that $B_{36}N_{36}$ cage is not suitable as a practical hydrogen storage material.

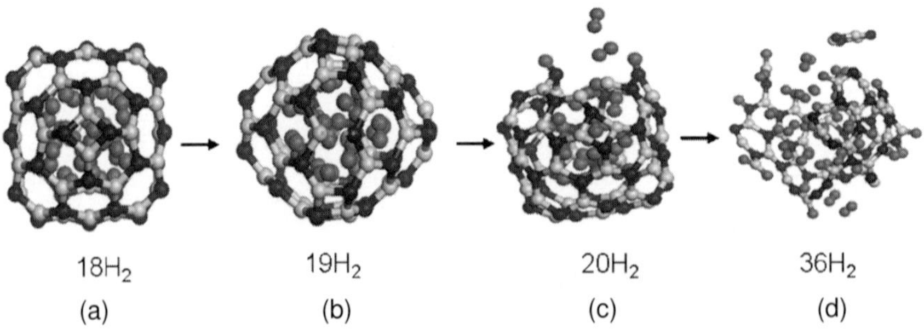

Figure 2. The stability of the cage structure as more H_2 molecules are embedded inside the cage.

3. Hydrogen storage in metal decorated C_{60}

Due to the light weight of C, carbon fullerenes, nanotubes and porous carbon have been considered for hydrogen storage. However, due to the weak interaction of hydrogen molecule with pure carbon based materials, researchers are trying to improve the binding of hydrogen through doping with transition metal atoms such as Sc, Ti and Ni. Recently two theoretical groups [35-39] have shown that metal atoms such as Sc and Ti coated on carbon fullerenes and nanotubes can bind hydrogen in molecular forms with a binding energy of the order of 0.5 eV/H_2 and with gravimetric density of up to 8 wt %. This result is based on the assumption that these metal atoms remain isolated on the nanostructured carbon substrate. We examined this possibility by carrying out theoretical calculations of the interaction of Ti with C_{60} and hydrogen [12] using similar theoretical procedures described in the above. We showed that Ti atoms prefer to cluster on the C_{60} surface and hence the ability of Ti decorated C_{60} to store hydrogen is severely reduced. Similar results are expected for any other transition metal atoms that are used to decorate C_{60}.

We later explored the possibility that there may be other metal atoms that can bind to a large number of hydrogen molecules and yet remain in isolated form on a C_{60} surface. We examined the capacity of $Li_{12}C_{60}$ to store hydrogen. Our rational for studying Li coated C_{60} was two fold: (i) Since C_{60} fullerene has a large electron affinity (2.66 eV) which is comparable to that of Cl (3.62 eV), Li atom would partially donate its valence electron to the fullerene resulting in strong bonding. The charge transfer from Li to the fullerene cage would leave the Li atom in a cationic state which can then bind hydrogen in molecular form due to the polarization mechanism [15, 16]. Since this bonding does not require charge transfer, the amount of hydrogen that can be stored on a $Li_{12}C_{60}$ is limited mainly by steric hindrance. Furthermore, the cohesive energy of Li is substantially smaller than those of transition metals and Li atoms bind more strongly to C than to itself. Thus, Li atoms are not likely to form clusters on the fullerene surface. In fact, experiments based on mass spectroscopy of Li decorated C_{60} fullerene have shown that $Li_{12}C_{60}$ is a very stable cluster [40].

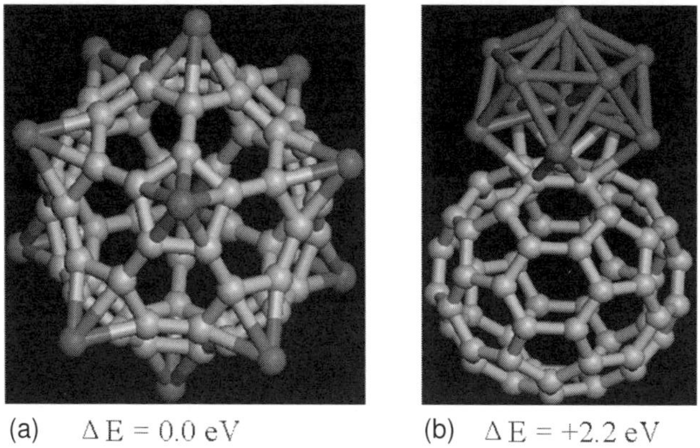

Figure 3. Two configurations of $Li_{12}C_{60}$: (a) Li atoms are kept isolated and (b) Li atoms are allowed to cluster. The relative energy ΔE is evaluated referring to configuration (a).

We showed that $Li_{12}C_{60}$ in the isolated configuration [Fig. 3(a)] is 2.2 eV lower in energy than the clustered configuration [Fig. 3(b)]. In the stable structure, 12 Li atoms remain isolated on the 12 pentagons resulting in a highly symmetric geometry. The average binding energy per Li atom with C_{60} in $Li_{12}C_{60}$ is 1.78 eV, which differs very little from that in LiC_{60}, namely, 1.80 eV. The Mulliken charge analysis shows that each Li atom in $Li_{12}C_{60}$ carries a charge of +0.5, very close to the Mulliken charge on Ni in Ni-C_{60} cluster [41]. Thus the bonding between Li atoms and C_{60} in $Li_{12}C_{60}$ is partly ionic and partly covalent. It is worth pointing out that the equilibrium geometry of $Li_{12}C_{60}$ is totally different from that of $Ti_{12}C_{60}$ where clustering of Ti was shown to lower the energy [12].

We found that up to 5 H_2 molecules can be bound to each of the 12 Li atoms in $Li_{12}C_{60}$. The resulting optimized structure is given in [Fig. 4(b)]. Here we note that hydrogen atoms remain molecular with a bond length of 0.753 Å. The total interaction energy of 60 H_2 molecules with $Li_{12}C_{60}$ is 4.5 eV which yields an average binding energy of 0.075 eV/H_2 molecule.

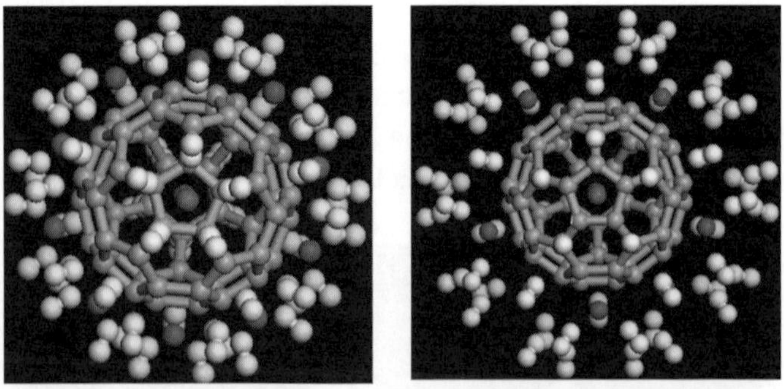

Figure 4. (a) Initial and (b) optimized geometry of $Li_{12}C_{60}(H_2)_{60}$.

While the above results are promising for an isolated $Li_{12}C_{60}$ cluster, one has to wonder about its potential as a hydrogen storage material. For example, is it possible to synthesize a cluster assembled material composed of $Li_{12}C_{60}$ clusters as building blocks where the structural identity of individual $Li_{12}C_{60}$ clusters remains? If so, do the clusters retain their original capacity to store hydrogen? To address these questions, we studied the interaction between two $Li_{12}C_{60}$ clusters. First, the geometry optimization of a $(Li_{12}C_{60})_2$ dimer was performed by starting with an initial configuration where the distance between two Li atoms was set to 2.70 Å, which is the equilibrium bond length of the Li_2 dimer [see Fig. 5(a)]. However, after full symmetry unrestricted geometry optimization, we found the structure of the $Li_{12}C_{60}$ dimer to be that in [Fig. 5(b)]. In the equilibrium configuration, the distance between the two Li atoms is 3.34 Å and the binding energy of the $Li_{12}C_{60}$ dimer is 0.40 eV. Thus, not only $Li_{12}C_{60}$ clusters do not coalesce, but also they maintain their structural identity.

Secondly, we started with another initial configuration where the Li atom capping the pentagonal face of a C_{60} interacts with the hexagonal face of the other $Li_{12}C_{60}$. The resulting optimized structure is shown in Fig. 6. This structure has a binding energy of 1.18 eV which is substantially larger than 0.40 eV corresponding to the structure in Fig. 5(b). Note that, in comparison, the binding energy of a Li atom to C_{60} is 1.68 eV. Interestingly, the integrity of the geometry of $Li_{12}C_{60}$ in the dimer remains, although there are minor changes in the bond lengths. This suggests that $Li_{12}C_{60}$ cluster can form the building blocks of a new kind of solid similar to that of crystals of C_{60} with one major exception. In the fulleride crystal, the hexagonal faces interact with each other while in a material composed of $Li_{12}C_{60}$, the interaction is between the hexagonal and pentagonal faces. Thus, the crystal structure of $Li_{12}C_{60}$ is likely to be different from than that of the fulleride.

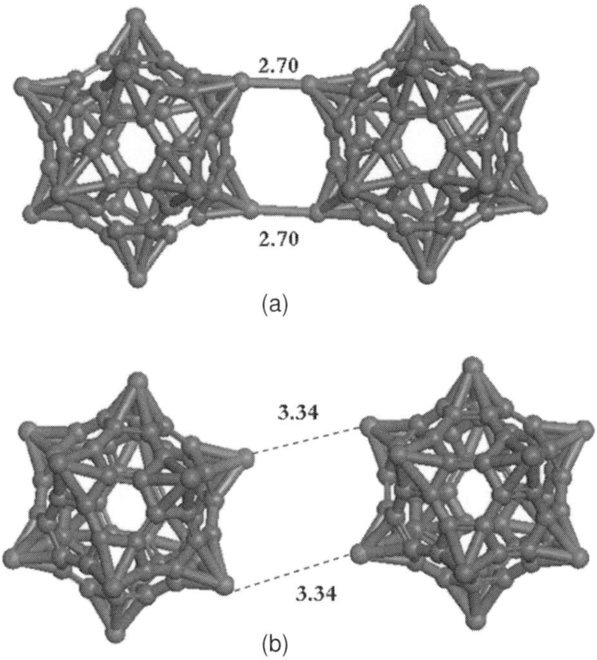

Figure 5. Initial geometry (a) and fully optimized geometry (b) for dimer with the interaction energy of 0.40 eV in the final configuration.

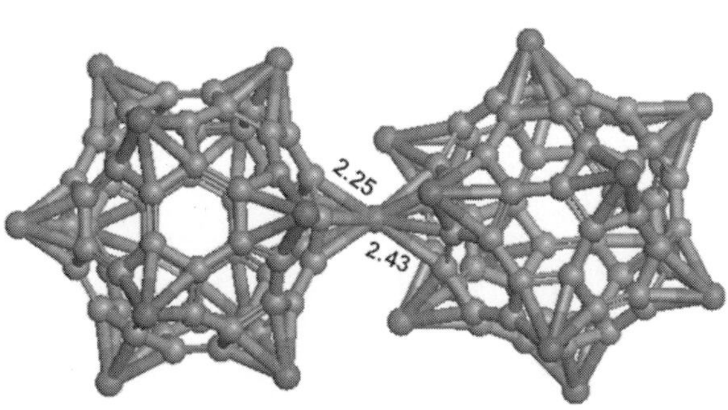

Figure 6. Optimized geometry for dimer with the hollow site configuration. The interaction energy is 1.18 eV.

When the $Li_{12}C_{60}$ dimer complex in Fig. 6 was allowed to interact with hydrogen, we found that the Li atom linking the two $Li_{12}C_{60}$ clusters is unable to bind H_2. This would lower the ultimate hydrogen capacity of a bulk material that contains $Li_{12}C_{60}$ as building blocks. In addition, the average binding energy of H_2 to $Li_{12}C_{60}$ is also small and hence the system may not be suitable for room temperature applications.

4. Hybrid nanostructures for hydrogen storage

As seen in the previous section, transition metal atoms can provide better binding for H_2 molecules, but they tend to cluster. Li atoms do not cluster, but they bind to H_2 molecules weakly. To see if one can use transition metals on a different substrate where their clustering can be avoided, we tried a different approach [14] by grafting the metal-Cp complex on silsesquioxanes (SQ). First we started with the Cp unit. When Sc atom is capped on a Cp ring, the binding energy is found to be 3.81 eV with a Sc-C bond length of 2.44 Å (see Fig.7). This complex is able to bind up to four H_2 molecules with an average binding energy of 0.69 eV/ H_2. The distance between H_2 and Sc is 1.995 Å. Note that the adsorption has little effect on the Cp geometry. All these results are in agreement with previous studies [35].

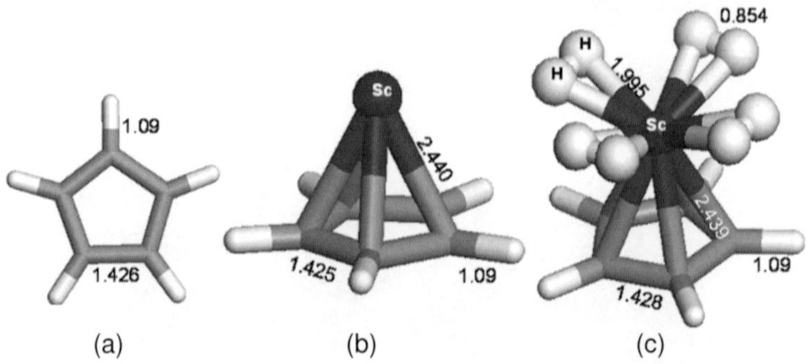

Figure 7. Geometry and bond length (in Å) for Cp, Cp-Sc, and Cp-Sc-4H_2 complex.

Next we studied the assembly of the Sc-Cp complex by using SQ as a matrix. This has the structural formula of $[RSiO_{3/2}]_n$, where the functional group R can be H, alkyl, alkylene, aryl, arylene, or their organo-functional derivatives. In Fig. 8, we show the geometry of $[HSiO_{3/2}]_8$, where the bond length of Si-O and Si-H is 1.640 and 1.475 Å, respectively, and the bond angle of H-Si-O is 109.5 degree, showing that all Si atoms are fully coordinated with sp^3 bonding.

The development of synthesis techniques enables R in $[RSiO_{3/2}]_n$ to be any molecular or nano structural unit. By introducing atoms or molecules into phenyl the system can be functionalized. Bent and Gunko [42] successfully synthesized hybrid structure of $[RSiO_{3/2}]_n$ with R= $-C_5H_5$ and n=8 and 10 using hydrolytic condensation of the silicon organic precursors. This experimental advance in assembling Cp with SQ sheds new hope for synthesizing materials for hydrogen storage. We introduced Sc as the ligand atom and found that it can adsorb four H_2 molecules in a nearly molecular state.

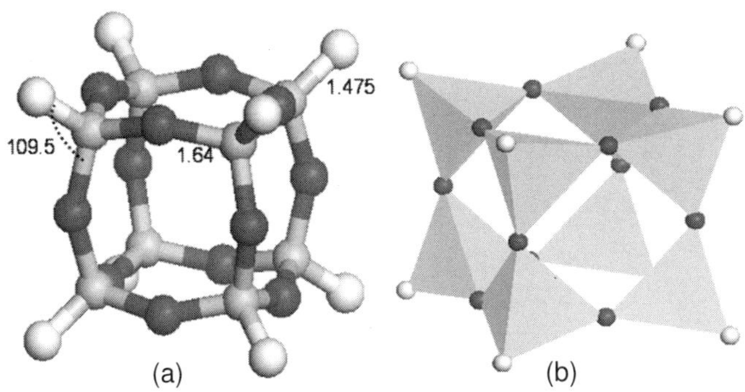

Figure 8. Geometry (a) and 4-coordinated bonding (b) of $H_8Si_8O_{12}$.

For more detailed information about hydrogen bonding in $Sc\text{-}C_5H_4\text{-}H_7Si_8O_{12}$ complex, we show in Fig. 9 the adsorption of one, two, three, and four hydrogen molecules. When one H_2 is introduced, it dissociates and binds atomically to Sc with a binding energy of 0.85 eV/ H. The distance between these two H atoms is 3.14 Å, and they are 1.832 Å away from Sc. When two, three, and four H_2 molecules are simultaneously introduced they bind nearly molecularly and the adsorption energy ranges from 0.56 to 0.66 eV $/H_2$. Note that all the hydrogen molecules have elongated bond lengths ranging from 0.836 to 0.864 Å. This is very similar to what was observed in an isolated Sc-Cp complex [35].

Next we go further from mono-grafted to multi-grafted structures as shown in Fig. 10. When going from one to four, six, and eight grafted complexes, we found that each Sc can still adsorb up to four H_2 molecules, and the corresponding hydrogen storage capacity increases from 1.5 to 3.7, 4.5, and 5.0 wt%. However, only minor changes occur in the average adsorption energy (0.64, 0.65, 0.65, and 0.64 eV / H_2), and in the average distance between Sc and hydrogen molecules (1.995, 1.996, 1.998, and 2.001 Å). The average H-H bond lengths, on the other hand, remain almost unchanged (0.851, 0.850, 0.851. and

0.850 Å). Therefore, we can see that assembling Sc-Cp units by grafting on SQ can not only prevent the clustering of Sc atoms but also can retain the favorable adsorption energy.

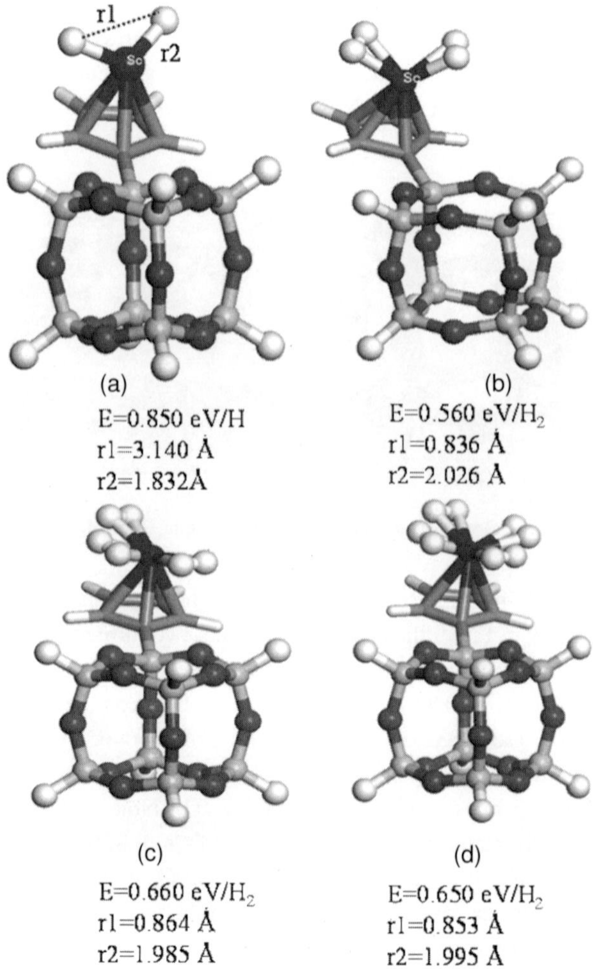

Figure 9. Hydrogen adsorption on mono-grafted structure. E is the adsorption energy, and r1 and r2 are the average distances of H-H and H-Sc.

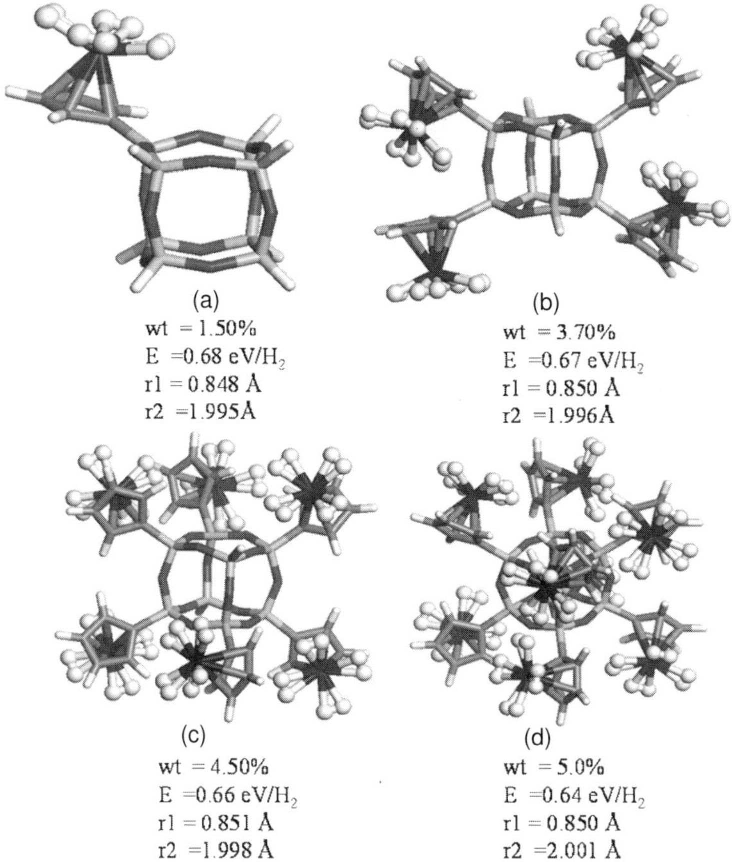

Figure 10. Hydrogen adsorption in one- (a), four- (b), six- (c), and eight- (c) Sc-Cp-grafted structures. E is the average adsorption energy, r1 is the average hydrogen molecule bond length, and r2 is the average distance between H_2 and Sc.

5. Hydrogen absorption in supra Cp structures

We have shown above that there are some advantages in using transition metal atoms as the adsorption centers for hydrogen molecules, and the Cp units can be used as the support for transition metal atoms. To find some ways to assemble these complex structures for hydrogen storage we used SQ in the above as the assembly linkage. Another way is to use supra Cp structures, which has been synthesized recently [43], consisting of six Cp units with one in the center, as shown in Fig. 11. We explored the hydrogen storing capacity of this complex.

Figure 11. Supra Cp structure.

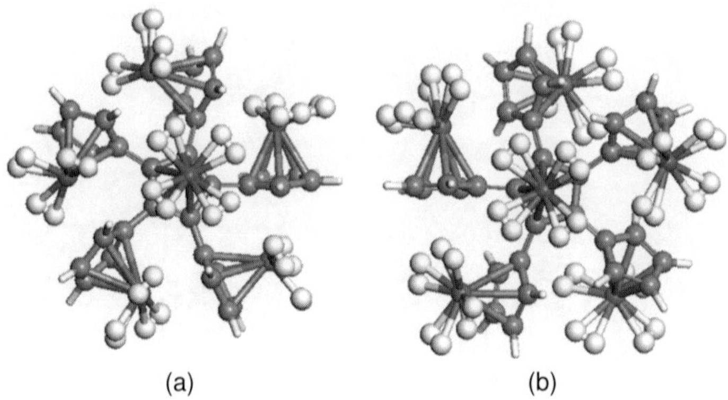

Figure 12. Hydrogen absorption in supra Cp structure.

To compare with previous results we used both Sc and Li as capping atoms. For supra Cp structure with Sc, we find that when three hydrogen molecules are introduced to each Sc site, the average binding energy is 0.808 eV / H_2, the weight percentage of stored hydrogen is 5.85%. If four hydrogen molecules are introduced, resulting weight percentage is 7.38%, and the binding energy becomes 0.36eV / H_2. This is much less than that in SQ structure discussed above. One of the main reasons is that the space available for H_2 absorption in supra Cp is less, accordingly the stress introduced by the adsorption results in a higher energy cost. If we replace Sc with Li, four H_2 molecules can be attached to each Li site and the weight percentage increases to 11.37%. However, the average binding energy is found to be 0.056eV/H_2, which is even less than the value of 0.075eV/H_2 in $Li_{12}C_{60}$ structure [13].

Next we discuss a ring structure composed of 12 Cp units linked by 12 Sc atoms [see Fig. 13(a)]. Frequency calculation indicates that the ring geometry is stable. When hydrogen molecules are introduced, we find that each Sc atom only takes two hydrogen molecules with the average binding energy of 0.44 eV / H_2, due to the limited space available for adsorption. The corresponding weight percentage is 3.6%. This shows that the way Cp units are assembled affects the amount of hydrogen that can be stored. In Fig. 14, we present the iso-surface plot of charge density for the supra Cp and the nano-ring.

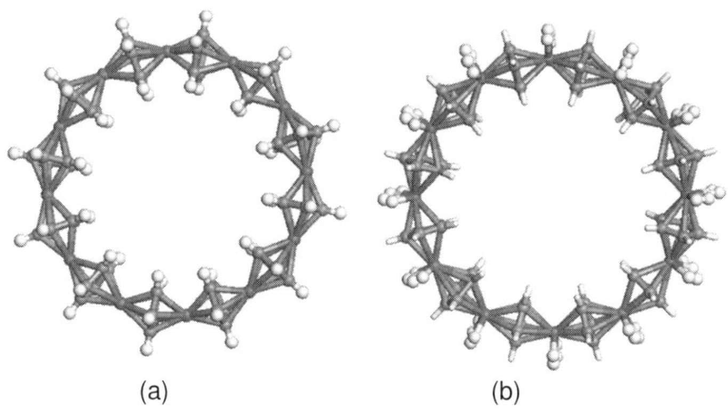

Figure 13. Cp ring (a) and hydrogen absorption (b).

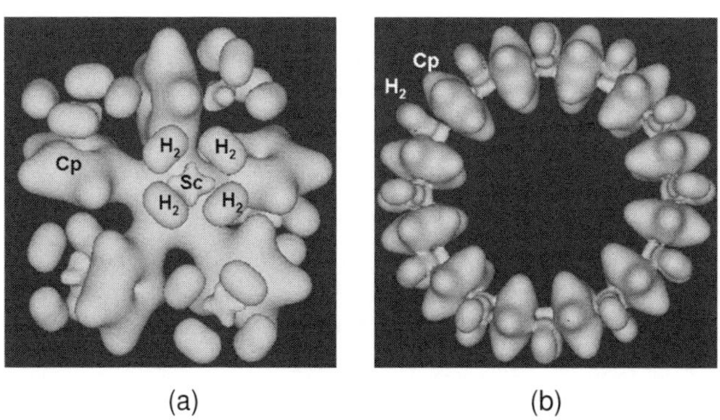

Figure 14. Charge density distribution in supra Cp (a) and Cp ring (b).

6. Summary

Using density functional theory we have explored the ability of nanomaterials such as BN nanocage, metal decorated C_{60}, metal grafted SQ, and supra assembly of Cp molecules to bind hydrogen in quasi-molecular form with binding energies that are intermediate between physisorption and chemisorption energies. We showed that positively charged metal atoms can bind to hydrogen through a charge polarization mechanism. While transition metal atoms supported on a C_{60} surface can bind hydrogen with binding energies that are ideal for applications under ambient thermodynamic conditions, they tend to cluster and thus undermine their hydrogen storage capacity. Li, on the other hand, does not cluster but it binds weakly to hydrogen, making Li decorated C_{60} unsuitable for room temperature applications. We showed that it is possible to dope metal atoms to molecules such as SQ or supra molecular assemblies of Cp rings and these nanomaterials can store hydrogen in quasi-molecular form and are suitable for applications under ambient conditions. Transition metal atoms in these complexes are bound strongly to the substrate and hence do not cluster.

Acknowledgments

The work is supported in part by a grant from the Department of Energy.

References

1. J. Alper, *Science* **299**, 1686 (2003).
2. R. D. Cortright, R. R. Davada, and J. A. Dumesic, *Nature* **418**, 964 (2002).
3. P. Chen, Z. Xiang, J. Z. Luo and K. L. Tan, *Nature* **420**, 302 (2002).
4. N. L. Rosi, J. Eckert, M. Eddaoudi, D. T. Vodak, J. Kim, M. O'Keefe, and O. M. Yaghi, *Science* **300**, 1127 (2003).
5. L. Schlappbach, and A. Zuttel, *Nature* **414**, 353 (2001).
6. see:http://www.sc.doe.gov/
7. see http://www.sc.doe.gov/bes/hydrogen.pdf
8. A. C. Dillion, K. M. Jones, T. A. Bekkedahl, C. H. Kiang, D. S. Bethune, and M. J. Heben, *Nature* **386**, 377 (1997).
9. C. Liu, Y. Y. fan, M. Liu, H. T. Cong, H. M. Cheng, and M. S. Dresselhaus, *Science* **286**, 1127 (1999).
10. H. M. Cheng, Q. H. Yang, and C. Liu, *Carbon* **39**, 1447 (2001).
11. Q. Sun, Q. Wang , and P. Jena, *Nano Letters* **5** (2005) 1273
12. Q. Sun, Q. Wang, P. Jena, and Y. Kawazoe, *J. Am. Chem. Soc.* **127**, 14582 (2005).

13. Q. Sun, P. Jena, Q. Wang, and M. Marquez, *J. Am. Chem. Soc.* **128**, 9742 (2006).
14. Q. Sun, Q. Wang, P. Jena, B. V. Reddy, and M. Marquez , *Chem. Mater.*, **19**, 3074 (2007).
15. F. L. Darkrim, P. Malbrunot, and G. P. Tartaglia, *Int. J. Hydrogen Energy* **27**, 193 (2002).
16. A. Zuttel, P. Sudan, Ph. Mauron, T. Kiyobayashi, Ch. Emmenegger, and L. Schlapbach, *Int. J. Hydrogen Energy* **27**, 203 (2002).
17. A. C. Dillon, K. E. Gilbert, J. L. Alleman, T. gennett, K. M. Jones, P. A. Parilla, and M. J. Heben in Proceedings of the U.S. DOE Hydrogen Program Review, 2001 (http:/www.eren.doe.gov/hydrogen/pdfs/30535am.pdf).
18. G.G. Tibbetts, C.P. Meisner, C. H. OLK, *Carbon* **39**, 2291 (2001).
19. M. Shiraishi, T. Takenobu, M. Ata, *Chem. Phys. Lett.* **367**, 633 (2003).
20. H.Kajiura, S.Tsutsui,K.Kadono,M.Kakuta,M.Ata, and Y.Murakami, *Appl. Phys. Lett.* **82**, 1105 (2003).
21. H. Dodziuk and G. Dolgonos, *Chem. Phys. Lett.* **356**, 79 (2002).
22. X. Blasé, A. Rubio, S. G. Louie, and M. L. Cohen, *Europhys. Lett.* **28**, 335 (1994).
23. N. G. Chopra, R. J. Luyken, K. Cherry, V. H. Crespi, M. L. Cohen, S. G. Louie, and A. Zettl, *Science* **69**, 966 (1995).
24. A. Loiseau, F. Williaime, N. Demoncy, G. Hug, and H. Pascard, *Phys. Rev. Lett.* **76**, 4737 (1996).
25. W. Han, Y. Bando, K. Kurashima, and T. Sato, *Appl. Phys. Lett.* **73**, 3085 (1998).
26. O. R. Lourie, C. R. Jones, B. M. Bartlett, P. C. Gibbons, R. S. Ruoff, and W. E. Buhro, *Chem. Mater.* **12**, 1808 (2000).
27. N. G. Chopra and A. Zettl, *Solid State Commun.* **105**, 297 (1995).
28. D. Shrivastava, M. Menon, and K. Cho, *Phys. Rev.* B **63**, 195413 (2001).
29. C. Tang, Y. Bando, X. Ding, S. Qi, and D. Goldberg, *J. Am. Chem. Soc.* **124**, 14550 (2002).
30. R. Ma, Y. Bando, H. Zhu, T. Sato, C. Xun and D. Wu, *J. Am. Chem. Soc.* **124**, 7672 (2002).
31. T. Oku, I. Naritab, A. Nishiwakic, and N. Koid, *Defect and Diffusion Forum.* **226-228**, 113 (2004).
32. G. Kresse and D. Joubert, *Phys. Rev. B* **59**, 1758 (1999).
33. G. Kresse and J. Furthmüller, *Phys. Rev. B* **54**, 11169 (1996).
34. S. Nose, J. Chem. Phys. 81, 511(1984).
35. Y. Zhao, Y-H Kim, A. C. Dillon, M. J. Heben, and S. B. Zhang, *Phys. Rev. Lett.* **94**, 155504 (2005).
36. T. Yildirim1 and S. Ciraci, *Phys. Rev. Lett.* **94**, 175501 (2005).
37. T. Yildirim1, J. Iniguez, and S. Ciraci, *Phys. Rev. B* **72**, 153403 (2005).

38. Y. Zhao, Y.-H. Kim, A.C. Dillon, J. M. Heben, and S.B. Zhang, *Phys. Rev. Lett.* **94**, 155504 (2005).
39. J. Niu, B.K. Rao, P. Jena, P. *Phys. ReV. Lett.* **68**, 2277 (1992).
40. U. Zimmermann, N. Maljnowski, A. Burkhardt, T.P. Martin, *Carbon* **33**, 995 (1995).
41. W.H. Shin, S.H. Yang, W.A. Goddard, J.K. Kang, *Appl. Phys. Lett.* **88**, 053111(2006).
42. M. Bent and Y. Gunko, *J. Organomet. Chem.* **690**, 463 (2005).
43. Y. Yu, A.D. Bond, P. W. Leonard, K. P. C. Vollhardt, and G. D. Whitener, *Angew. Chem.***118**, 1826 (2006).

Fuel Cells

ENHANCEMENT OF PROTONIC CONDUCTIVITY IN THE NEAR SURFACE REGIONS OF RADIATION INDUCED POLYMER ELECTROLYTE MEMBRANES [*]

B. TSUCHIYA*, S. NAGATA, K. SAITO, T. SHIKAMA[†]

Institute for Materials Research, Tohoku University,
2-1-1, Katahira, Aoba-ku, Sendai 980-8577, Japan

New protonic conduction processes of the perfluorosulfonic acid polymer electrolyte membranes by gamma-ray irradiation at the dose up to 530 kGy and room temperature in air have been found by a direct current (DC) resistance measurement. The conductivities between the polymer electrolyte and the electrode, made of platinum, at 300 and 373 K in vacuum were enhanced to be about two and one, respectively, order of magnitude higher than that of the unirradiated one. The new and original activation energies of the conductivities in the temperature range below and above 343 K were distinguished to be 0.12±0.05 and 0.84±0.03 eV, respectively, which corresponded to potential energy of hydrogen diffusion due to the radiation induced defects and the existing sulfonate group. It was also revealed by means of ultraviolet, visible and infrared optical absorption and hydrogen ion-exchange capacity measurements that the radiation induced defects such as fluorocarbon and peroxy radicals, and C=O including in carbonyl groups were related to the new proton conduction processes. The modification of the hydrogen absorption characteristics due to the radiation induced defects in the near surface regions induces the enhancement of the proton conductivity.

1. Introduction

Perfluorosulfonic acid (PFSA) membranes having a high protonic conduction characteristic at operating temperature around 373 K promise to be useful as electrolytes for fuel cells in some earth as well as space environments [1]. Several reports have recently been published on the degradations of the structure and mechanical and thermal properties for the fluorinated polymers, based on a tetrafluoroethylene backbone with ether-linked side-chains terminating in a sulfonate group, due to chain scissions, long-chain branches and cross-links and by ionizing radiations with X-ray and electrons [2-7]. On the other hand, our group has proposed the improvement of the protonic conduction, investigated by direct current (DC) resistance measurements, by

[*] This work is supported by a research grant from The MAZDA Foundation.

[†] Work partially supported by Takasaki Research Establishment of Japan Atomic Energy Agency (JAEA).

gamma-ray irradiation [8]. However, the correlation between the structure change and the proton immigration is as yet not fully understood.

In the present study, as part of an ongoing program to develop new electrolytes with further higher proton conductivity at lower operating temperature near room temperature and understand the proton behavior on the decomposed chains in the polymer, gamma-ray irradiation to the membranes was examined in air at room temperature and the radiation effects on the protonic conduction process were investigated by proton conductivity, optical absorption and ion-exchange capacity measurements.

2. Experiments

The PFSA membranes used in the present study were Aciplex-SF-1004® with dimension of 10x10x0.117 mm^3. The polymers were irradiated with 1.17 and 1.33 MeV gamma-ray from a cobalt-60 source, installed at the Takasaki Research Establishment of Japan Atomic Energy Agency (JAEA), at room temperature and atmospheric pressure. The resultant ionization doses to the polymers by the gamma-ray irradiation were 530 kGy.

The DC resistance measurements with carried out in air at humidity of 40 % and vacuum, evacuated under 6×10^{-5} Pa, by applying the voltage of -5 to +5 V. The electrode was Al plate with 8x8x0.005 mm^3. In order to investigate temperature dependence on the conductivity, the DC measurement was performed in the temperature range 300 to 393 K in vacuum. Also, the conductivity to the temperature was measured after heating to 393 K in vacuum.

The structure changes of the irradiated polymers were observed in air by two kinds of optical absorption measurements with in the ultraviolet and visible (UV-Vis) wavelengths of 190~760 nm and infrared (IR) wavenumbers of 500~3600 cm^{-1}. Each instrument were transmission spectroscopy for UV-Vis and Fourier transform-infrared (FTIR) one using the attenuated total refraction (ATR) technique with an incident angle of 45°for IR. The measurement range for FTIR is about 0.4～3 μm depth. In addition, the relation between the structure change and the hydrogen absorption was investigated by an ion-exchange capacity measurement with 1 mol/l NaCl and 0.01 mol/l NaOH liquids at room temperature. The ion-exchange capacity E [meq/g] was determined with following equation and chemical reactions of hydrogen ions H, absorbed in the membranes R, with NaCl and NaOH;

$$E = \frac{b \times c}{a} \tag{1}$$

$$R - H + NaCl \Leftrightarrow R - Na + HCl \quad (2)$$

$$HCl + NaOH \Rightarrow NaCl + H_2O \quad (3)$$

where *a*, *b* and *c* represent weight of the membrane (about 100 g), mole concentration (0.01 mol/l) and amount of NaOH, respectively.

3. Results and discussion

Fig. 1 shows proton conductivity for the perfluorosulfonic acid membranes, measured at 300 K in air by the DC resistance measurement, after gamma-ray irradiation at the several doses up to 414 kGy. The conductivity was calculated from the applied voltage and the measured current and dimension of the polymers. It can be seen in Fig.1 that the conductivity increases with increasing the dose. The conductivities at 300 K in air atmosphere rapidly increased until about 50 kGy, and achieved to be higher by about three orders of magnitude than that of the unirradiated one.

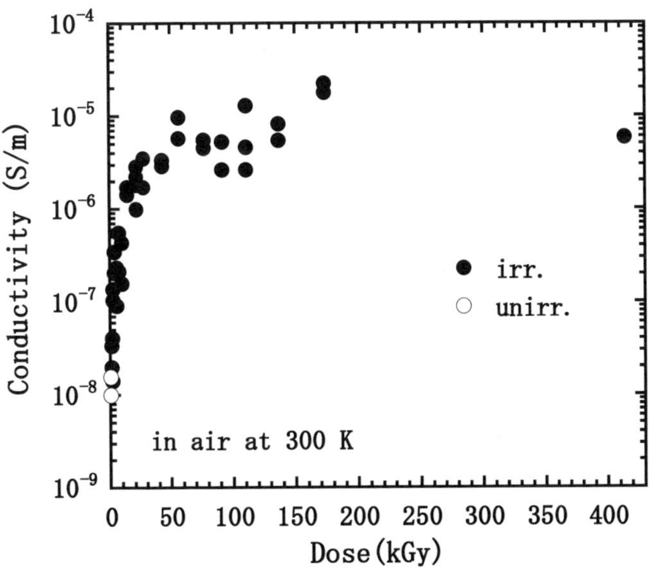

Figure 1. Dose dependence of the conductivity at 300 K in air for the gamma-ray irradiated perfluorosulfonic acid membranes.

Fig. 2 shows Arrhenius plots of conductivities in vacuum at a pressure of 6×10^{-5} Pa against the temperature 300 to 393 K for the perfluorosulfonic acid membranes before and after the irradiation at the doses of 5, 14 and 137 kGy. The absolute values of the conductivity at 300 K in vacuum for the unirradiated and irradiated membranes were about one and two, respectively, orders of magnitude lower than those in air. Even if it is in vacuum condition, the conductivities at 300 K for 5, 14 and 137 kGy were enhanced to be one and two orders magnitude higher than that for the unirradiated one. The temperature dependence of the conductivity for the unirradiated membrane has one slope only. The activation energy can be estimated to be about 0.84±0.03 eV from the slope, and is associated with proton behavior in sulfonate group (SO_3^-). On the other hand, the temperature dependence for irradiated membranes has two slopes in the temperature ranges below and above 343 K. The activation energies of the conductivities at lower and higher temperatures can be calculated to be about 0.12±0.05 and 0.84±0.03 eV, respectively. Their values indicate that there are two kinds of protonic conduction mechanisms, associated with the radiation induced new proton trapping sites and the existing sulfonate group. At the temperatures above 343 K the absolute values of the conductivities for 5 and 14 kGy became almost same with that for unirradiated one, while that for 137 kGy was about one order magnitude higher.

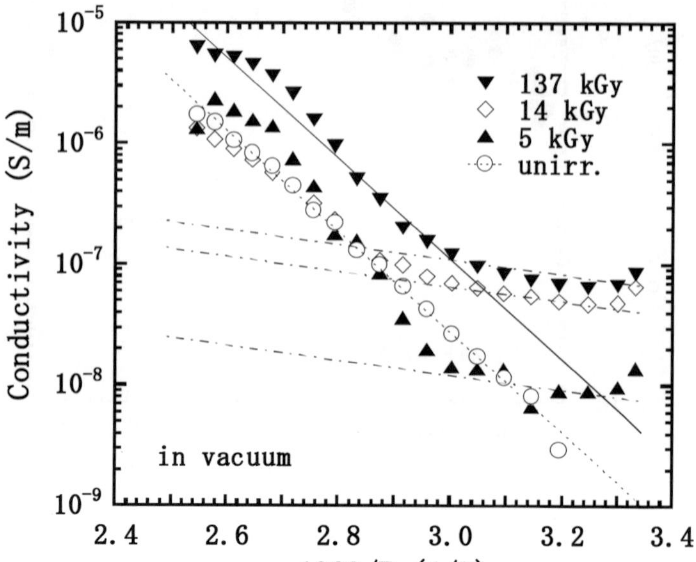

Figure 2. Arrhenius plot of the conductivity in vacuum for the membranes irradiated at 5, 14 and 137 kGy, as compared with that for the unirradiated one.

Fig. 3 shows temperature dependences of conductivities in vacuum before and after heating the perfluorosulfonic acid membrane at the dose of 14 kGy to 393 K. The conductivities after heating decreased at higher temperature and reached to that for the unirradiated one at lower one. Particularly, it was found that the new protonic conduction process at lower temperature disappeared by heating. Moreover, it was confirmed that the conductivity increased again when was exposure to air at room temperature for 20 days. The conductivity depends on humidity in the environments and, namely, contents of water and hydrogen in the membrane. The modification for the absorption characteristic of water on the topmost surface greatly contributes to the new conduction mechanism at lower temperature.

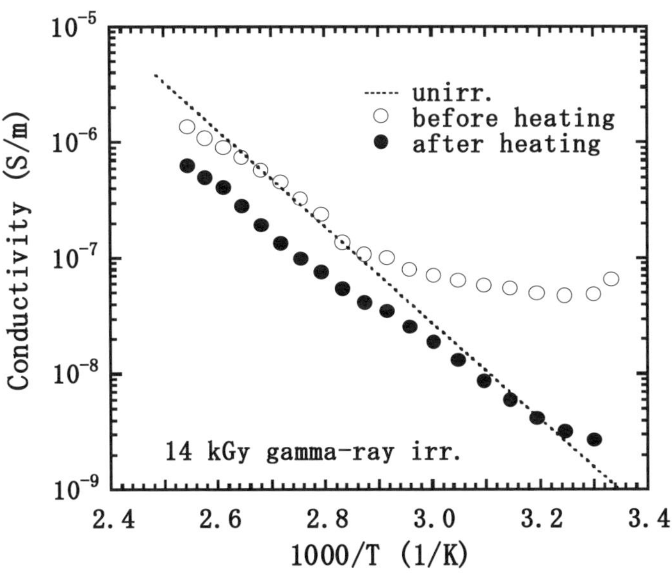

Figure 3. Arrhenius plot of the conductivity in vacuum for the membrane irradiated at 14 kGy before and after heating to 393 K.

In order to investigate the structure changes in the irradiated membranes, the UV-Vis and FTIR optical absorption measurements were performed at room temperature in air. Figs. 4 and 5. show typical UV-Vis and FTIR spectra in the wavelength regions of 190-300 nm and wavenumber of 500-1900 cm^{-1}. The UV-Vis optical absorption spectra of the unirradiated membrane in Fig. 4 exhibit three bands around 190, 215 and 275 nm. The intensities of the optical absorption increased with increasing the dose. The results almost coincide with those by X-ray irradiation, although the bands have been exhibited at 196, 230 and 273 nm [3]. Several absorption peaks at 190, 215 and 275 nm are associated with fluorocarbon (R-CF_2-C·F-CF_2-R or R-CF_2-CF_2-C·F_2) and peroxy (R-CF_2-CF(OO·)-CF_2-R or R-CF_2-CF_2OO·) radicals and C=O groups (R-COF) with unsaturated electron bonding, respectively. The free fluorocarbon radicals are due to the cross-linkage and chemical bond breaking of the chains by gamma-ray irradiation. They can react with oxygen at room temperature and form the peroxide free radicals. The some radicals may mainly provide the new proton immigration with the activation energy of 0.12 eV at temperatures below 343 K. On the other hand, the FTIR optical absorption spectra in Fig. 5 exhibit a new absorption bond at 1771 cm^{-1} which is close to bands (1773 [3], 1776 [5] and 1777 cm^{-1} [4]) associated with the appearance of C=O band in carboxyl group (R-COOH), at the dose above 110 kGy. The carboxyl group is produced by the hydrolysis of the COF group, preferred in near surface rejoins, with water [4, 5, 7]. The conduction due to proton in carboxyl group may provide the increase of the conductivity at higher temperatures above 343 K for the membrane irradiated at 137 kGy in Fig. 3. In contrast, there are no radiation induced change on the peak intensities at 509, 626, 1146 and 1201 cm^{-1} for the symmetric and anti-symmetric stretching vibrations manly involving C-F bonds of CF_2, 803 cm^{-1} for the C–S stretching vibration, 970 and 980 cm^{-1} for the C-O-C stretching vibration, 1060 and 1130 cm^{-1} for the symmetric and anti-symmetric stretching modes of SO_3^- group, around 1300 cm^{-1} for the C-C stretching vibration, around 1450 and 1730 cm^{-1} for the C=C or C-F stretching vibrations of CF=CF or CF=CF_2 groups which have been obtained in the present study [3-6, 9-11] and 1630 and 3450 cm^{-1} for the O-H stretching vibration of H_2O [10].

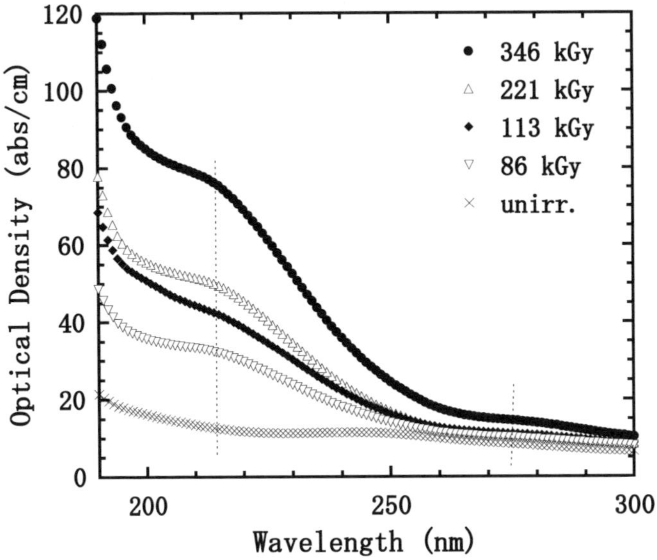

Figure 4. UV-Vis optical absorption spectra for the unirradiated and irradiated perfluorosulfonic acid membranes.

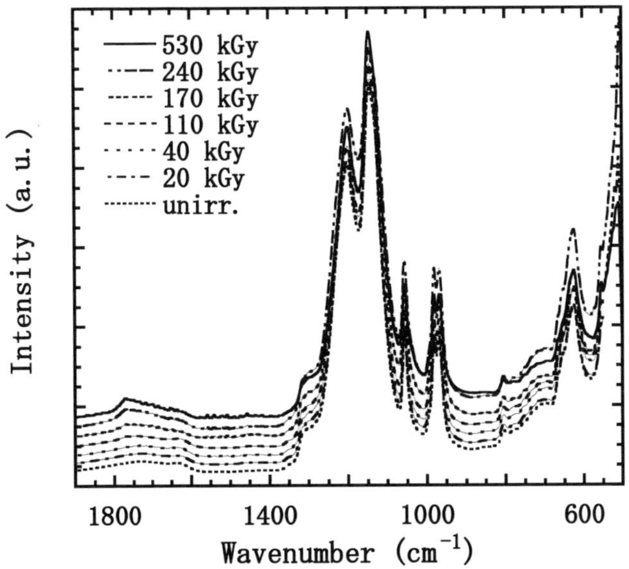

Figure 5. FTIR optical absorption spectra for the unirradiated and irradiated perfluorosulfonic acid membranes.

Fig. 6 shows changes in the ion-exchange capacity concerning with hydrogen ions as a function of dose. The unirradiated membrane has the specific ion-exchange capacity of 0.95 meq/g, namely 0.95 milli-mol sulfonic groups per g, which is close to 0.91 meq/g for Nafion® 117 [2]. The number of ion-exchange capacity gradually increased with increasing the dose. It seems that a number of the ion-exchange capacity corresponds to the amount of the radicals, COF and COOH groups, as compared with Figs. 4 and 5. The results indicate that the hydrogen absorption characteristics are dynamically enhanced by radiation, because the radiation induced defects are hydrophobic. The water or hydrogen absorbed with the radiation induced defects play an important role for new protonic conduction processes. Therefore, the productions of fluorocarbon and peroxy radicals and C=O groups including Carboxyl induces the increment of the proton conductivity.

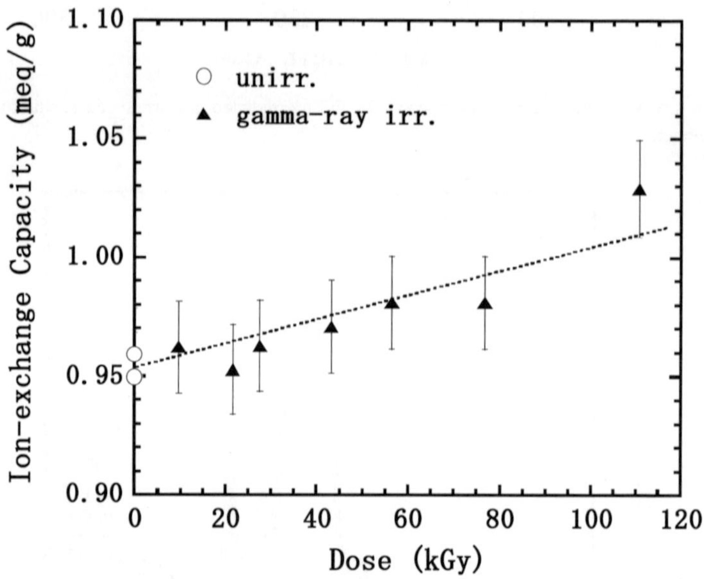

Figure 6. Change in hydrogen ion-exchange capacity of the gamma-ray irradiated perfluorosulfonic acid membranes as a function of the dose.

4. Conclusion

The proton conductivity of the perfluorosulfonic acid membranes greatly changed by the gamma-ray irradiation in air at room temperature. The conductivity at 300 K, measured in air using the electrical resistance method with direct current, increased as the dose increased, and reached to be about three orders magnitude higher than that for the unirradiated one. The conductivity at 300 K when putting the unirradiated and several irradiated membranes in vacuum evacuated under $6x10^{-5}$ Pa decreased to one to two orders magnitude. The conductivity at 373 K in vacuum for the membranes, irradiated at the dose below 14 keV, were almost same with that for the unirradiated one, whereas that for the one at 137 kGy was about one order magnitude higher. For the temperature dependence on the conductivity, Arrhenius plots had two kinds of slopes in the temperature ranges below and above about 343 K. The activation energies at lower and higher temperatures, obtained from the slopes, were determined to be 0.12±0.05 and 0.84±0.03 eV, respectively, and correspond to potential energy of hydrogen diffusion due to the radiation induced defects in near surface regions and the existing sulfonate group. However, the slope at lower temperature disappeared by heating to 393 K. It was observed form the UV-Vis and FTIR optical absorption spectra that the structure of the membranes was modified with the radiation induced defects such as fluorocarbon and peroxy radicals, C=O including in carbonyl groups. In addition, it was also confirmed by the ion-exchange capacity measurement concerning hydrogen ions that the hydrogen absorption characteristic which was significantly concerned to the reaction with H_2O was enhanced by the radiation induced defects. Therefore, the modification of the proton conduction corresponds to enhancement of the water or hydrogen absorption characteristics due to the decomposition of the polymer chains. The some radicals may probably dominate for the new protonic conduction with the activation energy of 0.12 eV. The carbonyl groups produced at high dose may mainly contribute to the increase of the conductivity around 373 K.

Acknowledgments

This work was supported by a research grant from The MAZDA Foundation of Japan.

References

1. H.P. Dhar, *J. Appl. Electrochem.* 23, 32 (1993).
2. M. Schulze, M. Lorenz, N. Wagner, V. Gülzow, Fresenius *J. Anal. Chem.* **365**, 106 (1999).
3. S.H. Almeida, Y. Kawano, *Polym. Degrad. Stab.* **62**, 291 (1998).
4. K. Lunkwitz, U. Lappan, U. Scheler, *J. Fluorine Chem.* **125**, 863 (2004).

5. D. Fischer, U. Lappan, I. Hopfe, K.-J.Eichhorn, K. Lunkwitz, *Polymer* **39**, 573 (1998).
6. M.M. Senna, H.A. Aly, Z.I. Ali, A.M. El-Naggar, *Polym. Degrad. Stab.* **71**, 53 (2001).
7. M.M. Nasef, H. Saidi, K.Z.M. Dahlan, *Radiat. Phys. Chem.* **68**, 875 (2003).
8. T. Adachi, S. Nagata, N. Ohtsu, B. Tsuchiya, K. Toh, T. Shikama, *J. Nucl. Mater.* **329-333**, 1499 (2004).
9. Z. Liang, W. Chen, J. Liu, S. Wang, Z. Zhou, W. Li, G. Sun, Q. Xin, *J. Membr. Sci.* **233**, 39 (2004).
10. M. Ludvigsson, J. Lindgren, J. Tegenfeldt, *Electrochim. Acta* **45**, 2267 (2000).
11. A. Gruger, A. Régis, T. Schmatko, P. Colomban, *Vib. Spectrosc.* **26**, 215 (2001).

NEW PEM FUEL CELL MEMBRANES FOR HIGHER TEMPERATURE, DRIER OPERATING CONDITIONS BASED ON THE HETEROPOLYACIDS

ANDREW M. HERRING,[*] NICCOLO V. AIETA, AND MEI-CHEN KUO

Department of Chemical Engineering, Colorado School of Mines
Golden, CO 80401, USA

JAMES L. HORAN, AND STEVEN F. DEC

Department of Chemistry and Geochemistry, Colorado School of Mines
Golden, CO 80401, USA

MATTHEW H. FREY, ANITHA GENUPUR, AND LUCY REN

3M Corporate Research Materials Laboratory, 3M
St. Paul, MN 55144, USA

STEVEN J. HAMROCK, MICHAEL A. YANDRASITS, AND GREGORY M. HAUGEN

3M Fuel Cell Components Program, 3M
St. Paul, MN 55144, USA

We are developing new proton exchange membranes for hotter and drier operating conditions in fuel cells. The materials we are developing are based on the interaction between the heteropoly acids (HPAs) and the proton donating groups in polymers. One set of materials are composite membranes taking advantage of the HPAs and perfluorosulfonic acid (PFSA) ionomers. These composite membranes have superior proton conductivity compared to the native ionomer under hotter and drier conditions and additionally appear to be more durable under the harsh oxidizing environment of the PEM fuel cell. In another set of materials HPAs are functionalized with monomers which are co-polymerized with monomers that donate protons and add structural features to the hybrid films. While we are yet to fabricate a "polyPOM" with superior proton conductivity these materials can be made to conduct protons as well as standard PFSAs without the need for sulfonic acids.

[*] Email: aherring@mines.edu

1. Introduction

Proton exchange membrane (PEM) fuel cells have many attractive attributes including high operating efficiencies, power densities, and system versatility. However, widespread commercial introduction of the PEM fuel cell is still hampered by issues such as durability, cost, the need for membrane hydration, and the low practical operating temperatures (*e.g.*, 80°C) achievable for this promising device. Higher temperature operation would deliver the benefits of relaxed fuel purity requirements and simpler heat exchange systems. The properties of current state of the art perfluorosulfonic acid (PFSA) ionomers, which are necessary to achieve a reasonable level of oxidative stability, drive the need for the low operating temperature and supplemental hydration of the PEM fuel cell.[1, 2] This is because sulfonic acid derived proton transport only delivers sufficient proton conductivity when the membrane is fully hydrated, which becomes less practical as the operating temperature approaches or exceeds 100°C. One strategy for preserving hydration at high temperature involves cell pressurization, but the energy cost of pressurization diminishes the overall system efficiency. Thus, there is a need for a low cost, durable PEM that will operate at temperatures >100°C on dry inlet gases without pressurization.[3]

2. PFSA/HPA composites

One approach to the improvement of PFSA ionomers is to form composites with inorganic particles.[4] We and others have shown that the proton conductivity and durability of the PFSA ionomers and fuel cell performance of membrane electrode assemblies (MEA) can be improved by the addition of heteropoly acids (HPAs).[5-11] The HPAs, a subset of the polyoxometallates, are an extensive class of structurally well-defined inorganic metal oxide clusters that contain a central heteroatom.[12] These superacidic inorganic oxides are synthetically versatile, exhibit redox catalyst activity, and have very high proton conductivity in the solid state. The HPAs are known to have strong interactions with the sulfonic acid groups of ionomers[13] into which they are doped, resulting in morphological changes, as compared with the undoped ionomers. Additionally, the HPAs may interact with catalyst layers in an MEA. In order to shed light on these complex interactions we report here the results of studies on the structural features of an HPA doped PFSA ionomer, using a variety of different HPAs.

We have shown that doping 12-sillicotungstic acid (HSiW) into PFSA membranes improves fuel cell performance under hot and dry operating conditions and can reduce the rate of F^- release from the fuel cell membranes by 50%.[6] 12-phosphotungstic acid (HPW) is not stable under the harsh fuel cell operating conditions. These two HPAs, HPW and HSiW, have the well known and commonly encountered Keggin structure. Keggin HPAs have the general formula $[X^{+n}M_{12}O_{40}]^{(8-n)-}$ in which a central heteroatom X (where X = B, Zn, Si, Ge, As, P, *etc*) is surrounded tetrahedrally by four groups of three MO octahedra (where M is commonly W or Mo), Figure 1. We have also studied the interaction of more complex HPA structures such as the Wells-Dawson anion, Figure 1. The PFSA ionomer chosen in this study is the polymer available from 3M, which is similar to the well known Nafion® material, in that it has a PTFE backbone, but differs by having a shorter side chain -O-$(CF_2)_4SO_3H$. In this study a polymer with an equivalent weight (EW) of 1000 was used (ion exchange capacity of 1.0meq/g). In general, experimental conditions where chosen that avoided contact between liquid water and the HPA doped membranes, to avoid HPA leaching. An account of this work, concerning only the interaction of HPW with the 3M ionomer, has been published.[14]

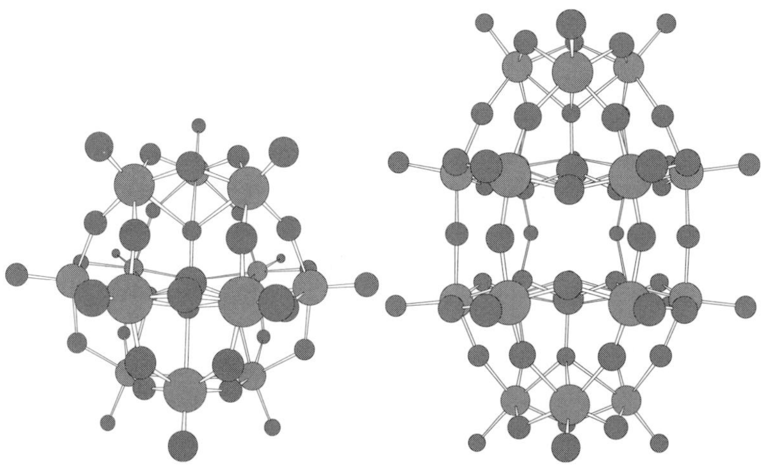

Figure 1. Idealized Keggin, left, and Wells-Dawson anions, right.

The common HPAs of the Keggin structure strongly interact with the 3M PFSA ionomer. This is manifested in the IR bands of the peripheral bonds in the HPA being shifted to lower energy, most likely by bonding with the sulfonic acid groups of the PFSA. Changes in the morphology of the PFSA polymer are observed in the SAXS patterns. Under dry conditions the HPAs exist as crystallites in sulfonic acid clusters, but under wet conditions the HPAs are distributed throughout the ionomer. HPAs assist proton transport under dry conditions. The interaction between these anions and this PFSA ionomer results in significant reduction in E_a, under lower RH. In fact the reverse trend of Ea with RH for the undoped PFSA membranes versus the HPA doped PFSA membrane may indicate a change in proton transport mechanism. The observation of the E_a for proton transport of ½ that of the undoped material a significant and has application for the design of materials for proton conduction under hotter and drier conditions than are currently possible.

We have continued this work by studying metal substituted HPA in composite membranes. Certain metal substituted HPA show dramatic improvements in both proton conductivity under hot and dry conditions and in fuel cell stability.

3. PolyPOMs

The HPA are water soluble and so will wash out of a PEM unless they are immobilized; however, because of their interest as catalysts and in biological systems, a large amount of data is available on immobilization of HPA. There are three basic methods of immobilizing HPA: 1. Electrostatically with a cationic substrate, although this will obviously reduce the proton conductivity of the HPA depending on the number of cations required to render the HPA insoluble. 2. Entrapment in a matrix, the problem being the lack of control in the entrapment process and the fact that HPA may be buried in the entrapping matrix and so may not be able to partake in proton conduction. 3. Covalently bonding a lacunary HPA to a functionality which may be of sufficient bulk to render the HPA insoluble (e.g., a polymerizable monomer). The third method offers the best opportunity to controllably assemble a material and ensure that the HPA moieties are optimally positioned for proton conduction.

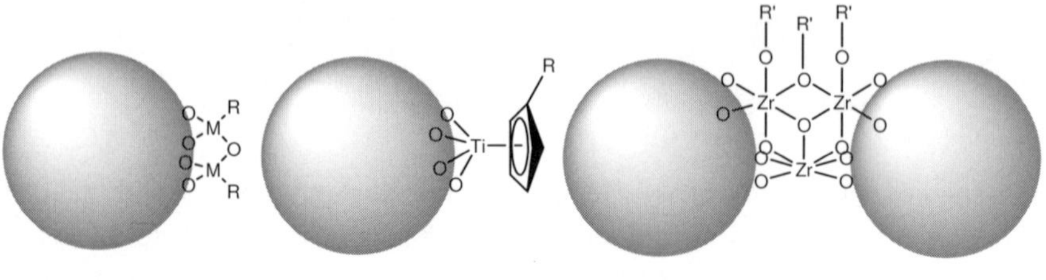

Figure 2. Possible linkages to HPA.

Three classes of hybrid HPA are known to be stable to hydrolysis: 1. Organometallic derivatives of the type RM (M = Si, Ge, Sn, Pb and R = alkyl or aryl). 2. Cyclopentadienyl-titanium derivatives. 3. Zirconium alkoxide or phosphate derivatives, all of which are illustrated in Figure 2. We have tested phenyl model compounds of all of these for stability by boiling them in 6M HCl or H_2O_2 solution. This study showed that only PhP-O-HPA moieties are stable under conditions likely to be encountered in a fuel cell. Never the less we continue to study model compounds of the type RSi-O-HPA due to the large diversity of available ethoxy- and chloro- silanes.

We functionalize the lacunary HPA, $[SiW_{11}O_{39}]^{8-}$ with vinyl, styrenyl, ethylstyrenyl or acrylate monomers. These are then co-polymerized with co-monomers to provide additional protons, cross-linking or other film attributes, Figure 3.

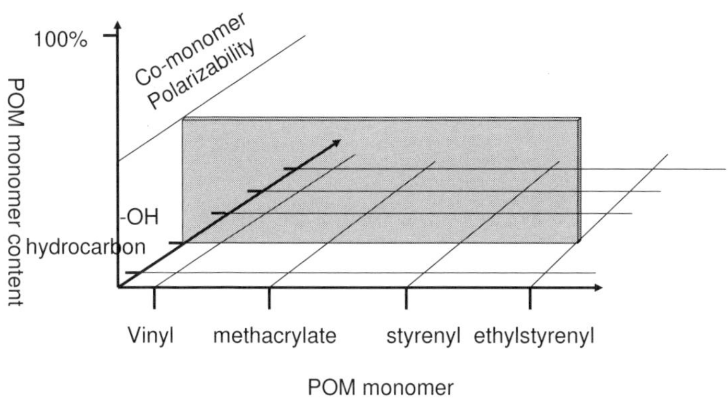

Figure 3. Poly POM design space.

To date we have primarily made polyPOM films with –OH moieties in the co-monmer. Not surprisingly the activation energy for proton conduction in these systems is much higher than in the PFSAs. The self-diffusion coefficients as measured by PFGSE NMR for a styrenylHPA monomer and polymer with 50wt% HPA are shown in Figure 4. Encouragingly we see that the diffusion coefficients of the polymer increase through 120°C. However, for practical proton conductivity these films still require the presence of liquid water. The proton conductivity of a polyPOM with 50wt% methylmetacrylate monomer is shown in Figure 5. Impressively the proton conductivity of the polyPOM at 80°C and 100%RH is 100 mS/cm equivalent to Nafion® 1100 EW under the same conditions. This conductivity is achieved without the use of sulfonic acids.

Unfortunately due to the high E_a of these materials for proton transport the proton conductivity rapidly falls with temperature and decreasing RH. The conductivity of the material in liquid water at room temperature is also *ca.* 100 mS/cm and on stirring only a small portion of this is lost due to theliberation of mobile HPA moieties. We have alos achieved similar results with vinyl polyPOMS.

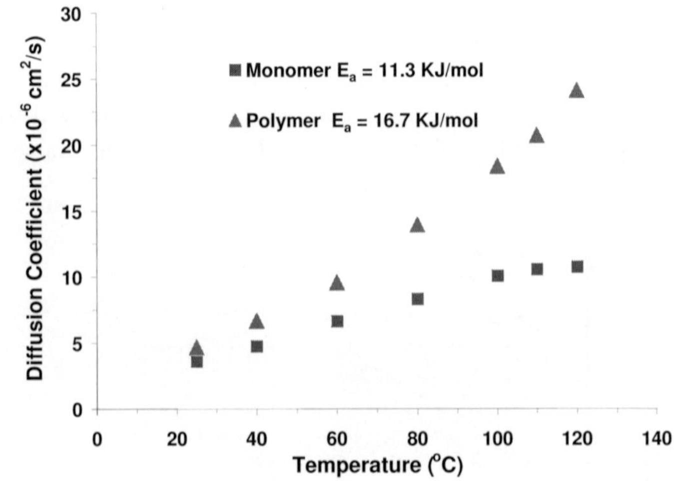

Figure 4. PFGSE H^+ self diffusion in PolyPOM50s Based Polymers and Monomers.

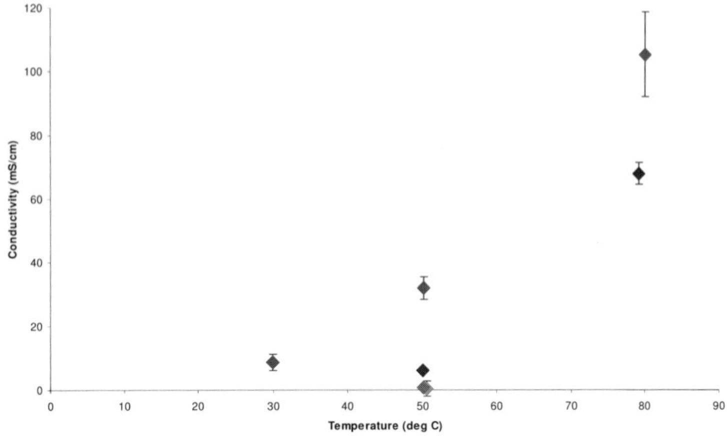

Figure 5. In-plane H^+ conductivity of a PolyPOM50m, ◆ 25%RH, ◆ 50% RH, ◆ 75% RH, ◆ 80% RH, ◆ 100% RH.

In Figure 6 we show the SAXS patterns of a series of a series of polyPOMs using methacryalte as the HPA monomer with various loadings and differing co-monomers. The peak at high q is attributed to the bonds of the HPA. It is interesting to note that the bonding changes depending on the constituents of the membrane. Some of the films show a peak at $q = 4nm^{-1}$ indicative of ordering of the HPA moieties. At low q there is evidence of polymer ordering but this is not true of all the films.

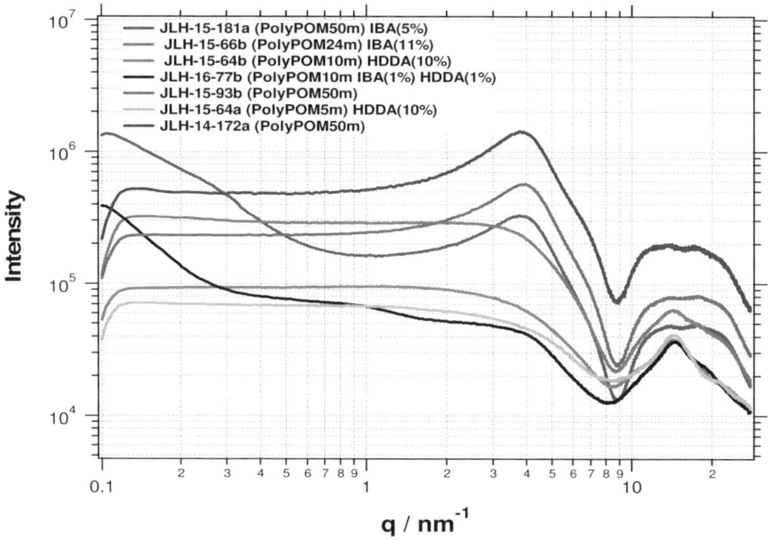

Figure 6. SAXS for a variety of polyPOM materials.

4. Conclusions

Our studies of HPAs continue to show that these inorganic moieties lend interesting properties to ionomer films. In practical films where the HPA is immobilized the film properties are not yet predictable.

Acknowledgments

This research was supported in part by the U.S. Department of Energy, EERE Cooperative Agreements No. DE-PS36-05GO95020 and DE-FG36-07G017006. DOE support does not constitute an endorsement by DOE of the views expressed in this presentation.

References

1. Herring, A. M., Fuel Cell Membranes. In *Encyclopedia of Chemical Processing*, Lee, S., Ed. Marcel Dekker: New York, 2006; Vol. 2, pp 1085-1097.
2. Hamrock, S. J.; Yandrasits, M. A., Proton Exchange Membranes for Fuel Cell Applications. *Polymer Reviews* **2006,** 46, (3), 219 - 244
3. Zhang, J.; Xie, Z.; Zhang, J.; Tang, Y.; Song, C.; Navessin, T.; Shi, Z.; Song, D.; Wang, H.; Wilkinson, D. P.; Liua, Z.-S.; Holdcroft, S., High temperature PEM fuel cells. *Journal of Power Sources* **2006,** 160, (2), 872-891.
4. Herring, A. M., Inorganic–Polymer Composite Membranes for Proton Exchange Membrane Fuel Cells. *Polymer Reviews* **2006,** 46, (3), 245 - 296
5. Li, M.; Shao, Z.-G.; Zhang, H.; Zhang, Y.; Zhu, X.; Yi, B., Self-Humidifying Cs[sub 2.5]H[sub 0.5]PW[sub 12]O[sub 40]/Nafion/PTFE Composite Membrane for Proton Exchange Membrane Fuel Cells. *Electrochemical and Solid-State Letters* **2006,** 9, (2), A92-A95.
6. Haugen, G. M.; Meng, F.; Aieta, N. V.; Horan, J. L.; Kuo, M.-C.; Frey, M. H.; Hamrock, S. J.; Herring, A. M., The Effect of Heteropoly Acids on Stability of PFSA PEMs Under Fuel Cell Operation. *Electrochemical and Solid-State Letters* **2006,** In press.
7. Ramani, V.; Kunz, H. R.; Fenton, J. M., Effect of particle size reduction on the conductivity of Nafion(R)/phosphotungstic acid composite membranes. *Journal of Membrane Science* **2005,** 266, (1-2), 110-114.
8. Ramani, V.; Kunz, H. R.; Fenton, J. M., Stabilized composite membranes and membrane electrode assemblies for elevated temperature/low relative humidity PEFC operation. *Journal of Power Sources* **2005,** 152, 182-188.
9. Ramani, V.; Kunz, H. R.; Fenton, J. M., Stabilized heteropolyacid/Nafion[trademark] composite membranes for elevated temperature/low relative humidity PEFC operation. *Electrochimica Acta* **2005,** 50, (5), 1181-1187.
10. Tazi, B.; Savadogo, O., Effect of Various Heteropolyacids (HPAs) on the Characteristics of Nafion HPAS Membranes and their H2 /O2 Polymer Electrolyte Fuel Cell Parameters. *Journal of New Materials for Electrochemical Systems* **2001,** 4, (3), 187-196.
11. Malhotra, S.; Datta, R., Membrane-Supported Nonvolatile Acidic Electrolytes Allow Higher Temperature Operation of Proton-Exchange Membrane Fuel Cells. *Journal of The Electrochemical Society* **1997,** 144, (2), L23-L26.
12. Cotton, F. A.; Wilkinson, G.; Murillo, C. A.; Bochmann, M., *Advanced Inorganic Chemistry*. 6th ed.; John Wiley & Sons: 1999.

13. Kim, Y. S.; Wang, F.; Hickner, M.; Zawodzinski, T. A.; McGrath, J. E., Fabrication and characterization of heteropolyacid (H3PW12O40)/directly polymerized sulfonated poly(arylene ether sulfone) copolymer composite membranes for higher temperature fuel cell applications. *Journal of Membrane Science* **2003,** 212, (1-2), 263-282.
14. Meng, F.; Aieta, N. V.; Dec, S. F.; Horan, J. L.; Williamson, D.; Frey, M. H.; Pham, P.; Turner, J. A.; Yandrasits, M. A.; Hamrock, S. J.; Herring, A. M., Structural and Transport Effects of Doping Perfluorosulfonic Acid Polymers with the Heteropoly Acids, H3PW12O40 or H4SiW12O40. *Electrochimica Acta* **2007,** In Press.

ALTERNATIVE MATERIALS TO Pd MEMBRANES FOR HYDROGEN PURIFICATION

THAD M. ADAMS AND PAUL S. KORINKO

Savannah River National Laboratory, Aiken SC 29803 USA

Development of advanced hydrogen separation membranes in support of hydrogen production processes such as coal gasification and as front end gas purifiers for fuel cell based system is paramount to the successful implementation of a national hydrogen economy. Current generation metallic hydrogen separation membranes are based on Pd-alloys. Although the technology has proven successful, at issue is the high cost of palladium. Evaluation of non-noble metal based dense metallic separation membranes is currently receiving national and international attention. The focal point of the reported work was to evaluate two different classes of materials for potential replacement of conventional Pd-alloy purification/diffuser membranes. Crystalline V-Ni-Ti and Amorphous Fe- and Co-based metallic glass alloys have been evaluated using both electrochemical and gaseous hydrogen permeation testing techniques.

1. Introduction

Hydrogen separation and purification has been identified as a bottleneck in the development of advanced hydrogen fuel technologies. Many techniques for hydrogen separation are in use or are currently being investigated, such as cryogenic separation, pressure swing adsorption, catalytic purification and selective diffusion. As a result of its high hydrogen permeability, good mechanical characteristics and highly catalytic surface, which dissociate hydrogen rapidly, palladium is still the membrane material of choice in many applications. Unfortunately, palladium and its alloys are extremely expensive, roughly twice the cost of gold, making them impractical for large-scale applications. Therefore, an economically feasible, palladium-based, commercial scale system would require a significantly reduced amount of palladium, which can be accomplished by techniques such as thin palladium membranes supported on porous substrates or highly permeable bulk substrates. The high cost of palladium has turned the attention of researchers to palladium-free membrane technologies, such as cermets and ceramics for high-pressure, high-temperature applications.

The current generation of gas purification/separation membranes is based on Pd/Pd-alloy used either independently or in conjunction with porous ceramic supports. Palladium/Palladium alloys have been known to possess the ability to dissolve a considerable volume of hydrogen and to demonstrate increasing permeability with increasing pressure differential and temperature. However, the major drawbacks to their industrial use are high cost for Pd, relatively low flux, and that during cycling above and below a critical temperature an irreversible change takes place in the palladium lattice structure which can result in significant damage to the membrane. Palladium coated ceramic membranes offer the potential for extended temperature range operations but suffer from the fatal flaw of "pinhole" short circuit paths. Any "pinholes" in the Pd-catalytic film on the surface of the ceramic substrate will allow for contaminant/intermediate species to pass directly through the membrane thus effectively reducing the purification factor of the membrane. Recent efforts in the hydrogen purification/separation membrane community have focused on the development and evaluation of non-palladium based membranes that offer a lower cost, high flux, and highly durable membranes to replace Pd-based systems. Group 5A metals such as V, Nb, and Ta are currently being evaluated by numerous researchers and show promising results with respect to hydrogen permeability [1-3]. However, these metals suffer from severe hydrogen embrittlement and thus are unacceptable for membranes. Japanese researchers have begun to evaluate alloying additions—Al, Ni, Co, and Mo—to vanadium in hope of decreasing the susceptibility to hydrogen embrittlement [4-5].

Crystalline Non-Noble Metal Membranes

The most interesting recent result has been the evaluation of Ni-Ti-Nb alloys for hydrogen permeation [6]. Ni-Ti has long been known as a shape memory alloy but it also possesses good hydrogen solubility and mechanical properties. The major drawback is that hydrogen diffusivity in Ni-Ti is considerably slower than either Pd/Pd-alloy, V, Nb, or Ta. In attempt to enhance the diffusivity, additions of Nb have been made to Ni-Ti alloys and permeation and mechanical stability have been evaluated. The limited study of a these ternary Nb-Ti-Ni alloy has shown permeation on an order equal to pure Pd and reasonable mechanical stability in hydrogen. Recent work on V-Ti-Ni and Ta-Ti-Ni alloys by the same authors has shown similar results. The permeabilites of the V- and Ta-alloys were not quite as high as either Pd or the previous studied Nb-alloys [7]. This was attributed to the inherent greater permeability of bulk Nb in comparison to V and Ta.

Amorphous Non-Noble Metal Membranes

The development of metallic glasses in bulk form has led to a resurgence of interest into the potential utilization of these materials for a variety of applications. Prior to this development, metallic glasses were produced exclusively in very thin sections by rapid solidification processing in most cases. The subsequent consolidation into something "bulk" frequently led to devitrification and a loss in the desirable properties characteristic of the glass. In fact, the unique properties of metallic glasses — strengths of 1-2 GPa, toughness of 30-70 MPa m-0.5, good environmental resistance and unique magnetic properties in some cases — have only been exploited in a few applications where thin sections are desirable (e.g., transformer sheet and magnetic strips for anti-theft tags). A potentially exciting application for these new bulk metallic glass materials is use as membranes for enhancing the efficiency of gas separations both in production processes and for fuel cell usage.

The current generation of gas separation membranes is based on Pd/Pd-alloy used either independently or in conjunction with porous ceramic supports. Palladium/Palladium alloys have been known to possess the ability to dissolve a considerable volume of hydrogen and to demonstrate increasing permeability with increasing pressure differential and temperature. However, the major drawbacks to their industrial use are the high cost for Pd/Pt, relatively low flux, and an irreversible structural change that occurs when the materials are cycled through a critical temperature range. This irreversible change takes place in the palladium lattice structure and can result in significant damage to the membrane. SRNL has previously worked with thin section (melt–spun ribbons) of metallic glass materials for membrane applications, however, with the relatively new ability to cast fully amorphous metallic glasses in bulk sections a new opportunity is opened for bulk metallic glasses as hydrogen membranes. The ability to readily cast metallic glass alloys will allow for easier fabrication of membranes—machine thin membranes from larger castings-- and will also ease mass production challenges in comparison to thin section (melt spun) metallic glass ribbons. Bulk metallic glass alloys are traditionally processed from multi-component system comprised of metallic species of varying atomic size. It is this vast difference in atomic sizes that results in slow diffusion/redistribution kinetics and allows for deep undercoolings to the point of freezing in the "liquid" structure to produce amorphous metallic alloys at relatively slow cooling rates (10-100 K/s). These metallic glass alloys have been shown to possess high permeation rates. For example the permeation rate for a --Zr-Al-Co-Ni-Cu BMG alloy --1.13 x 10-8 mol/m s Pa½--is comparable to permeation the rate measured for pure Pd metal. Furthermore, these metallic glass alloys have also been shown to possess high elastic toughness and excellent resistance to hydrogen degradation, i.e., structural changes. Both of these properties—high permeation and high elastic toughness-- potentially make these materials attractive for gas separation membranes.

The focal point of this work is to extend the Nb-Ti-Ni membrane development work with a direct replacement of Nb with V. Characterization of the resulting microstructure and measurement of the permeability of the novel V-based alloy is reported.

2. Experimental Approach

Electrochemical Permeation Testing

Arc melted buttons of approximately 25 gms each were prepared using a Centorr System VII arc melter system with a tungsten electrode. Arc melting was performed following evacuation to approximately 10^{-4}Torr and backfilled with argon. The V-Ti-Ni alloys were prepared using 99.7%V, 99.95%Ti, and 99.95% Ni raw materials supplied by Alfa-Aesar. The V-Ni-Ti alloy tested as part of this study contained the following alloy composition—53wt%V, 26wt%Ti, and 21wt%Ni. Characterization of the as-cast microstructure was performed using light optical microscopy on polished and etched samples. Scanning electron microscopy and energy dispersive x-ray spectroscopy—including X-ray dot mapping—using a Hitachi S3600 were performed to characterize the phase structure and an alloying element distribution. Disk approximately 12mm in diameter and 0.5-0.75mm in thickness were sectioned from the arc melted buttons and prepared via grinding on SiC papers to provide a 1200 grit finish.

Hydrogen permeation testing was conducted using a Devanathan and Stachurski type-electrochemical apparatus—Figure 1. Permeation testing of V-Ni-Ti alloy was conducted on foils approximately 0.6mm in thickness with an exposed surface area of 0.4 cm^2. The test solution consisted of 0.1M NaOH solution at room temperature. The solution was purged with nitrogen 24 hours prior testing as well as during the test. The electrochemical parameters included a charging current of $100\mu A/cm^2$ on the cathode side and an applied potential of -125mV versus a saturated calomel electrode on the anode side. In an attempt to truly measure the actual permeability of the V-Ni-Ti alloy both sides of the alloy disc were coated with a flash layer of Pd. As a means of comparison Pd foils approximately 0.25 and 0.5mm in thickness have been tested under identical conditions.

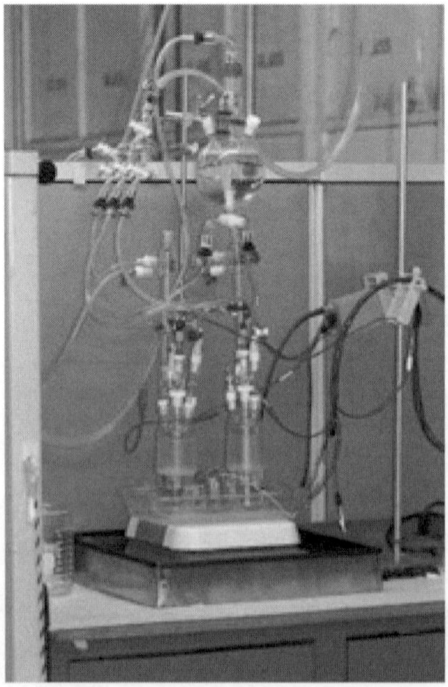

Figure 1. Devanathan-Stachurski Electrochemical Hydrogen Permeation Apparatus.

Analysis of the electrochemical data provides a measure of the hydrogen flux through the sample by measuring the steady-state current density I_p (A/cm^2) on the anodic side of the cell. This steady state current density can be converted to the steady state hydrogen permeation flux, J_∞, (mol/m^2s) via equation 1 below.

$$J_\infty = I_p /nF \tag{1}$$

The steady-state hydrogen permeation rate, V, (mol/m s), can be defined according to equation 2

$$V = J_\infty L = LI_p /nF \tag{2}$$

Where L is the sample thickness, Ip is the steady-state current density, n is the number of electrons transferred, and F is Faraday's constant.

Gas Permeation Testing

Hydrogen permeation testing was conducted using the permeation test rig shown in Figure 2. Samples, 19 mm diameter and 0.89 mm thick or disks were welded into 2.12" diameter Conflat (CF) flanges. Crevices were seal welded using electron beam welding to minimize the effects of virtual leaks. The sample assemblies were placed in a 1" OD vacuum system fabricated with 2.12" CF flanges. Copper gaskets were used to seal the samples. The samples were evacuated to at least 1 x 10-6 Torr for a period of at least four hours at room temperature. The samples were then heated to 100C for 8 to 16 hours to outgas the system and up to the final test temperature. A leak rate test was conducted by closing the appropriate valve. If the leak rate was not linear, the sample was evacuated for additional time, after an acceptable leak rate curve was obtained, the sample section valves were closed and the desired pressure of deuterium was introduced. It took approximately 2-3 minutes for the pressure to reach the target value. The pressure rise on the low pressure side of the system was monitored. The data were logged at either a ten to 30 second interval. The data were reduced to estimate the diffusivity and permeability.

Figure 2. SRNL Gaseous Permeation Test Rig.

The raw data were plotted as a function of time. The data exhibit three distinct regions, the background in-leakage region, a transition region, and a steady state region, nearly linear region. The diffusivity (D) was estimated by calculating the slope and the intercept of the linear region using a least squares method. These two variables were then used to determine the lag time (tl), i.e., the time at which the line crossed the y-axis at zero. Lag time, tl, time was used in the equation: tl = x2 / 6 D (1) to determine D. The permeability (Φ) was estimated from the slope (M) of the curve, the expansion volume (V), the sample area (A), and the test pressure (ΔP) as shown in Eq. 2. Φ = M * V * t / A √ΔP (2). The permeability is the product of the solubility (S) and the diffusivity as shown in Eq. 3. Φ = S * D (3)

3. Results and Discussion

Microstructure Analysis

Evaluation of the microstructure of the V-Ti-Ni alloy was performed using scanning electron microscopy combined with x-ray mapping of the element constituents. Previous work on Nb-Ti-Ni alloys being investigated for advanced hydrogen separation membrane use has attributed positive results to microstructures consisting of a large primary $Nb_{83}Ti_{13}Ni_4$ phase surrounded by eutectic (NiTi +NbTiNi); $Nb_{83}Ti_{13}Ni_4$ was postulated to be the high diffusivity phase while the eutectic structure contributes to the lack of susceptibility to hydrogen embrittlement [6]. SEM backscattered and secondary electron micrographs of the V-Ti-Ni alloy display a similar microstructure to the Nb-Ti-Ni alloy with a primary phase surrounded by interdendritic eutectic structure—Figure 3. X-ray mapping of the elemental constituents provided in Figure 4 shows the primary phase in the microstructure to be high in vanadium content. Additionally, the interdendritic eutectic is rich in Ni and Ti. Semi-quantitative chemical analysis of the composition of the vanadium rich primary phase indicates an approximate composition of $V_{75}Ti_{16}Ni_9$.

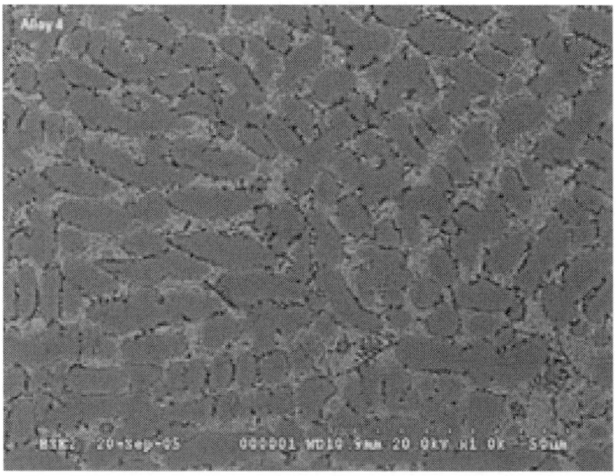

(a)

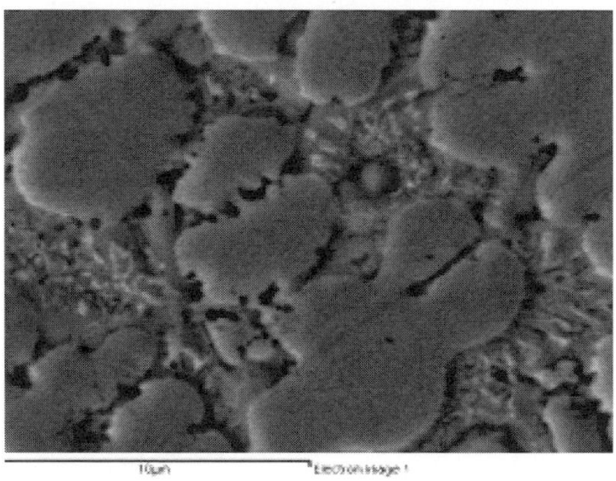

(b)

Figure 3. Scanning Electron Microscope Images of a V53-Ti26-Ni21 Alloy (a) Back-scattered and (b) secondary electron image.

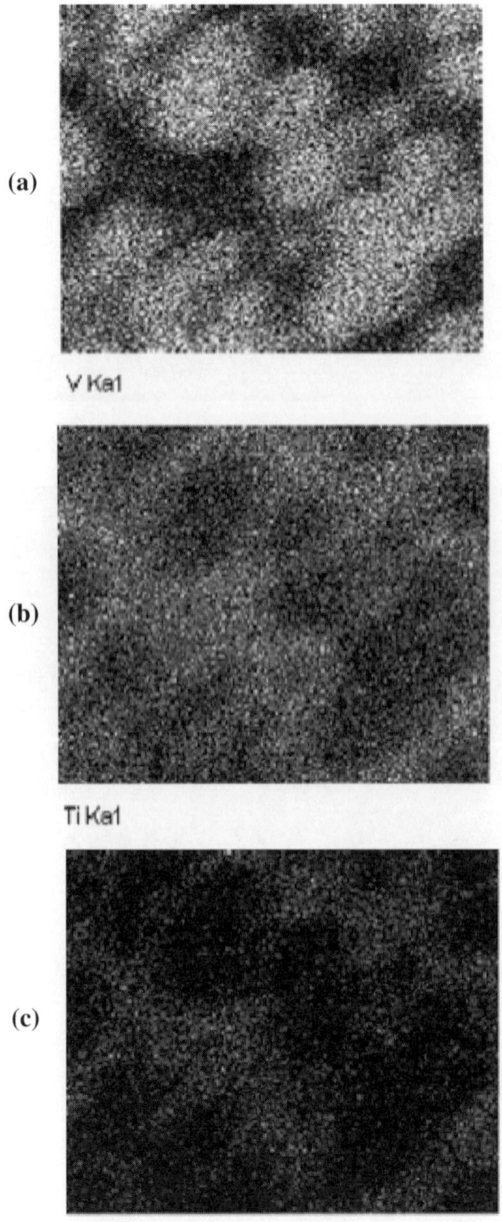

Figure 4. X-ray Mapping of Elemental Constituents within the micrsotructure of a V53-Ti26-Ni21 membrane alloy: (a) V-Kα1 map, (b)Ti Kα1 map, and (c) Ni-Kα1 map.

Electrochemical Hydrogen Permeation

Measurement of the steady state hydrogen permeation flux and rate was conducted and compared to measured values for pure palladium. Comparison of the results for the V-Ti-Ni all to palladium since Pd/Pd-alloys are the current dense metallic membrane materials of choice. The testing was conducted under similar condition—100μA/cm^2 charging current at 22°C-- using the apparatus previously shown in Figure 1. A single set of results are shown in Figure 5, graphed as current density versus time. Examination of the current density plots for the two alloys shows an almost order of magnitude higher steady state current density value for the Pd membrane when compared to the V-Ti-Ni alloy. This higher steady state current density translates into a larger steady state hydrogen flux through the Pd membrane. However, due to the significant difference in thickness between the two membrane materials—L_{Pd}=0.05mm and L_{V51}=0.635mm—the overall hydrogen permeation rate as calculated from equation 2 is larger by an order of magnitude for the V-Ti-Ni alloy. Table 1 displays the calculated steady-state permeation rates for both materials. Thus, from these initial low temperature results the V-Ti-Ni alloy appears to possess a hydrogen permeability greater than Pd under the same conditions. Finally, additional testing at higher cathodic charging currents showed increasing anodic current densities that appeared to saturate.

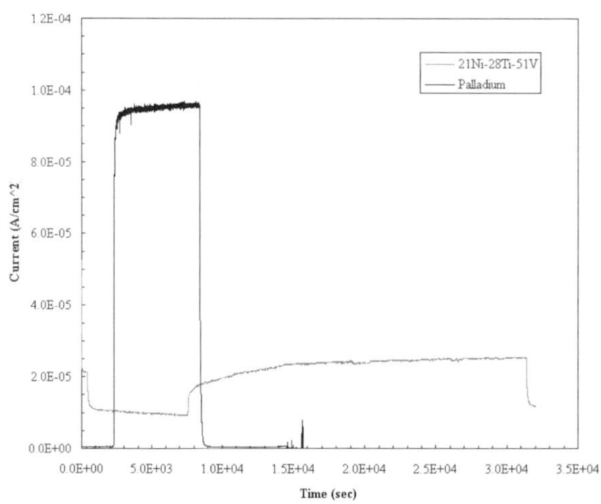

Figure 5. Measured Anodic Current Density during Electrochemical Hydrogen Permeation Testing for Pd and V-53-Ti26-Ni21 materials at 22°C.

Table 1. Steady -State Hydrogen Permeation Rate Measured for Pd and V-Ti-Ni Alloy

Alloy	Permeation Rate (mol H_2/m s)
Palladium	**$3.3 - 4.3 \times 10^{-10}$**
V53-Ti26-Ni21	**$1.0 - 3.7 \times 10^{-9}$**

Gaseous Hydrogen Permeation Testing

Measurement of the steady state permeation flux for both the crystalline V-Ti-Ni alloys (see microstructure of V51-Ti28-Ni21 and V54-Ti28-Ni18 alloys in Figure 6) and the amorphous metallic glass materials was conducted and compared to literature data for palladium membranes.

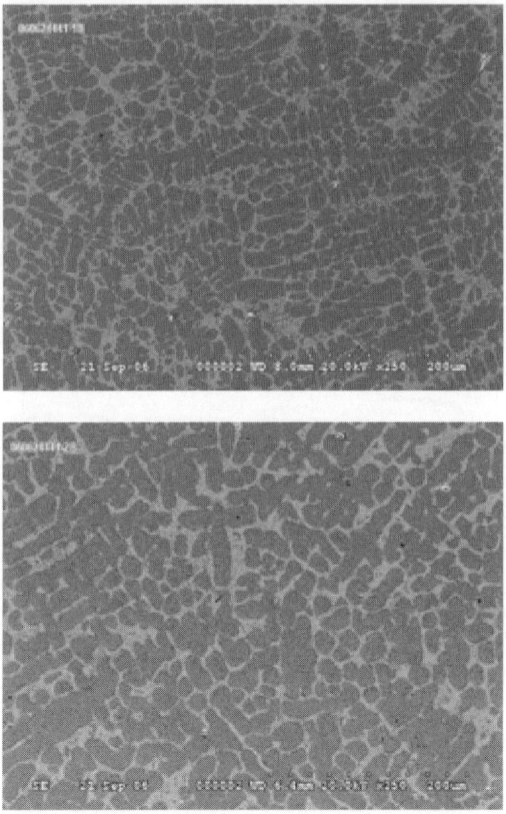

Figure 6. SEM-backscattered electron images of the microstructures of (a) V51-Ti28-Ni21 and (b) V54-Ti28-Ni18 alloys.

The testing was conducted under sub-atmospheric pressures but at values typically used at the Savannah River Site for hydrogen isotope purification. A typical raw data curve for a V51-Ti28-Ni21 alloy at a temperature of 400°C and a pressure of 10 torr is shown in Figure 7. The calculated permeabilities from test data collected at 400°C for this alloy as well as for a V54-Ti28-Ni18 alloy are provided in Table 2.

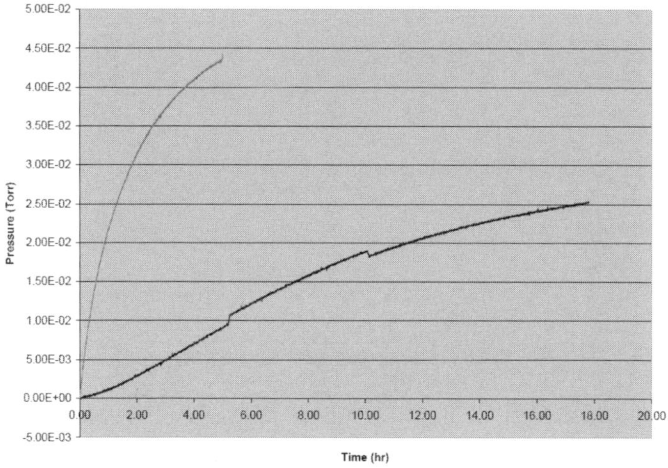

Figure 7. Hydrogen saturation test data for V51-Ti28-Ni21 Alloy.

Table 2. Permeability Values for V-Ti-Ni alloys tested at 400° C

Alloy	Permeability (mol H_2 m^{-1} s^{-1} $Pa^{-1/2}$)
V51-Ti28-Ni21	1.26×10^{-8}
V54-Ti28-Ni18	9.71×10^{-9}

Similar to Pd and Pd-alloy membrane materials the V51-Ti28-Ni21 alloy demonstrated a susceptibility to hydriding when cooled through a critical temperature range in the presence of hydrogen. During testing a system power failure allowed the sample to cool from the test temperature (400°C) to room temperature under hydrogen. During re-start of the system it was determined that the sample has failed and this failed sample was then examined using x-ray diffraction in order to determine the cause of failure. Results from the XRD analysis shown in Figure 8, clearly show the formation of vanadium- hydride phase in this sample which resulted in the subsequent failure of the membrane.

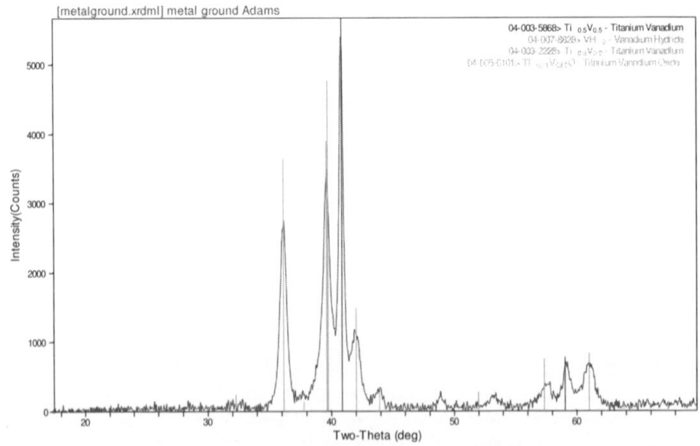

Figure 8. XRD data indicating formation of vanadium hydride on cooling to room temperature under H2 cover gas.

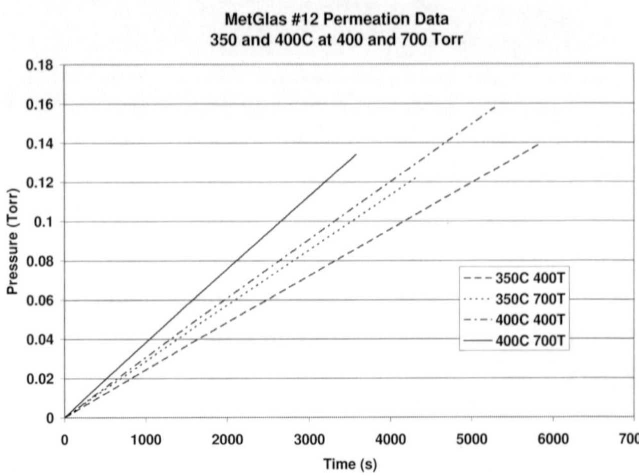

Figure 9. Permeation data for a typical MetGlas sample.

In addition to the V-based alloys, metallic glass materials have been tested using the same approach, temperature range and pressures. Due to concerns about crystallization of the material, the samples were fixtured in 0.75" diameter VCR fittings using silver plated nickel gaskets. The samples were verified leak tight to at least 2x10-9 sccm He. The commercial-off-the-shelf (COTS) metallic glass material tested exhibits a permeability and flux within two decades of Pd. The raw data are shown in Figure 9.

Table 3. Permeability data for COTS metallic Glass materials compared to Pd

Alloy	Permeability (mol H_2 m^{-1} s^{-1-} $Pa^{-1/2}$)
Palladium 350°C	1.8×10^{-8}
Palladium 400°C	2×10^{-8}
MetGlas 12 350°C	1.81×10^{-9}
MetGlas 12 400°C	1.94×10^{-9}

Due to the low thickness of this sample, approximately 25μm, no determination of a lag time was possible. Testing using mass spectrometer generally indicated hydrogen at near saturation levels within the first sampling frequency, which is limited to about 15 seconds for the instrumentation used. The data were analyzed using the standard data reduction method and the permeability of this alloy is indicated in Table 3.

With the relative promise of this material and its relatively high strength, testing of additional COTS materials is on-going and will be reported in future articles.

4. Conclusions

V-Ti-Ni alloys and Fe- /Co-Based metallic glasses have been evaluated with respect to hydrogen permeability for potential use in hydrogen purification membrane reactor application. Microstructural characterization of the V-Ti-Ni alloy using SEM has shown similar microstructural features to a previously evaluated Nb-Ti-Ni alloy; namely, the occurrence of a primary phase surrounded by interdendritic eutectic..

Hydrogen permeation rate for a V53-Ti26-Ni21 alloy was measured electrochemically and compares favorable to rates also measured for pure Pd. Subsequent, gaseous hydrogen permeation testing of similar V-Ti-Ni alloys once again demonstrated permeabilites on par with commercially available Pd/Pd-alloy membrane materials. Permeation testing of the Fe-/Co-based metallic glass alloys demonstrated permeabilities slightly lower than Pd/Pd-alloys however, the cost savings afforded by these materials (approximately 650X lower) warrants further study/evaluation of this class of materials.

References

1. R. E. Buxbaum and T. L. Marker, Journal of Membrane Science, 85, 29-38, (1993).
2. N. M. Peachey, R. C. Snow, and R. C. Dye, Journal of Membrane Science, 111, 123-133, (1996).
3. T. S. Moss,N. M. Peachey, R. C. Snow, and R. C. Dye, International Journal of Hydrogen Energy, 23, 99-106, (1998)
4. C. Nishimura, M. Komaki, S. Hwang, and M, Amano, Journal of Alloys and Compounds, 330-332, 902-906, (2002).
5. Y. Zhang, T. Ozaki, M. Komaki, and C. Nishimura, Scripta Materialia, 47, 601-606, (2002).
6. K. Hashi, K. Ishikawa, T. Matsuda, and K. Aoki, Journal of Alloys and Compounds, 368, 215-220, (2004).
7. K. Hashi, K. Ishikawa, T. Matsuda, and K. Aoki, Journal of Alloys and Compounds, 404-406, 273-278, (2005).

Safety and Education

STRUCTURAL-METALS CONSIDERATIONS FOR THE CONTAINMENT OF HIGH-PRESSURE HYDROGEN GAS

C. SAN MARCHI, B.P. SOMERDAY, K.A. NIBUR AND M. YIP

Sandia National Laboratories, 7011 East Ave, Livermore CA 94550

All engineering alloys are vulnerable to hydrogen-assisted fracture (hydrogen embrittlement) under some intersection of microstructural, mechanical, and environmental conditions, thus it is important to develop a comprehensive understanding of the service conditions for a given application. In addition, quantitative assessment of structural integrity aids the management of hydrogen-assisted fracture in structurally efficient components designed for high-pressure gaseous hydrogen. In this presentation, we describe two methods of assessing structural integrity: strength-based and fracture-control methodologies. The implications of these differing methods are briefly described in the context of hydrogen-assisted fracture. Strength-based methods lead to conservative designs based on indirect assessment of the hydrogen compatibility of materials, since hydrogen generally does not affect strength even in materials susceptible to hydrogen-assisted fracture. Fracture-control methods explicitly address failure mechanisms associated with existing defects, in particular crack extension in fatigue can be evaluated. These fracture control methods incorporate quantitative evaluation of materials properties (ie, resistance to hydrogen-assisted crack propagation) for establishing efficient structural design for hydrogen service. In general, engineering tools exist for quantitative assessment of structural integrity in high-pressure gaseous hydrogen environments using fracture control methodologies and these tools are being exploited in the engineering community.

1. Introduction

Mechanical properties of structural metals measured in hydrogen environments can be strongly influenced by testing protocols, thus an understanding of the thermodynamics and kinetics of hydrogen transport in metals is important for interpreting test results and formulating comprehensive strategies for assessment of structural integrity. Although this topic is much too extensive to be treated here in any depth, briefly, hydrogen dissociates on metal surfaces, forming atomic hydrogen, which then dissolves into the microstructure of metal. Atomic hydrogen has the unique feature of being relatively mobile in metals at low temperature, thus affecting deformation and fracture processes in the metal. The role of hydrogen in a particular material under a given set of environmental and mechanical conditions is determined by thermodynamics (how much?) and kinetics (when?) and this can vary substantially for different classes of materials. By way of example, consider the precipitation-strengthened, austenitic stainless steel A286. The tensile properties of A286 are unchanged, when tensile straining is concurrent with hydrogen exposure; however, if hydrogen is pre-charged into the metal to simulate long-time exposure then subsequently subjected to tensile straining, a significant reduction in ductility is measured [1, 2]. The origin of this difference can be interpreted in the context of hydrogen transport: in stable austenitic stainless steels, there is no appreciable hydrogen diffusion on the time scale of days at room temperature [2]. In other words, so-called "slow strain rate tests" in hydrogen gas are not adequate for this class of materials. Hydrogen in low-alloy ferritic steels, in comparison, diffuses as much as eight orders of magnitude faster than austenitic stainless steels, thus the effects of hydrogen can be observed on time scales appropriate to slow strain rate tests. A brief discussion of testing approaches is presented in Ref. [2] in the context of different materials classes.

To some extent appropriate testing protocols will be determined by the design space (intersection of environmental, mechanical and materials variables), as well as the data required to support the specific design methodology. In this discussion, we focus on aspects of component design and fitness-for-service (FFS) assessment of cracks and crack-like flaws, emphasizing the importance of fracture mechanics in designing to accommodate hydrogen-assisted fracture. All structural metals are susceptible to hydrogen-assisted fracture, depending on numerous microstructural, mechanical and environmental conditions (which contributes much confusion about materials selection). The concept of accommodating hydrogen-assisted fracture is central to the design of robust hydrogen systems that allow for efficient use of materials while maintaining a high-level of confidence in the structural integrity of the component.

2. Structural Integrity

2.1. *Strength-based methodology*

Approaches to evaluating structural integrity can generally be classified in two broad categories: strength-based and fracture-control methodologies. The strength-based approach addresses plastic collapse: the condition when the average applied stress exceeds the yield strength or tensile strength of the material. For tubular structures, one common formulation for design against plastic collapse can be expressed as:

$$t = \left(\frac{PD}{2S}\right)\frac{1}{F} \tag{1}$$

where P is the design pressure, S is the yield strength of the material, t is the wall thickness, D is the outside diameter of the pipe and F is a design factor (< 1), which is based on a number of criteria and typically varies from less than 0.3 to as high as 0.72.

In considering fatigue, strength-based approaches use so-called S-N curves, i.e., plots of stress amplitude (or strain amplitude) versus cycles to failure. These tests make use of smooth specimens, thus the test methodology is biased toward crack initiation. In real components, small flaws and stress risers exist that can facilitate initiation of cracks. Safety factors are necessary to account for stress concentrations and flaws.

Many structures are designed based on the strength-based approach, particularly for components subject to constant pressure, and these designs tend to be very conservative. Although many other factors are generally considered and the design equations may change (e.g., for thick-walled structures such as high-pressure tubing), the essence of the strength-based approach is that if the stresses in the component are kept low, the structure will not fail. The disadvantage of the strength-based approach is that the structure may be very inefficient or impractical, particularly for high-pressure applications. Additionally, since the margin against crack propagation (an observed failure mode in hydrogen) is not explicitly assessed with strength-based methods, the safety and reliability of hydrogen containment structures cannot be quantified with respect to hydrogen-assisted fracture.

2.2. *Strength-based assessment of hydrogen compatibility*

In general, the strength-based approach does not account for hydrogen-assisted fracture. Many structural metals, for example, do not show degradation of strength in gaseous hydrogen environments, but experience significant reduction in ductility and transitions to more brittle modes of fracture. These latter properties reflect a material's resistance to crack propagation, but there are no quantitative criteria in the strength-based approach to account for these changes. 17-7PH stainless steel is an instructive example: this steel shows essentially no loss of strength when tested in high-pressure hydrogen, but it is considered "extremely embrittled" by hydrogen [3]. The inadequacy of standard smooth bar tensile tests for the assessment of hydrogen compatibility was apparently recognized in studies funded by NASA in the late 60s and early 70s. Walter,

Chandler and co-workers advocated for the notched tensile strength as a strength-based metric for hydrogen compatibility. Although the notched tensile strength is sensitive to the hydrogen environment, it remains a comparative evaluation with arbitrary categories that cannot be used in quantitative design of components. For robust system design, experimental determination of crack propagation properties that can be used in the design process is highly desirable.

2.3. *Fracture-control methodology*

During the past decade FFS assessments have gained international acceptance in the nuclear and petrochemical industries. One aspect of the FFS approach formalizes quantitative evaluation of the structural integrity of engineering components containing flaws [4]. There are several formal documents that address FFS, including British Energy R-6, British Standards Institute BS 7910, and American Petroleum Institute (API) RP-579. There are many elements of these documents, but for the purposes of this presentation, we are primarily interested in the principles engendered in the failure assessment diagram (FAD), which allows for a quantitative evaluation (Figure 1) of crack propagation and plastic collapse in a flawed structure [4, 5].

The principles of FFS can be applied to any structural component, provided the relevant failure modes are addressed. Cracks and crack-like flaws are of particular concern in pressure-bearing components. Linear elastic fracture mechanics uses the stress-intensity factor K to quantify the stress field at a crack tip, which is used with the FAD to assess a material's resistance to crack propagation under constant load. Unlike many petrochemical applications, however, gaseous fuel infrastructure experiences substantial fatigue cycles, e.g., natural gas and hydrogen fuel tanks. Therefore, fatigue analysis that explicitly addresses crack propagation is necessary to complement the analysis of crack propagation under constant load. The American Society of Mechanical Engineers (ASME) provides a methodology for evaluating crack propagation in fatigue using fracture mechanics data, which is described in Article KD-4 in Section VIII, Division 3 of the Boiler and Pressure Vessel Code (BPVC). Briefly, fatigue crack growth rates generally follow the Paris relationship (Figure 2), a power law of the form:

$$da/dN = C\Delta K^m \tag{2}$$

where a is the depth of the crack, N is the number of stress (pressure) cycles, and ΔK is the difference of stress-intensity factors at peak and minimum loads, while C and m are constants. The stress-intensity factor, and thus ΔK for a given pressure cycle, can be calculated for specific component features and dimensions. Thus, given a specific flaw or crack, the propagation of the flaw can be quantitatively evaluated for given pressure cycles. This relatively straightforward approach can be used to predict fatigue life (i.e., number of cycles to failure) and to define inspection intervals, as demonstrated below.

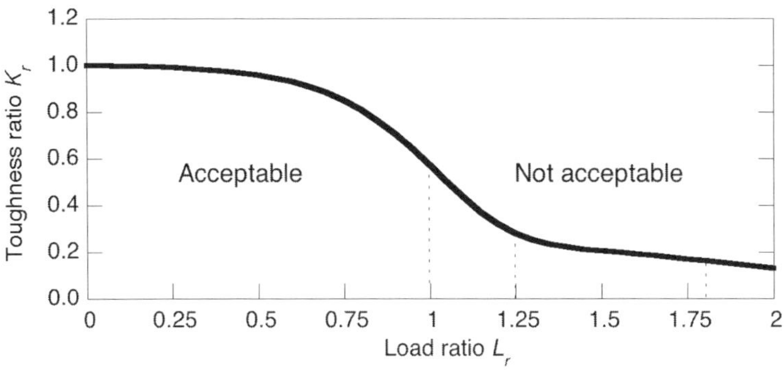

Figure 1. API RP-579, FAD level 2; dotted lines represent material-specific cut-off values.

2.4. *Fracture-based assessment of hydrogen compatibility*

ASME has taken the position that design methods using fracture control are necessary for high-pressure hydrogen pressure vessels [5]. The ASME BPVC Committee has recently adopted article KD-10 in Section VIII, Division 3, which requires comprehensive design for hydrogen storage tanks, using failure assessment diagrams (from API RP-579) and fatigue life analysis (Article KD-4). Article KD-10 also provides comprehensive guidance on testing protocols for measuring the required crack propagation properties of alloys in high-pressure hydrogen gas using established methodology (e.g., existing ASTM testing standards) modified to account for the particular physics of hydrogen.

Determination of the fracture properties in high-pressure hydrogen is imperative to comprehensive FFS evaluation for hydrogen service. High-pressure hydrogen is known to significantly reduce the fracture toughness of common pressure vessel steels [6] and accelerate the propagation of cracks in fatigue [7] (Figure 2). Assessment of hydrogen-assisted fracture, however, must be interpreted judiciously since the testing protocols can influence results. For example, the lack of a cracking response does not necessarily establish a lower bound for sustained-load fracture in high-pressure hydrogen. Tests in our own laboratory have shown that specimens loaded to K higher than the measured threshold for crack propagation may not crack in some cases [8]. This observation is not fully understood, but appears to be related to several phenomena including crack branching and surface-limited transport of hydrogen. Testing protocols may need to be modified in the future as our understanding develops.

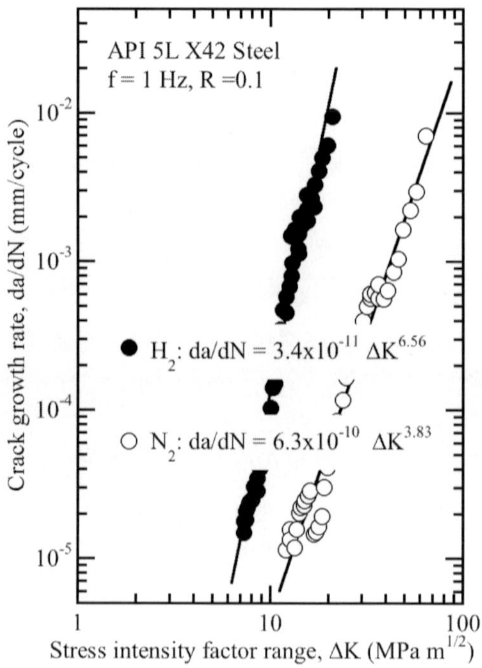

Figure 2. Crack growth rates in fatigue for maximum pressure of 6.9 MPa from Ref. [7].

Frequency is another variable that requires further study, as hydrogen-assisted fatigue appears to be strongly sensitive to time scales [9]. This is likely due to surface and transport kinetics of hydrogen; consequently, upper bounds for testing frequency must be established, which are representative of the load cycles that can be expected for service conditions.

3. Design Problem

To illustrate key concepts of the two design strategies for hydrogen systems, we consider a simple tubular structure for operation in hydrogen gas: 300 mm (~12 in) inside diameter "pipe" with a maximum operating pressure of 6.9 MPa

(1000 psi). The material of construction is assumed to be carbon steel that can be characterized by API 5L X42 carbon steel with specified minimum yield strength (SMYS) of 290 MPa (42 ksi). Reports in the literature give tensile and fracture properties for X42 carbon steel measured in 6.9 MPa hydrogen gas as shown in Table 1.

Table 1. Mechanical properties of API 5L X42 carbon steel from Ref. [7].

Testing environment	Yield strength (MPa)	Tensile strength (MPa)	Reduction of area (%)	Fracture toughness (MPa $m^{1/2}$)
Air	311	490	52	180
6.9 MPa H_2	338	476	41	107

3.1. *Strength-based analysis*

Using equation 1 with $F = 0.4$, we determine the required wall thickness to be 9.2 mm, while for $F = 0.72$, the minimum wall thickness is 5 mm. There is no way to account for hydrogen-assisted fracture in the strength-based approach, but it should be clear from the data in Table 1 that hydrogen has a substantial effect on tensile ductility (reduction of area) and fracture toughness. These reductions in resistance to crack propagation are anticipated in most steels and, based on conservative engineering intuition, drive F lower.

3.2. *Fracture-control analysis*

For the purpose of this illustration, we assume a semi-infinite flaw along the full length of an idealized tubular structure, Figure 3. Relationships from Anderson [10] are used to calculate the stress-intensity factor (K) as a function of crack depth (a), pressure, and component dimensions. We consider two wall thicknesses similar to those calculated above: 10 mm and 5 mm, corresponding to maximum hoop stress of 36 and 72% of the SMYS respectively for maximum pressure of 6.9 MPa. In the context of static loads these are conservative designs since flaws through a quarter of the thickness (a/t = 0.25) result in stress-intensity factors of < 25 MPa $m^{1/2}$ compared to the fracture toughness of 107 MPa $m^{1/2}$ in 6.9 MPa hydrogen gas. With respect to the level 2 FAD (Figure 1), K_r is likely to be < 0.25, well within the acceptable region, although the exact value will depend to some extent on other structural considerations and which definitions of K_r and L_r are used [4].

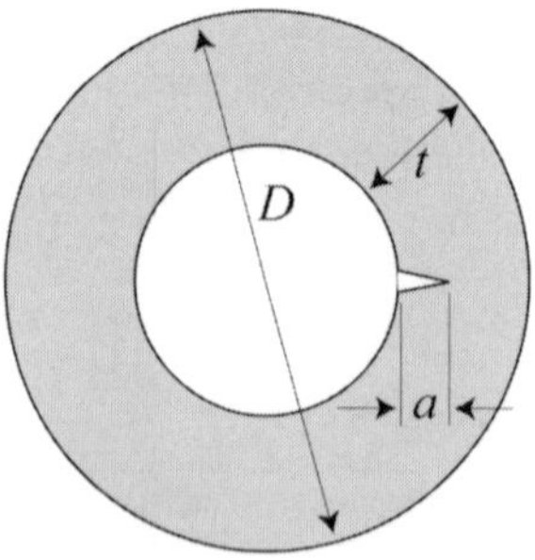

Figure 3. Schematic of idealized tubular structure.

The analysis of crack propagation in fatigue tells a somewhat different story. Using the data in Figure 2 and the method outlined above with pressure cycling between 6.9 and 0.69 MPa ($R = 0.1$), the evolution of crack extension is predicted as shown in Figure 4. Hydrogen increases the rate of crack growth by more than an order of magnitude compared to cracks growing in nitrogen (Figure 2), resulting in a substantial reduction of the fatigue life (i.e., number of cycles to failure). For the cases evaluated in Figure 4, the initial ΔK is highest for case 1, thus the crack grows comparatively rapidly reaching critical crack length for fracture under constant load in about 2000 cycles. For the thicker wall design (case 3), the crack grows at a slower rate because ΔK is lower initially. The power-law form of the Paris law (Equation 2) implies that even small reductions of ΔK can have substantial impact. The initial ΔK can also be reduced and the fatigue life extended if the initial flaw sizes can be reduced; for example, by reducing the initial flaw size by half ($a/t = 0.05$, case 2), the fatigue life is increased by almost an order of magnitude. In other words, design and FFS can be strongly influenced by methods for non-destructive examination (NDE) of existing flaws. Of course, the inspection intervals and design life (i.e., number of allowable cycles) are not equal to the number of cycles to failure, but some fraction of this, for example as proposed in Article KD-10 [5].

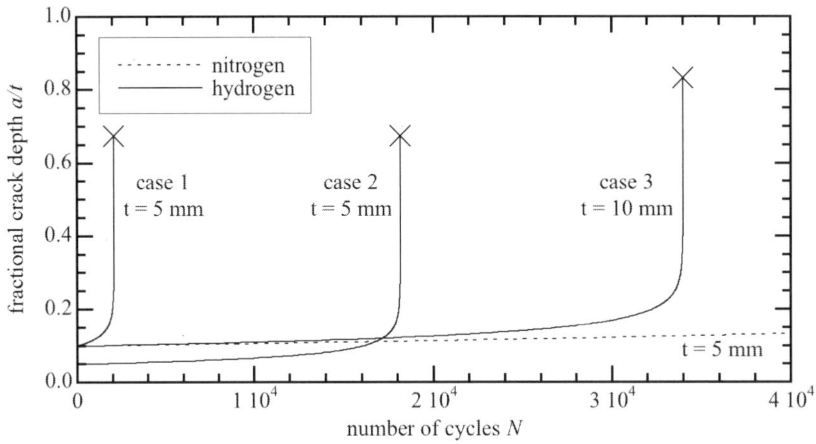

Figure 4. Crack extension curves for API 5L X42 tubular structure with a semi-infinite crack along the length of the structure: 300 mm inside diameter, 10 MPa maximum gas pressure. The "X" denotes the critical crack depth for fracture under constant load.

4. Summary

Comprehensive structural integrity assessment has been gaining acceptance in industries that require large capital investment (e.g., nuclear and petrochemical). However, the principles of FFS assessments can be implemented for any component design. This brief description was motivated by an apparent need in the engineering community to use comprehensive structural analysis in the design of components for hydrogen service. Strength-based analysis is limited in its ability to address hydrogen-assisted fracture, since material strength is often unaffected by hydrogen. In comparison, a fracture-based approach (such as API RP-579) can better address the "embrittling" effects of hydrogen. In addition, a comprehensive structural-integrity methodology for components that will experience crack propagation in fatigue is imperative for materials exposed to hydrogen. In general, the engineering tools exist to address the design of components for hydrogen service, and these are not limited to the example given here. However, some effort is necessary to adapt materials testing protocols to the unique physics of hydrogen in metals (and more generally in materials) and to acknowledge the limitations of various testing methodologies by specific class of material. "Materials compatibility" with hydrogen depends on the design, and service environment, and there is no all-encompassing test method for establishing structural integrity in hydrogen environments.

Acknowledgments

Sandia is a multiprogram laboratory operated by Sandia Corporation, a Lockheed Martin Company, for the United States Department of Energy under contract DE-AC04-94AL85000.

References

1. AW Thompson, *Hydrogen in Metals*. IM Bernstein and AW Thompson, editors. Metals Park OH: American Society of Metals (1974) p. 91.
2. C San Marchi and BP Somerday, 2007 SAE World Congress, Detroit MI.
3. RP Jewitt, RJ Walter, WT Chandler and RP Frohmberg, NASA CR-2163, March 1973.
4. TL Anderson and DA Osage, *Int J Pressure Vessel Piping* **77,** 953 (2000).
5. MD Rana, GB Rawls, JR Sims and E Upitis, 2007 ASME Pressure Vessels and Piping Division Conference, San Antonio TX.
6. AW Loginow and EH Phelps, *Corrosion* **31,** 404 (1975).
7. HJ Cialone and JH Holbrook, *Metall Trans* **16A,** 115 (1985).
8. KA Nibur, BP Somerday and C San Marchi, 2008 ASME Pressure Vessels and Piping Division Conference, Chicago IL.
9. RJ Walter and WT Chandler, *Effect of Hydrogen on Behavior of Materials*. AW Thompson and IM Bernstein, editors. New York: AIME (1976) p. 273.
10. TL Anderson, *Fracture Mechanics: Fundamentals and Applications*. Boca Raton FL: CRC Press (1995).

A NATIONAL AGENDA FOR HYDROGEN CODES AND STANDARDS

CHAD BLAKE

National Renewable Energy Laboratory, 1617 Cole Boulevard, Golden, CO, 80401

This paper provides an overview of hydrogen codes and standards with an emphasis on the national effort supported and managed by the U.S. Department of Energy (DOE). With the help and cooperation of standards and model code development organizations, industry, and other interested parties, DOE has established a coordinated national agenda for hydrogen and fuel cell codes and standards. With the adoption of the Research, Development, and Demonstration Roadmap and with its implementation through the Codes and Standards Technical Team, DOE helps strengthen the scientific basis for requirements incorporated in codes and standards that, in turn, will facilitate international market receptivity for hydrogen and fuel cell technologies.

1. Introduction

Large quantities of hydrogen have been used safely as a chemical feedstock and industrial gas for many years. Standards, codes, and regulations governing its storage, distribution, and use at industrial sites are well established. The use of hydrogen as an energy carrier for consumer markets is expected to grow over the next decade, and the development and promulgation of codes and standards for this use are essential to establish a market-receptive environment for commercial hydrogen products and systems.

Hydrogen standards are typically written under a consensus process by technical committees representing a cross-section of interested parties and issued in the U.S. by organizations such as the American Society of Mechanical Engineers (ASME) for pressure vessels, pipelines, and piping; the Compressed Gas Association (CGA) for pressure vessel operation and maintenance; and the Underwriters Laboratory for product certification. In the U.S., the American National Standards Institute (ANSI) facilitates the development of national standards by accrediting the procedures of standards developing organizations (SDOs) such as those mentioned above [1].

For hydrogen energy use in the U.S., the International Code Council (ICC) and the National Fire Protection Association (NFPA) are the two principal model code development organizations. The ICC develops and publishes a family of model codes; the most relevant for hydrogen energy are the International Fire Code (IFC), International Fuel Gas Code (IFGC), International Building Code (IBC), and International Mechanical Code (IMC) [2]. The NFPA develops and publishes both standards and codes [3]. For hydrogen energy, the most widely used of these are NFPA 55 (Standard for the Storage, Use, and Handling of Compressed Gases and Cryogenic Fluids in Portable and Stationary Containers, Cylinders, and Tanks), NFPA 52 (Vehicular Fuel Systems Code), and NFPA 30A (Code for Motor Fuel Dispensing Facilities and Repair Garages). The NFPA is currently incorporating all of its provisions for hydrogen into a single document, NFPA 2, Hydrogen Technologies, which is scheduled for publication in 2010.

The adoption and enforcement of codes and standards in the U.S. takes place under the jurisdiction of some 44,000 entities that include city, county, and state governments, as well as special districts such as port and tunnel authorities. Regulations make use of existing standards, either by incorporating appropriate sections of the standards (incorporation by transcription), or by referring to those sections (incorporation by reference). The extremely decentralized enforcement of codes and standards means that the permitting process for hydrogen fuel facilities can be very cumbersome.

The federal government plays a limited role in the development, adoption, and enforcement of codes and standards, but federal safety regulations are incorporated in the Code of Federal Regulations (CFR). Those that apply to hydrogen are embodied primarily in 49 CFR (1995) and 29 CFR (1996), under the jurisdictions of the Department of Transportation (DOT) and Occupational Safety and Health Administration (OSHA), respectively. The DOT regulates the transportation of hydrogen. The OSHA regulates the safe handling of hydrogen in the work place. OSHA regulations are intended to provide worker safety for the industrial use of hydrogen [5].

While most industrialized countries have adopted regulations, codes, and standards (RCS) that govern the use of hydrogen, many of these countries also support the development of international standards to facilitate international trade and commerce. For hydrogen energy, the key international SDOs are the International Organization for Standardization (ISO) and the International Electrotechnical Commission (IEC). Information about domestic and international hydrogen RCS, current activities of relevant ISO and IEC technical committees, including draft standards under preparation or review, can be found at www.fuelcellstandards.com, a website supported by the U.S. Department of Energy (DOE). Another useful source of information on hydrogen safety, codes and standards is the Hydrogen Safety Report, a monthly newsletter published by the National Hydrogen Association (NHA) at www.hydrogensafety.info, also supported by DOE.

2. DOE Program for Hydrogen Codes and Standards

For the past decade, the Office of Hydrogen, Fuel Cells and Infrastructure Technologies in DOE has sponsored a collaborative national effort by government and industry to prepare, review, and promulgate codes and standards needed to expedite hydrogen infrastructure development and to help enable the emergence of hydrogen as a significant energy carrier. In addition, DOE has worked to harmonize national and international standards, codes, and regulations that are essential for the safe use of hydrogen by consumers in the U.S. and throughout the world. The National Renewable Energy Laboratory (NREL) provides technical and programmatic support to DOE for this effort.

DOE has also launched a comprehensive research, development, and demonstration (RD&D) effort to obtain the data needed to establish a scientific basis for requirements incorporated in hydrogen codes and standards. This RD&D is planned, conducted, and evaluated in collaboration with industry through the U.S. FreedomCAR and Fuel Partnership formed to examine and advance pre-competitive research and development of technologies to enable high volume production of affordable hydrogen fuel cell vehicles and the national hydrogen infrastructure necessary to support them. The codes and standards activities of the Partnership are conducted through the Codes and Standards Technical Team that adopted a Roadmap to guide the RD&D.

2.1 National Templates

Over the past several years, a coordinated national agenda for hydrogen and fuel cell codes and standards has emerged through DOE leadership and the support and collaboration of industry and key standards and model code development organizations (SDOs and CDOs). For example, hydrogen is recognized as a fuel gas, and hydrogen applications are incorporated in the 2003 and 2006 editions of the ICC model codes. Provisions for the safe use of hydrogen are included in ICC's International Building, Residential, Fire, Mechanical, and Fuel Gas Codes. Also, NFPA has incorporated hydrogen safety requirements into its family of codes and standards, as noted above. The consolidation of all hydrogen safety requirements into a single document (NFPA 2) will be a major step toward development of a national hydrogen code.

A key to the success of the national hydrogen and fuel cell codes and standards development efforts to date has been the creation and implementation of national templates through which DOE, NREL, and the major SDOs and CDOs coordinate the preparation of critical standards and codes for hydrogen and fuel cell technologies and applications. The national templates have helped the DOE to create and maintain a coordinated national agenda for hydrogen and fuel cell codes and standards. DOE leadership has coincided with the emergence of heightened national and international interest in hydrogen energy in general and in codes and standards in particular.

The national templates have been accepted by the major SDOs and CDOs in the U.S., the FreedomCAR and Fuel Partnership, key industry associations, and many state and local governments as the guideposts for the coordinated development of standards and model codes. All of the relevant major SDOs and CDOs in the U.S. are part of this national effort: the American National Standards Institute (ANSI), American Society of Mechanical Engineers (ASME), American Society of Testing and Materials (ASTM), Compressed Gas Association (CGA), CSA America, International Code Council (ICC), National Fire Protection Association (NFPA), Society of Automotive Engineers (SAE), and Underwriters Laboratories (UL). Industry participants include the FreedomCAR-Fuel Partnership (Chrysler, Ford Motor Company, General Motors, BP, Chevron, ConocoPhillips, ExxonMobil, Shell Hydrogen); other industry members, such as Ballard Power Systems, General Electric, Plug Power, Hydrogenics, UTC Power; and industry associations, such as the American Petroleum Institute (API), National Hydrogen Association (NHA), and the US Fuel Cell Council (USFCC). Other federal agencies involved include the Department of Transportation (DOT) and the National Institute of Standards and Technology (NIST). Other organizations participate on an as-need basis.

The objectives of the national templates are to:

- Establish by a consensus of the national codes and standards development organizations the CDO or SDO that will have the lead in the development of codes and standards for establishing safety requirements for specific components, subsystems and systems (as shown in the templates) and the organizations that will work collaboratively with (or in support of) the lead organization
- Minimize duplication of efforts in the codes and standards development

- Establish "boundaries" and interfaces among standards for components, subsystems, and systems and identify harmonization requirements across such standards
- Identify codes and standards development needs and gaps and identify the organizations that should have responsibility for addressing the gaps.

Implementation of the national templates is coordinated through the National Hydrogen and Fuel Cells Codes and Standards Coordinating Committee, created by DOE, NREL, NHA, and USFCC. The committee conducts monthly conference calls to update participants on current activities and to discuss key issues. In addition, the committee meets quarterly to coordinate codes and standards development and prevent duplication of effort, identify critical deficiencies and gaps in hydrogen codes and standards development that could have an adverse impact on market acceptance, determine a collaborative strategy and action plan to address critical gaps and deficiencies, and identify specific opportunities for organizations to work together in developing codes and standards. The minutes of conference calls and proceedings of meetings are posted at www.hydrogenandfuelcellsafety.info.

DOE supports implementation of the templates through subcontracts with a number of SDOs and CDOs designated for lead roles on the templates. It should be noted that significant work to implement the templates is being done by organizations not funded by DOE. While the templates were not intended to specify which organizations should receive DOE funding, they have helped to solidify the roles of the organizations identified as having a lead role in developing a particular standard.

In summary, the templates continue to function as the seminal documents that help to create a more unified national approach to the development of hydrogen and fuel cell codes and standards. The templates and the National Hydrogen and Fuel Cells Codes and Standards Coordinating Committee that was formed to manage the templates have created a "virtual national forum" for SDOs, CDOs, industry, government, and interested parties to address codes and standards issues, both immediate and long-term.

2.2 Research, Development, and Demonstration for Codes and Standards

The RD&D Roadmap helps guide DOE activities that will provide data required for SDOs to develop performance-based codes and standards for a commercial hydrogen fueled transportation sector in the U.S. The Roadmap reflects the experience and priorities of the members of the FreedomCAR and Fuels Partnership, which include the DOE, energy companies (BP, Chevron, ConocoPhillips, ExxonMobil, Shell Hydrogen), and the automotive companies (Chrysler, Ford, General Motors) belonging to the U.S. Consortium for Automotive Research (USCAR). The contents of the Roadmap are reviewed and revised by the Partnership as needed to reflect changing needs and opportunities.

By evaluating specific needs for RD&D, assessing the status of on-going RD&D, and revising the Roadmap as needed, the Partnership will ensure new U.S. projects are efficiently leveraged and coordinated with those undertaken internationally. Through the International Partnership for the Hydrogen Economy (IPHE), DOE works with individual countries as well as contributing to global RD&D efforts. Information requirements of international SDOs are considered to help align RD&D projects with needs for code and standard development.

The Roadmap includes an assessment of existing hydrogen and fuel cell codes and standards and those that are in the process of being established domestically and internationally and identifies information needs and gaps related to those codes and standards for a hydrogen-based transportation system. The Codes and Standards Technical Team (CSTT) of the Partnership reviews RD&D projects to address gaps and to provide documented research to SDOs on a continuing basis.

The Roadmap is organized into four Focus Areas:

- Hydrogen Behavior
- Hydrogen-fueled Vehicles
- Hydrogen Fuel Infrastructure
- Fuel-Vehicle Interface

The technical goal for each of these Focus Areas is to gather sufficient information and validating experience on technology applications so that the responsible SDO or CDP can proceed with better data upon which to base requirements incorporated in its codes and standards. Each Focus Area is subdivided into key Target Areas, which identify important information needs for which information is required by SDOs and CDOs to fully develop codes and standards. Completion of RD&D for the individual technical Target Areas, in conjunction with information distribution, is expected to result in the subsequent development of safe, performance-based codes and standards. The Roadmap will be implemented over the next five years as proposed in the timeline for the Focus Areas.

3. Conclusion

Two key needs for hydrogen safety in consumer applications are the incorporation of data and analysis from RD&D into the codes and standards development process and the adoption and enforcement of these codes and standards by state and local code officials. DOE supports a comprehensive program to address both these aspects of hydrogen safety. For the first, DOE is working with the automobile and energy industries to identify and address high priority RD&D to establish a sound scientific basis for requirements that are incorporated in hydrogen codes and standards. The high priority RD&D needs are incorporated and tracked in a Roadmap adopted by the Codes and Standards Technical Team (CSTT) of the FreedomCAR and Fuel Partnership. DOE and its national laboratories conduct critical RD&D and work with key standards and model code development organizations to help incorporate RD&D results into the codes and standards process. To address the second aspect, DOE has, for example, launched an initiative to facilitate the permitting process for hydrogen fueling stations (HFS). A key element of this initiative will be a web-based information repository, a toolkit that includes information fact sheets, networking charts to encourage information exchange among code officials who have permitted or are in the process of permitting HFS, case studies of likely near-term HFS configurations, and a database of requirements incorporated in key codes and standards. The information repository will be augmented by workshops for code officials and station developers in jurisdictions that are likely to have HFS in the near future.

The national templates have guided DOE's effort to accelerate the development of key standards and model codes for hydrogen and fuel cell applications. With the help and cooperation of SDOs, CDOs, industry, and other interested parties, DOE has established a coordinated national agenda for hydrogen and fuel cell codes and standards. With the adoption of an RD&D Roadmap by the Partnership and through its implementation by the CSTT, the DOE will help strengthen the scientific basis for requirements incorporated in these codes and standards that, in turn, will facilitate international market receptivity for hydrogen and fuel cell technologies.

References

1. ANSI, Overview of the U. S. Standardization System, 2nd edition, 2007
2. See http://www.iccsafe.org (accessed August 8, 2007)
3. See http://www.nfpa.org (accessed August 8, 2007)
4. See http://www.hydrogen.gov/regulations (accessed August 8, 2007)

5. See 29CFR Part 1910.103 and http://www.osha.gov (accessed August 8, 2007)
6. See http://www.cganet.com/isotc197 (accessed August 8, 2007)
7. See http://www.csa-america.org (accessed August 8, 2007)

PRELIMINARY PERFORMANCE ASSESSMENT OF COMMERCIALLY-AVAILABLE HYDROGEN SENSORS

NATHAN D. MARSH AND THOMAS G. CLEARY

Fire Research Division, National Institute of Standards and Technology, 100 Bureau Drive, Stop 8664, Gaithersburg, MD 20899, USA

As part of an effort to develop standard test methods for the performance of commercial hydrogen sensors, we employed the Fire Emulator / Detector Evaluator, an instrumented flow system designed to study the response of fire detectors (smoke, heat, gas), in a preliminary study to evaluate the performance of a representative selection of commercially-available hydrogen sensors. These sensors depend on a variety of sensing technologies including metal-oxide semiconductors, electrochemical cells, catalytic bead pellistors, thermal conductivity sensors, and sensors employing a combination of technologies. They were evaluated both for their response to hydrogen concentrations up to half the lower flammability limit, and their response to nuisance gases (CO, CO_2, NOx, hydrocarbon gas and vapor—all potentially present in hydrogen dispensing and storage areas), as well as dynamic changes in environmental conditions by varying temperature, humidity, and flow velocity. These performance evaluations provide guidance for the development of a test method designed to assess real-world performance of hydrogen gas sensors. The ultimate goal is to develop standard test methods to be employed by product certification agencies.

1. Introduction

The hydrogen economy envisions wide application of energy delivery solutions based on hydrogen fuel cells or combustion systems. The public's acceptance of these new energy delivery systems will rely to some extent on the perceived and actual safe application of the technologies. To this end, reliable detection of an accidental hydrogen gas release and mitigation of the hazard through designed safety systems is a key component of hydrogen powered systems in commercial, residential, and transportation uses. In anticipation of this emerging market, inexpensive hydrogen gas sensors based on a range of sensing technologies are becoming increasingly available. There is a need to characterize sensors in conditions relevant to their end-use application.

Currently acceptance standards applied to hydrogen sensors follow the existing UL 2075 "Standard for Safety Gas and Vapor Detectors and Sensors" and the relevant flammable gas standards in the US such as NFPA 52 and 55. The International Organization for Standardization (ISO) Technical

Committee 197 has formed a working group (WG 13: Hydrogen Detectors) to focus on an international standard, ISO/CD 26142 "Hydrogen detection apparatus". In the ISO standard there is a recognized need to test sensor performance in terms of sensitivity, response time, recovery time, environmental changes (temperature, humidity, pressure) and nuisance sources, i.e. substances which may trigger a false alarm. However, the standard recommends a static test chamber, which is limited in its ability to expose sensors to dynamic and repeated changes in the environment and gas composition; in particular, it is unclear how one would test sensor recovery time in such a system.

In this work, we are interested in testing performance under conditions representative of real-world challenges. We therefore considered where hydrogen sensors might ultimately be deployed. As automotive applications appear to be an early adopter of hydrogen technology, current and near future use of hydrogen sensors might take place in hydrogen filling stations, which often are part of or adjacent to traditional gasoline filling stations, and residential or commercial garages. All of these spaces may be outdoors, although sheltered, neither heated nor air conditioned, and experiencing relatively high concentrations of automobile exhaust including CO, CO_2, and unburned hydrocarbons.

To this end, we acquired a representative sample of seven sensors, from four manufacturers, employing four different sensing technologies. These sensors were first calibrated and tested for exposure to hydrogen in a benchtop calibration flowcell. They were then tested in our Fire Emulator / Detector Evaluator (FE / DE) [1] an apparatus previously used for extensive studies of fire detectors [2-4]. The FE / DE is easily modified for the evaluation of hydrogen sensors, with the primary difference that we use only the gas exposure part system, and do not use any of the smoke generation options. The sensors were again tested for hydrogen exposure, as well as CO, CO_2, propene (propylene, C_3H_6), condensing water vapor, and temperature variation. These environmental changes and gas compositions were also tested in conjunction with hydrogen exposure to determine whether any synergistic or obfuscating effects were significant.

2. Procedure

2.1. *Calibration Flow Cell*

The calibration cell consists of a chamber 0.1 m cross-section and 0.6 m long. Mixed gas from two mass flow controllers is introduced into one end of the chamber, where it must pass through a 0.1 m long section of honeycomb flow straightener, past the sensor, and out of the chamber.

Sensors were placed one at a time in the calibration flow cell, powered by the appropriate DC voltage from regulated DC power supplies, the sensors' outputs connected to analog voltage inputs (0 V to 10 V) of a data acquisition system. Because some sensors have 4 mA to 20 mA current-loop outputs, in these cases the loop was completed with a 500 Ohm resistor and the resulting 2 V to 10 V drop across the resistor measured. Other sensors exhibited high-frequency oscillations, visible on an oscilloscope; these sensor outputs were filtered by connecting a 1 μF capacitor across the output terminals, resulting in a simple RC filter.

Sensors were exposed to between 500 μL/L and 6500 μL/L hydrogen in air, as determined by the mixing of a calibrated bottle of 2 % hydrogen in air with additional air via the mass flow controllers. The dimensions and flowrates used resulted in gas velocities from 15 cm/sec to 25 cm/sec.

2.2. *Fire Emulator / Detector Evaluator*

A schematic of the FE/DE is shown in Figure 1. A variable-speed fan draws room air and passes it through a series of 9 annular finned heating elements (5 kW each for a total maximum heat input of 45 kW) resulting in air velocity at the test section between 0.02 m/s to over 2 m/s and an available rate of temperature rise of 0.5 °C/s, up to maximum of about 80 °C. The flow is conditioned before it reaches the 0.5 m × 0.3 m test section by passing through a 10 cm long aluminum honeycomb with 5 mm rectangular openings. CO, CO_2, or other gas blends may be metered into the flow just downstream of the heater

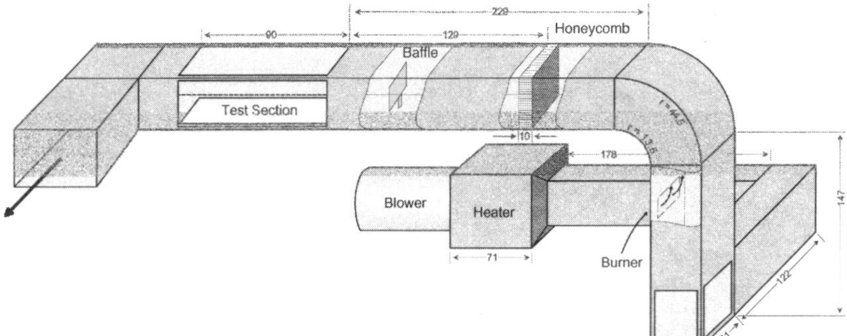

Figure 1. Schematic of the FE/DE

via electronic mass flow controllers. A laboratory steam generator can inject low-pressure steam, also just downstream of the heater, to humidify the air from ambient room to saturated conditions at elevated temperature. Water, CO, CO_2, and hydrocarbon gas concentrations at the test section are monitored by non-dispersive infrared (NDR) analyzers. Temperature and gas analysis are recorded in the same data acquisition system as the sensors.

Sensors, summarized in Table 1, were installed three or four at a time in the test section of the FE/DE, powered and monitored in the same way as in the calibration cell. Sensors were exposed to the following challenges:

- Temperature rise from 25 °C to 50 °C followed by a return to 25 °C
- 100 % relative humidity with condensing water vapor
- Carbon monoxide (120 μL/L to 250 μL/L) and carbon dioxide (2000 μL/L)
- Propene (130 μL/L)
- Hydrogen (250 μL/L)
- Hydrogen (250 μL/L) with temperature rise from 25 °C to 50 °C followed by a return to 25 °C
- Hydrogen (250 μL/L) with 100 % relative humidity and condensing water vapor
- Hydrogen (250 μL/L) with carbon monoxide (50 μL/L) and/or carbon dioxide (600 μL/L)
- Hydrogen (250 μL/L) with propene (120 μL/L)

These tests were carried out with an air flow rate of 12 cm/sec to 25 cm/sec, with the lower velocity used for chemical exposures and the higher velocity used for temperature and moisture exposure.

Table 1. Summary of Tested H2 Sensors

Sensor	Tech	Range (vol fraction)	
A	TCD	0.0 % to 100 %	
B	MOS	0.0 % to 2.0 %	
C	MOS	0.0 % to 2.0 %	
D	CAT	0.0 % to 2.5 %	
E	Multi	0.4 % to 5.0 %	Film resistor and MOS capacitor, Pd/Ni film
F	MOS	0.0 % to 0.20 %	Includes molecular sieve
G	CAT	0.1 % to 4.0 %	Includes molecular sieve

TCD: Thermal Conductivity Detector; MOS: Metal Oxide Semiconductor; CAT: Catalytic Bead Pellistor; Multi: Multiple integrated technologies

3. Results

Figure 2 shows typical results to a sensor test, in this case exposure to 1) 50 µL/L CO; 2) 50 µL/L CO and 250 µL/L H_2; and 3) exposure to H_2 alone. Most notable from this test is that while Sensor B does respond with a limited false positive to CO exposure, this response is not added to that for hydrogen when both gases are present. None of the other sensors in this test had any response to CO exposure.

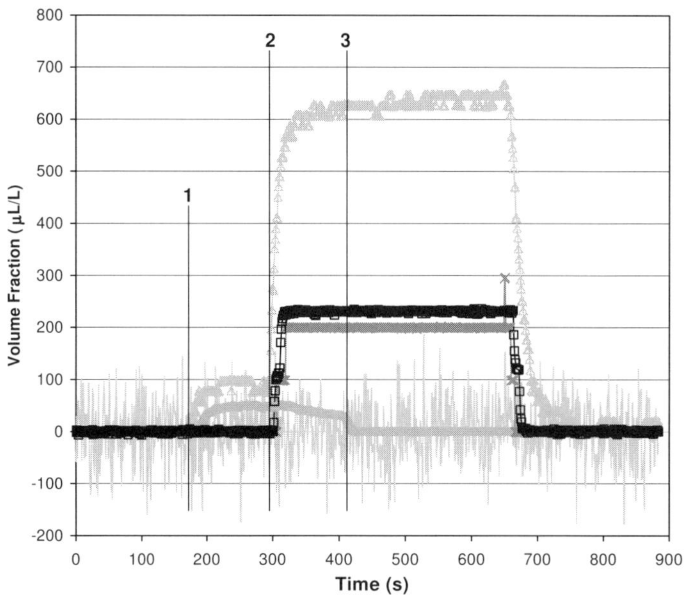

Figure 2. Typical result of an exposure test. Circles: CO; no symbol: Sensor A (TCD); light triangles: Sensor B (MOS); medium Xs: Sensor C (MOS); dark squares: Sensor D (CAT). 1) 50 µL/L CO; 2) 50 µL/L CO and 250 µL/L H_2; and 3) 250 µL/L H_2

The performance of the sensors tested here can be summarized as follows:

- Sensor A (TCD) was not sensitive enough to detect H_2 anywhere, even up to 7000 µL/L in the calibration cell. It was however sensitive to condensing water vapor, reading the equivalent of 3000 µL/L H_2 at 25 °C and 100 % relative humidity.
- Sensor B (MOS) experienced the most cross-sensitivity, responding to temperature, humidity, CO/CO_2 and propene. It also read consistently high in the presence of H_2. In general, cross sensitivities appear to be linear combinations, i.e. no synergistic effects.

- Sensors C and F (both MOS) experienced some cross-sensitivity. In Sensor C there appears to be a synergistic effect with humidity and H_2: it appears to be sensitive to humidity **only** in the presence of H_2.
- Sensor D (CAT) is cross-sensitive to everything except CO/CO_2. It is extremely sensitive to hydrocarbons. It is also inversely temperature sensitive: increasing the temperature by 25 °C reduces the baseline by a voltage equivalent to 200 μL/L. (Reducing the temperature by the same amount raises the baseline—essentially producing a false positive.)
- Sensors E (Multi) and G (CAT) were not sensitive to any challenge gases or conditions. However they were also not sensitive enough to detect 250 μL/L of H_2 in the FE / DE.

Table 2 summarizes the response of the different sensors to hydrogen and the various challenges, all expressed in equivalent volume fraction of hydrogen. Reported actual values are from the independent monitoring instruments, so for example the 130 μL/L reported as the actual fraction of propene is the value recorded from the NDIR hydrocarbon analyzer.

Table 2. Responses of Hydrogen Sensors (all values in μL/L)

	A	B	C	D	F	Actual
H_2	0	590	200	200	350	250
+ 25 °C	0	20	0	-200	0	+ 25 °C
H_2O	3000	120	0	300	0	condensing
CO / CO_2	0	190	0	0	10	120, 2000
C_3H_6	0	930	600	2500	70	130
H_2 + 25 °C	0	640	300	-200	320	250, + 25 °C
		510[a]	100[a]			
$H_2 + H_2O$	3000	740	300	300	390	250, condensing
$H_2 + CO/CO_2$	0	610	200	220	310	250, 50, 600
$H_2 + C_3H_6$	0	1210	900	2800	390	250, 120
Uncertainty[b]	1000	10	100	100	10	

[a] High value only on T rise with hydrogen present; low value for hydrogen added at high T
[b] Type B uncertainty based on sensor manufacturer's documentation

It is worth noting the relatively disparate responses of the sensors to an ostensibly uniform flow of hydrogen. We attribute this discrepancy to four possible sources. First, although the flow in the FE/DE has been observed to be relatively uniform [1-4], the profile of hydrogen itself has not been measured in the FE/DE, and is in fact the subject of ongoing work. Hydrogen's buoyancy and high diffusivity may in fact lead to its non-uniform accumulation in unexpected ways. Second, the volume fraction of hydrogen in the FE/DE was at the low end of the sensor calibration range; thus any non-linear response in one or more sensors to low concentration of hydrogen could account for disagreements between the sensors. Third, the effect of flow conditions on the sensors is unknown; differences in velocities both during the calibration procedure and between the calibration cell and the FE/DE could have an affect on the sensors' detection efficiency. Finally, the sensors were calibrated using dry compressed air and a calibration-grade hydrogen-air mixture, while the FE/DE uses room air, which in addition to having a relative humidity of 40% to 60% may contain other trace gases or vapors to which some sensors are more sensitive than others.

Sensors were also evaluated for response time in the calibration cell. Response times were characterized as being the time from the initiation of hydrogen flow to reach 95 % of the maximum reading (activation), and the time from the cessation of the hydrogen flow to reach 5 % of the maximum reading (relaxation). We make three general observations for the sensors tested here.

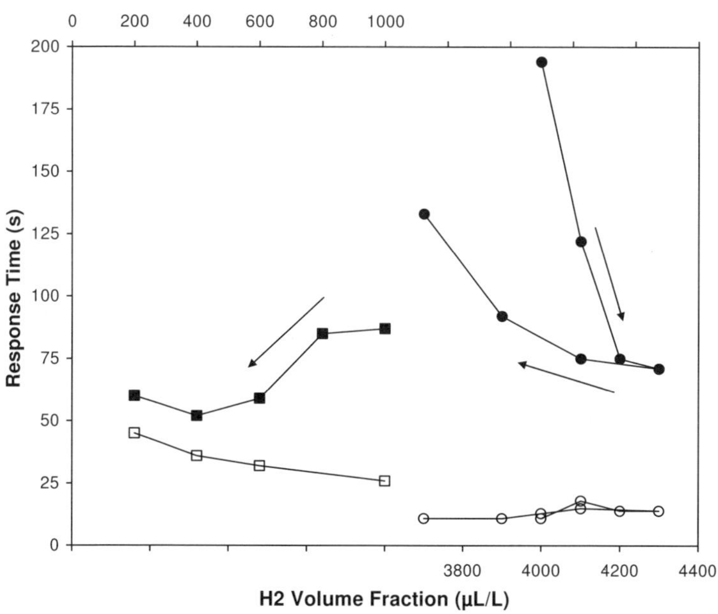

Figure 3. Response times of sensors. Squares: Sensor F (MOS); Circles: Sensor E (Multi); Filled symbols: response to hydrogen flow initiation; open symbols: response to hydrogen flow cessation. Arrows indicate order of tests. Times are not corrected for the response time of the calibration cell.

First, activation times were generally on the scale of 1 min to 3 min, with the inherent response time of the calibration cell (i.e. the response time of the system for a sensor with instantaneous response) being less than 10 s. Second, relaxation times were generally much faster than activation times. Third, the effect of concentration on activation times was not consistent between different sensors: in some sensors, increasing concentration increases activation time, while in others it decreases activation time. Figure 3 shows the uncorrected response times for Sensors E and F.

4. Conclusions

The FE/DE was modified to test an array of commercially-available hydrogen sensors that may be used for leak detection in hydrogen dispensing and storage facilities. Sensor cross-sensitivities to heat, moisture, and various gases in low concentrations were measured. In the presence of hydrogen, cross-sensitivities appeared to be additive in some cases and synergistic in other cases. The extent to which the observed cross-sensitivities would lead to nuisance alarms or missed alarms is unknown. Further testing at the desired hydrogen alarm concentrations needs to be performed. Sensor response times were on the order of one to three minutes, with relaxation times observed to be faster. Ultimately, performance evaluations need to consider dynamic environmental and concentration changes to assess temporal sensor performance.

References

1. M. Anderson, A. Chernovsky, T. Cleary, and W. Grosshandler, "Particulate Entry Lag in Spot-Type Smoke Detectors," *Proceedings of the 6th International Symposium on Fire Safety Science*, International Association for Fire Safety Science, 779 (2000).
2. T. Cleary, M. Anderson, J. Averill, and W. Grosshandler, "Evaluating Multisensor Fire Detectors in the Fire Emulator / Detector Evaluator," *Proceedings of the 8th International Conference on Fire Science and Engineering*, (Interflam '99), Interscience Commusications, 453 (1999).
3. T. Cleary, W. Grosshandler, and A. Chernovsky, "Smoke Detector Response to Nuisance Aerosols," *Proceeding of the 11th International Conference on Automatic Fire Detection* (AUBE '99), Joachim Agst Verlag, 32 (1999).
4. Grosshandler, W.L., "Toward the Development of a Universal Fire Emulator/ Detector Evaluator," *Fire Safety Journal* **29**, 113 (1997); also in *Proceeding of the 10th International Conference on Automatic Fire Detection* (AUBE '95), Mainz-Aachen, 368 (1995).

PANEL SUMMARY

SCOTT W. JORGENSEN

Chemical and Environmental Sciences Lab, GM Research and Development, 30500 Mound Rd, Warren, MI 48090, USA

R. CHAHINE

Hydrogen Research Institute, Universite Du Quebec a Trios-Riviers, 3351 Des Forges Bldg, Trios-Riviers, Quebec G9A 5H7, Canada

J. P. MEYERS

Materials Science and Engineering, The University of Texas at Austin, 1 University Station, Austin, Texas 78712, USA

G. D. PARKS

Research and Development, ConocoPhilips, 344A PL BTC, Bartlesville, OK 74004, USA

A. A. PUNDT

Universitat Gottingen, Friedrich-Hund-Platz 1, 37077 Gottingen, Germany

Y. FILINCHUK

European Synchrotron Radiation Facility, Swiss-Norwegian Bea Lines, 6 Rue Jules Horowitz, 38043 Grenoble, France

The main focus of this symposium, materials advances in the areas of hydrogen production, storage and fuel cells, reflects the world wide research focus on these key areas required to form the technical foundation for a hydrogen economy. The sessions on tank engineering materials, safety, education and standards also reflects the growing emphasis on these issues which will become the focus once the initial technical hurdles in science and or engineering are surmounted. Better materials are very much needed in each of these areas either to improve performance, and durability or decrease costs. The conference summarized the progress made to date in these field and highlighted the progress that remains to made to make hydrogen a dominant energy carrier in all aspects of the world economy. This review follows the flow of hydrogen through a constructed hydrogen economy to summarise the papers and discussion at the ISHE, referencing the barriers and progress globally.

Because, like electricity, hydrogen is a secondary power source that must be generated from primary sources such as fossil fuels, nuclear energy, tidal energy, geothermal heat, solar energy or its' derivatives wind and biomass, the hydrogen economy necessarily starts with hydrogen production and possibly subsequent transportation to users. A few interesting papers were given showing advances in hydrogen production from solar energy either directly by photochemical or indirectly via photobiological processes. This is in line with the plenary talk that stressed the fact that "solar is the only energy source with sufficient capacity to fulfill the energy needs of the future" and urging that "we should be developing H_2 [technologies] not only for cars but for the big picture". In general, the progress described was more evolutionary than revolutionary perhaps due to the maturity of the energy supply industry, but these works describe significant progress on several fronts related to hydrogen production.

Production and Delivery

In the short term economics and existing infrastructure dictates that bulk hydrogen production will largely originate from fossil fuels. Production of hydrogen using an improved water-gas shift catalyst was described. In these experiments and calculations showed that subsurface copper promoted platinum activity and made the catalysts less susceptible to poisons. Other research focuses on the use of oxygen-permeable membranes in partial oxidation (POX) reactors. Results with planar membranes showed reasonable oxygen fluxes and high CO and hydrogen selectivities. Additional research on oxygen transport membranes is aimed at improving steam reforming of methane.

Electrolysis, like SMR, could serve for distributed or large scale production. Nano-composite electrodes for natural gas-assisted steam electrolysis were described, aimed at improving electrode activity and stability.

One advantage a secondary power source has is that it may be able to access energy in waste streams. Although hydrogen production form the reaction of water with aluminum has been studied extensively for decades, recently researchers have increased efficiency by using the reaction to produce high pressure hydrogen. Combined with the use of waste aluminum as a feedstock, this process is somewhat more economical and efficient than similar processes. In reference to the hydrogen economy this is not a global solution for hydrogen supply, but does illustrate the general concept of extraction of energy that would otherwise be lost in waste, a principle that may play a role in the larger suite of hydrogen generation techniques.

In contrast, photoelectrochemical hydrogen could be exercised on a large scale if the difficulties in cost effective production with appropriate land and water use were solved. Researchers from multiple locations discussed hydrogen production using direct photoelectrochemical water splitting. While none of the photocatalytic materials are near commercialization, progress is being made on efficiency and corrosion control. One technique discussed uses mesoporous transition metal oxides for photocatalytic hydrogen production, but most required ultraviolet light to affect water splitting. An approach less dependent on new materials is the use of concentrated sunlight to thermally decompose water. Here, the use of catalyst-coated monolith reactors to facilitate decomposition and "trap" oxygen formed was described. By cycling multiple systems using solar heat to release oxygen and prepare for continued hydrogen production a continuous hydrogen stream is possible. Other thermochemical cycles to facilitate thermal water splitting include the modified sulfur-ammonia cycle, and the sulfur-iodine cycle - the later used either directly or with use of oxygen transport membranes to improve hydrogen production. Of course an alternative approach is to use sunlight to power biological hydrogen production. Work with green algae seeks to increase the H_2 yields, optimize adsorption of sunlight, and explore cost-effective reactor designs.

Only one presentation dealt with hydrogen delivery- a review of challenges and recent developments in the field that have been funded by the US DOE. While possibly less glamorous, delivery either by pipeline or vehicle, or alternately delivery of another energy source with subsequent on-site hydrogen-generation is a key link in the hydrogen economy and must be properly developed.

Storage

Once created and delivered, hydrogen will frequently need to be stored. This is a requirement for use in vehicles. Many feel that improvement in this area is a major barrier to launching the hydrogen economy, though several auto makers have fielded vehicles using existing storage technology and some have demonstrated 300 mile range is possible for vehicles that represent a portion of world fleet. Nonetheless, greater storage capacity and lower cost would surely facilitate faster and deeper penetration into this key sector of energy use.

An accordingly large share, (more than 50%) of the oral presentations at the symposium, were dedicated to hydrogen storage. There was a profusion of screening and modeling results showing good synergy with experiments in the solid hydrides. As with the global research to date, the focus was largely directed toward high wt% or specific mass of storage techniques while less attention was given to volumetric density which can be an equally critical factor for onboard storage in some applications.

An increasing number of one-component systems, like $LiBH_4$ and $Ca(BH_4)_2$ have been shown to be in principle reversible, although at high temperature (~600°C) and hydrogen pressure (~200 bar). There were few such presentations at the ISHE, and this follows the trend of a continued but decreasing activity in the relatively mature field of known, simple, materials.

By contrast, a new trend in research on hydrogen storage systems is to modify and combine known light hydrides in order to improve their hydrogen storage properties; and several papers on this topic were presented. New developments center on doping existing compounds (introducing chemical substitutions) and making them react with other H-rich solids. These attempts to modify properties often bring very interesting results. Properly chosen a mixture of two hydrides desorbs hydrogen at lower temperature than a one-component system, sometimes accompanied by lesser amounts of biproducts, e.g. diborane or ammonia, in the desorbed hydrogen gas. The following systems were presented at the symposium:

- Borohydrides ($LiBH_4$, $Ca(BH_4)_2$) + binary hydrides (LiH, MgH_2);
- Borohydrides ($LiBH_4$, $NaBH_4$) + amides ($LiNH_2$, $NaNH_2$);
- Aminoborane (NH_3BH_3) + binary hydrides (LiH, NaH, CaH_2);
- Borohydride ($NaBH_4$) + alanate ($NaAlH_4$).

In some cases, the initial compounds form hydrogen-rich intermediate phases, which release hydrogen at lower temperature than the starting compounds. A number of such new phases were presented both by oral presentations and posters. These presentations were a mix of pure and applied research.

From the perspective of laying a foundation for a hydrogen economy, an experimental study of a two-component system might ideally provide information about the reactivity of the components, hydrogen desorption temperature and H-capacity. In addition the research should isolate the presence of new intermediate phases that may contain meaningful amounts of hydrogen, and thus may be used for H-storage on their own, and describe the existence of new decomposition products, which may reabsorb hydrogen. This data should be also supported by the thermodynamic information, showing that the reaction enthalpies are not excessively high. In general such a complete study is of greater scope than any one researcher's area of interest, or perhaps that of their funding body. Fortunately, when several works from around the world are taken together, as may be seen in major conferences, this complete picture begins to emerge.

In the ISHE it was possible to connect PCT diagrams which characterize the bulk properties of a system, with diffraction studies that help to determine which phases are involved in reactions when more than two components are present in the mixture. The measurement of PCT diagrams tends to go first, as they demonstrate whether the "destabilization" is achieved. Diffraction study can clarify the mechanism of such "destabilization", i.e. the reaction mechanism. In particular, the dehydrogenation process has to be analyzed for new intermediate phases, and for new hydrogen-poor (or hydrogen-free) decomposition products. The latter may be tested as starting compounds in rehydrogenation processes. New compounds may show different properties and maybe even reversibility!

Even if only two starting components are used, a system becomes complicated (multi-component) when hydrogen desorption begins. As a number of intermediate and decomposition products are involved, the system becomes multi-component, and thus it appears essential to know which components are involved in the crucial steps of hydrogen release. In several works an *in-situ* diffraction study of hydride mixtures was important in the identification of new phases and gives a sequence of intermediate compounds.

Thus, there were several reports searching not only for new H-rich substances, but also for new hydrogen-poor, relatively unstable phases composed of light elements, which may appear for the first time as decomposition products, but later may turn out to store hydrogen reversibly. Different catalysts can be tested at this stage. Finally, the properties of the system can be improved, for example by using nanoengineering.

The new accomplishments using the "hydrogenography" approach were presented in the study of solid solution and two phase metal. It would be interesting to see this or other very high throughput methods used in the light hydride systems. Thin films of multiple light hydrides can be deposited at various concentrations and studied by diffraction. A crystalline powder, as the most common form of hydrogen storage materials, can be studied by *in-situ* powder diffraction, both at variable temperature and hydrogen pressure. Such scan reveals all possible transformations of the new material or a continuum of mixtures aiming to destabilize the hydride. Such measurements can now be routinely done at some synchrotron beam lines and neutron facilities and upgrading them to support combinatorial work could hasten the discovery process. Inelastic and quasi-elastic studies of silicon and boron-containing hydrides show the power of these techniques.

Probing dynamics by experimental techniques, inelastic X-ray scattering, infra-red and Raman (vibrational) spectroscopies and NMR is also effective and instructive. These methods provided a link between structure and properties, providing fundamental information that helps to reduce the "gap" between theory and experiment. Dynamics is something hidden from routine crystallographic studies, which often represent structures as "static". The presentation on the dynamics of amino borane illustrated the knowledge that can be gained pairing theory and spectroscopy. Theoreticians, on the other side, will have difficulties modeling a system that is ill-defined. Given the lack of information about dynamics, many calculations aim to reproduce the structure only. Experimental information on structure dynamics provides reference points for density functional theory (DFT) calculations, thus helping to avoid the most common pitfalls of structure prediction. It was also shown how complete intermediate information is required to properly calculate reaction enthalpies and predict reversibility at pressures and temperatures acceptable for use in the hydrogen economy.

There was discussion on several occasions on the "gap of realities" between theory and experiments. One point raised in this connection was the value of a theoretician directly participating with experimental groups. This will help to apply theory directly and continuously to experimental activities, with immediate feedback in both directions. Theory often brings attention to subtle but important features, which experimentalists may not naturally notice, or even cannot directly measure. Likewise without rapid data to test predictions, theorists can only slowly refine models and may spend considerable effort on predictions made with incomplete models that yield less accurate predictions. This synergy seems to be growing world wide and will surely benefit all research in the area of hydrogen storage.

For the last 5 years research on metal-organic frameworks has shown considerable progress; most recently culminating in a series of materials storing 7 to 7.5 wt% excess hydrogen storage at 77K reviewed here. This work includes a MOF with 32 g per liter volumetric storage capacity and hope of exceeding 40g/L. The enthalpy of hydrogenation tends to be low in these materials so they fill rapidly; for example new data showing that MOF-74 initially exhibits an 8.8 kJ/mol adsorption enthalpy, which drops to half that value with 2 wt% hydrogen uptake, was discussed. Possible applications of these materials will be more likely if research can produce higher volumetric capacity and higher absorption enthalpy.

Spectroscopic techniques nicely compliment synthesis and capacity work. One paper reviewed recent work on neutron powder diffraction, where MOFs loaded with different deuterium pressure were studied to directly pin point the location of hydrogen in these porous systems. For the first time the absorbed hydrogen molecules were located in the organic linker, and this highlights their importance. It would be interesting to test this approach on the other systems that store weakly bound hydrogen.

There was also an interesting triangle of presentations on Metal-H_2 (dihydrogen) complexes. One was a reaffirmation of the theory of multiple (up to 6 molecule binding) of hydrogen in metal assisted organics (so called soft chemisorption, e.g. C_2H_4-Ti); a second was announcement of the synthesis on the titanium complex at the picogram scale (hopefully soon to be independently confirmed); and thirdly a review on metal-H_2 complexes by the original discoverer, which showed that of the more than 600 compounds found so far only ~ 2 % contain 2 hydrogen molecules and the remaining only one molecule of hydrogen. It also mentioned the difficulties in obtaining high capacity materials in the condensed phase. These three papers together, point to the previously mentioned gap between prediction and successful creation of materials. Extensive theoretical predictions suggest each Sc atom in a $Sc_{12}C_{60}$ cluster may take 4 hydrogen molecules reaching 7 wt% of hydrogen storage capacity, but the clustering of these Sc atoms may affect material stability. The hydrogen absorption enthalpy for $Li_{12}C_{60}$ is predicted to be 6.4 kJ/mol based on a baseline structure. However, recent experimental results show completely different metal-fullerene structures displaying 0.2-0.5 wt% uptake of H_2 at 77K. At present it is unclear if the predictive models need refinement or if new synthesis techniques are required. It is clear that close interaction between experiment and theory groups is needed to resolve the situation.

On balance significant progress made to date over a very short time both reflects and validates the level of effort, the creative and deductive thought, and the amount of funding directed at hydrogen storage. Such funding must be sustained to drive this technology to a point where it is widely applicable in the hydrogen economy.

Fuel Cells

The final step in the hydrogen flow through a hydrogen based economy is consumption for useful work. While internal combustion engine technology has been known for some time, the major research effort is on fuel cells. Fuel cells also date to the 19th century, but application has only been likely with recent improvements in power density. While increasingly sophisticated stationary and mobile fuel cells have been fielded over the last 10 years, durability and cost have inhibited wide scale production in mobile applications.

During the ISHE, good progress toward the various international targets was reported. Durability of the new membranes can provide 40 to 50x reduction in fluoride evolution and last as much as 20x longer than their predecessors. New R&D efforts are aiming to increase the operation temperature and humidity requirements in order to reduce the cooling requirements and simplify the BOP.

Progress was also reported in modeling the reaction and transportation processes on fuel cell catalysts and through membranes, using multiple paradigms as well as starting from first principle quantum mechanics to 'train' a reactive force field that can be applied for large scale molecular dynamics simulations. It is expected that the model would "enable the conception, synthesis, fabrication, characterization, and development of advanced materials and structures for fuel cells".

Education and Safety

As illustrated by some of the latter papers, there is a role for governments around the world in educating the public and facilitating the transition. In addition, it will eventually be highly important to understand hydrogen embrittlement of metals, and the compatibility of materials in a moderate to high pressure hydrogen environment. Depending on the storage mode, this material may need to tolerate cryogenic temperatures or elevated temperatures. Work in these areas is not as intense as those discussed above, but as the hydrogen economy nears and applications increase, it may be expected that structural materials, standards, and education issues will grow in both public awareness and concomitantly in technical importance.

Hydrogen Economy beyond the Transportation Sector

Finally, as in every hydrogen conference, there was discussion of the issue of the DOE Hydrogen storage targets for onboard vehicle and the oft encountered comments of "being difficult to achieve". In all fairness these targets started as being US targets but similar targets were adopted years later by other national and international H_2 programs. Are they too high? We have to achieve high targets if we are seeking a world wide market application of FCVs, not only for personal mobility but for the transport of goods and other uses. By many accounts, mass production of FCVs will not happen before many years. By 2025, according to the recent report of the Air Resources Board of the state of California which has in the past set a tone for America and the rest of the world when it comes to pollution. So the questions arose, is an FCV the only hydrogen application out there, and given that several hydrogen technologies suffer from slow kinetics, how do we accelerate it? The answer is an application in demand where the targets are easier to hit. For example, cordless electric applications ranging from laptop computers to power tools where there is a real demand for longer run times. These applications are mostly going to lithium ion batteries. The Li battery market is estimated at $5 billion/year with double digit annual increase. So Instead of directing the quasi-totality of the R&D efforts on developing storage materials for the ultimate FCV application, which require that we meet a set of technical targets that is difficult to achieve, some felt we should invest efforts on these other applications where the targets are easier to hit than the corresponding DOE hydrogen targets. Battery targets are often an order of magnitude lower than H_2 storage targets, for example $900-1500/kWh compared to $2-8/kWh of storage and similar differences in life and energy density. Even accounting for the fuel cell mass and cost these are still a morel likely entry point. This might be where the hydrogen moves from a chemical industry technology to a wider field of application. The availability of such a system will create a real market and the crucially important supply chain where technological progress and innovations happen most rapidly on the trip up the experience curve. A strong consumer demand and rapid market 'kinetics' coupled with aggressive R&D could then open an automotive fuel cell / hydrogen-storage market. With the huge talent within the hydrogen scientific community there is reason to feel a hydrogen economy is possible.

SCIENTIFIC PROGRAM

Sunday, November 11

5:00 - 8:00 PM Registration, Submission of Manuscripts
6:00 - 8:00 PM Reception

Monday, November 12

INAUGURAL SESSION

9:00 - 9:10 AM Introductory Remarks
Puru Jena, Symposium Chair, Virginia Commonwealth University, USA

9:10 - 9:20 AM Welcome Address
John B. Fenn, Nobel Laureate, Virginia Commonwealth University, USA

SESSION A: Key Note
Chairman: Constantina Filiou, European Commission, Netherlands

9:20 - 10:05 AM "Progress and Challenges of a Hydrogen Economy"
Mildred S. Dresselhaus, Massachusetts Institute of Technology, USA

10:05 - 10:35 AM **COFFEE BREAK**

10:35 - 11:35 AM "DOE Hydrogen Program: Production, Delivery and Fuel Cells: Technologies, Challenges, Infrastructure Costs, and Material Needs"
Mark D. Paster, Department of Energy, USA

11:35 - 1:30 PM **LUNCH**

SESSION B: Production I
Chairman: U. (Balu) Balachandran, Argonne National Lab., USA

1:30 - 2:00 PM	"Materials Issues for Photoelectrochemical Water Splitting: Chalcopyrite Thin-Films and III-V Nitrides" **John A. Turner**, National Renewable Energy Laboratory, USA
2:00 - 2:30 PM	"Hydrogen Production via Water Splitting in Solar Reactors: The Hydrosol Process" **Athanasios G. Konstandopoulos**, Aerosol & Particle Technology Laboratory, Greece
2:30 - 3:00 PM	"Development of Photocatalysts for Solar Hydrogen Production" **Akihiko Kudo**, Tokyo University of Science, Japan
3:00 - 3:15 PM	"A Cu/Pt Near-Surface Alloy for Watr-Gas Shift Catalysis Studied by STM, XPS, TPD, and DFT" **Ronnie T. Vang**, Jan Knudsen, Joachim Schnadt, and Flemming Besenbacher. *Interdisciplinary Nanoscience Center (iNANO and Department of Physics and Astronomy), University of Aarhus, Denmark.*
3:15 - 3:45 PM	**COFFEE BREAK**

Session C: Storage I (Molecular)
Chairman: George Thomas, Department of Energy, USA

3:45 - 4:15 PM	"7.5 wt % Hydrogen Storage in Metal Organic Frameworks" **Omar M. Yaghi**, University of California, USA
4:15 - 4:45 PM	"Henry's Law and Isoteric Heats in Physisorbents" **Channing Ahn**, California Institute of Technology, USA
4:45 - 5:15 PM	"Novel Organometallic Fullerene Complexes for Vehicular Hydrogen Storage" **Anne C. Dillon**, National Renewable Energy Laboratory, USA

5:15 - 5:45 PM | "Engineered Nano-Materials for High Capacity Hydrogen Storage"
Taner Yildirim, NIST, USA

5:45 - 6:00 PM | "Design of materials for storing hydrogen in quasi-molecular form"
Qiang Sun[1,2], Qian Wang[1], and Puru Jena[1], *Physics Department, Virginia Commonwealth University, and Department of Advanced Materials and Nanotechnology, Peking University, China*

6:00 - 8:00 PM | **DINNER**

8:00 - 10:00 PM | **Poster Session I**

Tuesday, November 13

Session D: Fuel Cells I
Chairman: Gary Sandrock, Department of Energy, USA

8:30 - 9:00 AM | "Materials Challenges in Proton Exchange Membrane Fuel Cells"
Biswajit Choudhury, E. I. du Pont Nemours & Company, USA

9:00 - 9:30 AM | "New PEM Fuel Cell Membranes for Higher Temperature, Drier Operating Conditions Based on the Heteropolyacids"
Andrew M. Herring, Colorado School of Mines, USA

9:30 - 10:00 AM | "Simulation of Reaction and Transport Processes in Fuel Cell Catalysts and Membranes"
William A. Goddard, III, California Institute of Technology, USA

10:00 - 10:15 AM | "Alternative Materials to Pd Membranes for Hydrogen Purification"
Paul S. Korinko and Thad Adams, Savannah River National Laboratory, USA

10:15 - 10:45 AM | **COFFEE BREAK**

Session E: Storage II (Nano-materials)
Chairman: Shengbai Zhang, NREL, USA

10:45 - 11:15 AM	"Carbide-Derived Carbons for Hydrogen Storage" **Gleb Yushin**, Drexel University, USA
11:15 - 11:45 AM	"Storage of Molecular Hydrogen in Carbon Based Systems" **Sa Li**, Virginia Commonwealth Univesity, USA
11:45 - 12:15 PM	"Hydride Chemistry in Nanoporous Scaffolds" **John J. Vajo**, HRL Laboratories, USA
12:15 - 12:30 PM	"High Density H_2 Storage on Nanoengineered Scaffolds of Carbon Nanotubes" **Carter Kittrell**, A.D. Leonard, S. Chakraborty, H. Fan, W.E. Billups, R.H. Hauge, H.K. Schmidt, M. Pasquali, J.M. Tour, *Department of Chemistry, Rice University, USA*
12:30 - 2:00 PM	**LUNCH**

Session F: Production II
Chairman: Michelle V. Buchanan, Oak Ridge National Laboratory, USA

2:00 - 2:30 PM	"H_2 Binding and Reactivity on Transition Metal Complexes underlying Biomimetic H_2 Production and New Materials for H_2 storage" **Gregory J. Kubas**, Los Alamos National Laboratory, USA
2:30 - 3:00 PM	"Materials Issues in Photobiological Production" **Anastasios Melis**, University of California, Berkeley, USA
3:00 - 3:30 PM	"Hydrogen Production from Hydrocarbons by using Oxygen Permeable Membranes" **Hitoshi Takamura**, Tohoku University, Japan

3:30 - 3:45 PM "Direct Production of Pressurized Hydrogen from Waste Aluminum without Compressor"
T. Hiraki[a], N. Okinaka[a], H. Uesugi[b] and T. Akiyama[a], *[a]Center for Advanced Research of Energy Conversion Materials, Hokkaido University, Japan, [b]Waseda University, Japan*

3:45 - 5:45 PM FREE TIME/NETWORKING

6:00 PM RECEPTION/DINNER: Jefferson Hotel*
***Buses leave OMNI at 6:00 PM for Jefferson Hotel**

SPEAKER: Ambassador Reno L. Harnish,
Principal Deputy Assistant Secretary, U.S. Department of State

Wednesday, November 14

Session G: Storage III (Chemical Hydrides)
Chairman: Maciej Gutowski, Heriot-Watt University, UK

8:30 - 9:00 AM "Indirect, Reversible Hydrogen Storage in Metal Ammine Salts: Recent Progress and Prospects"
Claus H. Christensen, Technical University of Denmark, Denmark

9:00 - 9:30 AM "Alkali Aminoboranes for Hydrogen Storage"
Ping Chen, National University of Singapore, Singapore

9:30 - 10:00 AM "Structure and Dynamics of Ammonia Borane"
S. Thomas Autrey, Pacific Northwest Laboratory, USA

10:00 - 10:15 AM "Molecular Simulation of Structural Changes of Ammonia Borane"
Gregory K. Schenter, Chris Mundy, Shawn M. Kathmann, Vencislav Parvanov, Nancy J. Hess, Wendy J. Shaw, Herman M. Cho and Thomas Autrey, *Pacific Northwest National Laboratory, USA*

10:15 - 10:45 AM COFFEE BREAK

Session H: Storage IV (Complex Hydrides)
Chairman: Karl Johnson, University of Pittsburgh, USA

10:45 - 11:15 AM	"Characterization of Complex Metal Hydrides by High Resolution Solid State NMR" **Robert C. Bowman**, Jet Propulsion Laboratory, NASA, USA
11:15 - 11:45 AM	"Hydrogenography: A combinatorial thin film approach to identify the thermodynamic properties of metal hydrides" **Bernard Dam**, Vrije Univerity, Netherlands
11:45 - 12:15 PM	"First-principles engineering of advanced hydrogen storage materials" **Vidvuds Ozolins**, University of California, Los Angeles, USA
12:15 - 12:30 PM	"Development of Metal Hydrides for High-Pressure MH Tank" **T. Matsunaga*,** T. Shinozawa, K. Washio, D. Mori, M. Ishikiriyama**,** *Higashifuji Technical Center, Toyota Motor Corporation, Japan*
12:30 - 2:00 PM	**LUNCH**

Session I: Fuel Cells II
Chairman: Peter Edwards, Oxford University, UK

2:00 - 2:30 PM	"Materials Challenges in Solid Oxide Fuel Cells" **Subhash C. Singhal**, Pacific Northwest National Laboratory, USA
2:30 - 3:00 PM	"The Development of Nano-Composite Electrodes for Natural Gas-Assisted Steam Electrolysis for Hydrogen Production" **Raymond J. Gorte**, University of Pennsylvania, USA
3:00 - 3:30 PM	"Near-surface alloys and Core-shell nanocatalysts for reactions involving hydrogen" **Manos Mavrikakis**, University of Wisconsin, USA

3:30 - 3:45 PM	"Hybrid Inorganic-Organic Polymer Composites for Polymer-Electrolyte Fuel Cells" **Andrea Ambrosini**, Cy H. Fujimoto, Christopher J. Cornelius, *Sandia National Laboratories, Albuquerque, USA*
3:45 - 4:15 PM	**COFFEE BREAK**

Session J: Storage V (Complexhydrides)
Chairman: Vitalij Pecharsky, Ames Laboratory, USA

4:15 - 4:45 PM	"Reaction Mechanism and Kinetics of Reactive Hydride Composites" **Martin Dornheim**, GKSS Research Centre Geesthacht, Germany
4:45 - 5:15 PM	"Single- and Double-Cations Borohydrides for Hydrogen Storage Applications" **Shin-ichi Orimo**, Tohoku University, Japan
5:15 - 5:45 PM	"Tetrahydroboranates: The New Hydrogen Storage Materials" **Andreas Borgschulte**, EMPA Materials Science and Technology, Switzerland
5:45 - 6:00 PM	"Storage of Compressed Hydrogen in Multi-capillary Arrays" N. K. Zhevago, *Kurchatov Institute, Russia* and **Dan Eliezer**, *Ben Gurion University, Israel.*
6:00 - 8:00 PM	**DINNER**
8:00 - 10:00 PM	**Poster Session II**

Thursday, November 15

Session K: Safety & Education
Chairman: B. S. Shivaram, University of Virginia, USA

8:30 - 9:00 AM — "Structural-Materials Considerations for Hydrogen Gas Containment"
Chris San Marchi, Sandia National Laboratory, USA

9:00 - 9:30 AM — "A National Agenda for Hydrogen Codes and Standards"
Chad Blake, National Renewable Energy Laboratory, USA

9:30 - 10:00 AM — "Educating Key Audiences about Fuel Cell Technologies"
Robert Remick, NREL, USA

10:00 - 10:15 AM — "Hydrogen behavior and coloration of tungsten oxide films prepared by magnetron sputtering and pulsed laser deposition"
S. Nagata[1], A. Inouye[2], S. Yamamoto[2], B. Tsuchiya[1], T. Shikama[1], [1]*Institute for Materials Reseach, Tohoku University, Japan,* [2]*Japan Atomic Energy Agency, Takasaki, Japan*

10:15 - 10:45 AM — COFFEE BREAK

Session L: Storage –VI
Chairman: Ragaiy Zidan, SRNL, USA

10:45 - 11:15 AM — "Hydrogen Storage and Delivery Using Liquid Carriers"
Guido Pez, Air Products and Chemicals Inc, USA

11:15 - 11:45 AM — "Hydrogen Storage Materials – Playing the Odds"
W.I.F. David, Oxford University, UK

11:45 - 12:15 PM — "Probing Structure, Bonding, and Dynamics in Hydrogen StorageMaterials by Neutron-Scattering Techniques"
Terrence J. Udovic, NIST Center for Neutron Research, USA

12:15 - 12:30 PM "Thermodynamics of Doped Complex Metal Hydrides"

J. Karl Johnson[1,2], Sudhakar V. Alapati[3], Bing Dai[1], David S. Sholl[2,3]
[1]Department. of Chemical Engineering, University of Pittsburgh, Pittsburgh, PA 15261, USA; [2]National Energy Technology Laboratory, Pittsburgh, PA 15236; [3]Department of Chemical Engineering, Carnegie Mellon University, Pittsburgh, PA 15213, USA

12:30 - 2:30 PM **LUNCH**

2:30 - 4:00 PM **Panel Discussion:**

Chair: Scott W. Jorgensen, General Motors, USA
Richard Chahine, Univ. du Quebec a Trois Rivieres, Canada
Jeremy P. Meyers, University of Texas, USA
George D. Parks, Conoco-Phillips, USA
Astrid A. Pundt, University of Goettingen, Germany
Yaroslav Filinchuk, European Synchrotron Radiation Facility

ORGANIZATION

Chairman: Puru Jena (U.S.A)

INTERNATIONAL ADVISORY BOARD

Frank DiSalvo (Cornell University, USA)
Mildred Dresselhaus (M.I.T, USA)
Peter Edwards (University of Oxford, U.K)
Constantina Filiou (JRC, Netherlands)
Ronald Griessen (Vrije Universiteit, Netherlands)
Maciej Gutowski (Heriot-Watt University, U.K)
Craig Jensen (University of Hawaii, USA)
Thomas Klassen (Helmut-Schmidt-University, Germany)
Nathan Lewis (California Institute of Technology, USA)
Laurie Mets (University of Chicago, USA)
Jens Norskov (CAMP, Denmark)
Shin-ichi Orimo, (Tohoku University, Japan)
Louis Schlapbach (EMPA, Switzerland)
Omar Yaghi (University of California at LA, USA)

NATIONAL PROGRAM COMMITTEE

Michelle Buchanan (Oak Ridge National Laboratory)
Anne Dillon (National Renewable Energy Laboratory)
Peter Eklund (Pennsylvania State University)
Karl Johnson (University of Pittsburgh)
Scott Jorgensen (General Motors)
Vitalij Pecharsky (Ames Laboratory)

LOCAL ORGANIZING COMMITTEE

Gang Chen (Virginia Commonwealth University)
Anil K. Kandalam (McNeese State University)
Sa Li (Virginia Commonwealth University)
Qiang Sun (Virginia Commonwealth University)
Qian Wang (Virginia Commonwealth University)
Mary Willis (Virginia Commonwealth University)

PARTICIPANTS

ALEXANDER ABRAMOV
CHEMISTRY, SCHOOL OF EPS
WILLIAM H. PERKIN BLDG.
HERIOT-WATT UNIVERSITY
EDINBURGH EH144AS
UK
Tel: +44(0)793 916-30-60
ava3@hw.ac.uk

CHANNING AHN
SENIOR RESEARCH ASSOC.
DIV. OF ENGINEERING & APPLIED SCIENCE
CALIFORNIA INSTITUTE OF TECHNOLOGY
1200 E. CALIFORNIA BLVD, MS 238-78
PASADENA, CA 91125
Tel: (626) 395-2174
Fax: (626) 795-6132
cca@caltech.edu

TOMOHIRO AKIYAMA
PROFESSOR
CAREM
HOKKAIDO UNIVERSITY
KITA 13 NISHI8 KITA-KU
SAPPORO 060-8628
JAPAN
Tel: +81-11-706-6842
Fax: +81-11-726-0731
takiyama@eng.hokudai.ac.jp

ANDREA AMBROSINI
SENIOR MEMBER OF TECHNICAL STAFF
FUELS AND ENERGY TRANSITIONS
SANDIA NATIONAL LABORATORIES
PO BOX 5800, MS 0734
ALBUQUERQUE, NM 87185-0734
Tel: (505) 284-1340
Fax: (505) 844-7786
aambros@sandia.gov

KIKUO ARIMOTO
MANAGER
RESEARCH & TECHNICAL CENTER
KURARAY AMERICA, INC.
11500 BAY AREA BLVD.
PASADENA, TX 77507
Tel: (281) 474-1557
Fax: (281) 474-1572
kikuo.arimoto@kurarayamerica.com

S.THOMAS AUTREY
SCIENTIST
CHEMICAL & MATERIALS SCIENCES DIV.
PACIFIC NORTHWEST NATIONAL LAB
PO BOX 999, K2-57
RICHLAND WA 99352
Tel: (509) 375-3792
Fax: (509) 375-6660
tom.autrey@pnl.gov

MARK S. BAILEY
MATERIALS SYNTHESIS
WILDCAT DISCOVERY TECHNOLOGIES
6985 FLANDERS DRIVE
SAN DIEGO, CA 92121
Tel: (858) 550-1986
Fax: (858) 638-7533
mbailey@wildcatdiscovery.com

U. (BALU) BALACHANDRAN
MANAGER CERAMICS SECTION
ENERGY SYSTEMS DIVISION
ARGONNE NATIONAL LABORATORY
9700 S. CASS AVENUE
ARGONNE, IL 60439
Tel: (630) 252-4250
Fax: (630) 252-3604
balu@anl.gov

VINCENT BERUBE
DEPARTMENT OF PHYSICS
MASSAHUSETTS INSTITUTE OF TECHNOLOGY
77 MASSACHUSETTS AVE, RM 7-008
CAMBRIDGE, MA 02139
Tel: (617) 253-7080
Fax: (617) 253-3484
vberube@mit.edu

VINAY BHAT
RESEARCHER
MATERIALS SCIENCE & TECHNOLOGY DIV.
ORNL
1 BETHELVALLEY ROAD
OAK RIDGE, TN 37830
Tel: (865) 574-0798
Fax: (865) 574-4450
bhatvv@ornl.gov

CHAD BLAKE
SENION PROJECT LEADER
HYDROGEN TECHNOLOGIES AND SYSTEMS
NATIONAL RENEWABLE ENERGY LAB
1617 COLE BOULEVARD
GOLDEN, CO 80401-3393
Tel: (303) 275-3843

JAMES G. BLENCOE
CHIEF SCIENTIST
HYDROGEN DISCOVERIES, INC.
200 BRASHEARS ROAD
HARRIMAN, TN 37749
Tel: (865) 882-4984
jblencoe@hydrogendiscoveries.com

ANDREAS BORGSCHULTE
GROUP LEADER 'STABILITY & KINETICS"
HYDROGEN AND ENERGY
EMPA
UEBERLANDSTRASSE 129
DUEBENDORF 8600
SWITZERLAND
Tel: +41 44 823 46 92
Fax: +41 44 823 4022
andreas.borgschulte@empa.ch

MARK BOWDEN
INDUSTRIAL RESEARCH LTD.
PO BOX 31-310
LOWER HUTT
NEW ZEALAND
m.bowden@irl.cri.nz

ROBERT C. BOWMAN
MTS
JET PROPULSION LABORATORY
MAIL STOP 79-24
4800 OAK GROVE DRIVE
PASADENA, CA 91109-8099
T: (818) 354-7941
F: (818) 393-4878
robert.c.bowman-jr@jpl.nasa.gov

FLORIAN BUCHTER
HYDROGEN AND ENERGY
EMPA
UEBERLANDSTRASSE 129
DUEBENDORF 8600
SWITZERLAND
Tel: +41 44 823 40 82
Fax: +41 44 823 40 22
florian.buchter@empa.ch

ANTHONY K. BURRELL
LOS ALAMOS NATIONAL LABORATORY
J514
LOS ALAMOS, NM 87545
Tel: (505) 667-9342
burrell@lanl.gov

JACOB W. BURRESS
CHIEF RESEARCH ASSISTANT
PHYSICS
UNIVERSITY OF MISSOURI-COLUMBIA
223 PHYSICS BLDG – UMC
COLUMBIA, MO 65211
Tel: (573) 882-1147
jacobburress@mizzou.edu

KRISTEN CASALENUOVO
PHYSICS DEPARTMENT
VIRGINIA COMMONWEALTH UNIV.
RICHMOND, VA 23220
Tel: (434) 420-4825
kcasanovo@gmail.com

RICHARD CHAHINE
DIRECTOR
HYDROGEN RESEARCH INSTITUTE
UNIVERSITÉ DU QUÉBEC À TROIS-RIVIÈRES
3351 DES FORGES BLDG.
PO BOX 500
TROIS-RIVIÈRES, QUÉBEC G9A 5H7
Tel: (819) 376-5139
Fax: (819) 376-5164
Richard.chahine@uqtr.ca

PHILIP A. CHATER
SCHOOL OF CHEMISTRY
UNIVERSITY OF BIRMINGHAM
EDGBASTON B15 255
BIRMINGHAM, UK
Tel: +441214144382
pac079@bham.ac.uk

GANG CHEN
PHYSICS Department
VIRGINIA COMMONWEALTH UNIVERSITY
1020 W. MAIN ST.
RICHMOND, VA 23284
gchen@vcu.edu

LI CHEN
PHYSICS DEPARTMENT
RENSSELAER POLYTECHNIC INSTITUTE
110 8TH ST.
TROY, NY 12180
Tel: (518) 225-5825
chenl@rpi.edu

PING CHEN
ASSISTANT PROFESSOR
PHYSICS AND CHEMISTRY
NATIONAL UNIVERSITY OF SINGAPORE
10 KENT RIDGE CRESCENT
SINGAPORE 117542
Tel: 65-65162982
Fax: 65-67776126
phychenp@nus.edu.sg

ZHONGFANG CHEN
ASSOCIATE RESEARCH SCIENTIST
DEPARTMENT OF CHEMISTRY
UNIVERSITY OF GEORGIA
1004 CEDAR ST.
ATHENS, GA 30602
Tel: (706) 621-2903
Fax: (706) 542-0406
chen@chem.uga.edu

BISWAJIT CHOUDHURY
SR. RESEARCH SCIENTIST
FUEL CELLS
E.I. DUPONT DE NEMOURS & CO.
CRP 701/214, 4417 LANCASTER PIKE
WILMINGTON, DE 19803
Tel: (302) 999-2726
Fax: (302) 999-2395
biswajit.choudhury@can.dupont.com

CLAUS CHRISTENSEN
PROFESSOR
CENTER FOR SUSTAINABLE & GREEN CHEMISTRY
TECHNICAL UNIVERSITY OF DENMARK
BUILDING 206, DK-2800 LYNGBY
DENMARK
Tel: +45 45 25 24 02
chc@kemi.dtu.dk

YINGYING CUI
INORGANIC CHEMISTRY LABORATORY
UNIVERSITY OF OXFORD
SOUTH PARKS ROAD
OXFORD OX1 3QR
UK
Tel: 00441865272643
Fax: 00441865272690
ying.cui@chem.ox.ac.uk

AUDE CUNI
PROJECT MANAGER
CLAUDE-DELORME RESEARCH CENTER
AIR LIQUIDE
1 CHEMIN DE LA PORTE DES LOGES
JOUY EN JOSAS 78354
FRANCE
Tel: 33 1 39 07 60 73
Fax: 33 1 39 07 61 13
aude.cuni@airliquide.com

BERNARD DAM
ASSOCIATE PROFESSOR
SOLID STATE PHYSICS
VRIJE UNIVERSITEIT
DE BOELELAAN 1081
NL-1081 HV AMSTERDAM
THE NETHERLANDS
Tel: +31-205987917
Fax: +31-205987992
b.dam@few.vu.nl

BILL DAVID
STFC SENIOR FELLOW
ISIS FACILITY
RUTHERFORD APPLETON LABORATORY
CHILTON
OXON OX11 0QX
UK
Tel: +44 1235 445179
Fax: +44 1235 445383
bill.david@rl.ac.uk

DANIEL E. DEDRICK
SENIOR MEMBER OF TECHNICAL STAFF
THERMAL & FLUID SCIENCE & ENGINEERING
SANDIA NATIONAL LABORATORY
PO BOX 969 MS9409
LIVERMORE, CA 94610
Tel: (925) 294-1552
Fax: (925) 294-3870
dededri@sandia.gov

ANNE C. DILLON
SENIOR SCIENTIST
NREL
1617 COLE BLVD.
GOLDEN, CO 80401
T: (303) 384-6607
F: (303) 384-6655
anne_dillon@nrel.gov

HIMASHINIE V. DIYABALANAGE
POSTDOCTORAL RESEARCH ASSOCIATE
MATERIALS, PHYSICS & APPLICATIONS DIV.
LOS ALAMOS NATIONAL LABORATORY
MS J514
LOS ALAMOS NM 87545
Tel: (505) 606-1625
himashinie@lanl.gov

OLEKSANDR DOLOTKO
POST DOCTORAL ASSOCIATE
AMES LABORATORY
IOWA STATE UNIVERSITY
AMES, IA 50011-3020
Tel: (515) 294.9158
Fax: (515) 294.9579
dolotko@ameslab.gov

MARTIN DORNHEIM
DEPARTMENT HEAD
DEPARTMENT OF NANOTECHNOLOGY
GKSS-RESEARCH CENTRE GEESTHACHT
MAX-PLANCK-STR. 1
21502 GEESTHACHT GERMANY
T: +49-4152-872604
F: +49-4152-872636
martin.dornheim@gkss.de

MILDRED S. DRESSELHAUS
INSTITUTE PROFESSOR
ELECTRICAL ENG. & COMPUTER SCIENCE
DEPARTMENT OF PHYSICS
MIT
ROOM 13-3005, MIT
MASSACHUSETTS AVENUE
CAMBRIDGE, MA 02139
T: (617) 253-6864
F: (617) 253-6827
millie@mgm.mit.edu

DAN ELIEZER
PROFESSOR
THE ERIC SAMSON CHAIR FOR ADVANCED MATERIALS AND PROCESSING
DEPARTMENT OF MATERIALS ENG.
BEN GURION UNIV. OF THE NEGEV
CHIEF SCIENTIST ADVISER C.EN CO.
BEER SHEVA 84105, ISRAEL
Tel: +972 8 6461467
Fax: +972 8 6472931
deliezer@bgumail.bgu.ac.il

HANI EL-KADERI
PROFESSOR
CHEMISTRY DEPARTMENT
VIRGINIA COMMONWEALTH UNIVERSITY
RICHMOND, VIRGINIA 23284-2006
Tel: (804) 828-7505
Fax: (804) 828-8599
helkaderi@vcu.edu

JOHN B. FENN
NOBEL LAUREATE
CHEMISTRY DEPT
VIRGINIA COMMONWEALTH UNIVERSITY
BOX 842006
1001 W. MAIN ST., OLIVER HALL
RICHMOND, VA 23284-2006
Tel: (804) 828-1298
jbfenn@vcu.edu

YAROSLAV FILINCHUK
BEAMLINE SCIENTIST
EUROPEAN SYNCHROTRON RADIATION FACILITY
SWISS-NORWEGIAN BEAM LINES
6 RUE JULES HOROWITZ
38043 GRENOBLE
FRANCE
Tel: +33 47 688 2775
Fax: +33 47 688 2694
yaroslav.filinchuk@esrf.fr

CONSTANTINA FILIOU
SCIENTIFIC OFFICER
INSTITUTE FOR ENERGY
EUROPEAN COMMISSION – DG JRC
WESTERDUINWEG 3
PETTEN
NL-1755LE THE NETHERLANDS
Tel: +31-224-565171
Fax: +31-224-565623
constantina.filiou@jrc.nl

STANISLAW FILIPEK
PROFESSOR
PHYSICAL CHEMISTRY OF SOLIDS
INSTITUTE OF PHYSICAL CHEMISTRY
UL.KASPRZAKA 44
WARSAW 01-224
POLAND
Tel: +48 22-343-3334
Fax: +48 22 343 3333
smf@ichf.edu.pl

GERD GANTEFÖR
UNIVERSITY OF KONSTANZ
FACULTY OF PHYSICS
KONSTANZ 78457
GERMANY
T: (49) 7531-88-2067
F: (49) 7531-88-3091
gerd.gantefoer@uni-konstanz.de

QINGFENG GE
ASSOCIATE PROFESSOR
CHEMISTRY AND BIOCHEMISTRY
SOUTHERN ILLINOIS UNIVERSITY
CARBONDALE, IL 62901
qge@chem.siu.edu

GUTSEV L. GENNADY
PHYSICS DEPARTMENT
FAMU
TALLAHASSE, FL 32307
Tel: (850) 599-3783
Fax: (850) 599-3577
Gennady.gutsev@famu.edu

WILLIAM A. GODDARD III
PROFESSOR
CHEMISTRY DEPT.
CALIFORNIA INSTITUTE OF TECHNOLOGY
139-74 CALTECH
PASADENA, CA 91125
Tel: (626) 395-2731
Fax: (626) 585-0918
wag@wag.caltech.edu

RAYMOND J. GORTE
PROFESSOR
CHEMICAL & BIOMOLECULAR ENGINEERING
UNIVERSITY OF PENNSYLVANIA
311 TOWNE BLDG, 220 SOUTH 33RD. ST.
PHILADELPHIA, PA 19104
T: (215) 898-4439
F: (215) 573-2093
gorte@seas.upenn.edu

ANDREJ GRUBISIC
GRADUATE STUDENT
DEPARTMENT OF CHEMISTRY
JOHNS HOPKINS UNIVERSITY
3400 N. CHARLES ST.
BALTIMORE, MD 21218
Tel: (410) 516-4675
Fax: (410) 516-8420
agrubisic@jhu.edu

HAKAN GUNOYDIN
MATERIALS SCIENCE &
ENGINEERING
HENRI SAMUELI SCHOOL OF
ENGINEERING & APPLIED SCIENCE
UNIVERSITY OF CALIFORNIA – LA
PO BOX 951595
LOS ANGELES, CA 90095-1595
Tel: (310) 267-5538
Fax: (310) 206-7353

MACIEJ GUTOWSKI
PROFESSOR
CHEMISTRY-SCHOOL OF EPS
HERIOT-WATT UNIVERSITY
RICCARTON CAMPUS
EDINBURGH EH14 4AS
UK
Tel: 0 131 451 3083
Fax: 0 131 451 3180
m.gutowski@hw.ac.uk

NEIL J. HENSON
TECHNICAL STAFF MEMBER
THEORETICAL CHEMISTRY AND
MOLECULAR PHYSICS
LOS ALAMOS NATIONAL
LABORATORY
PO BOX 1663, Mail Stop B268
LOS ALAMOS, NM 87544
Tel: (505) 667-7795
Henson@lanl.gov

ANDREW M. HERRING
ASSOCIATE PROFESSOR
CHEMICAL ENGINEERING
CSM
GOLDEN, CO, 80401
T: (303) 384-2082
F: (303) 273-3730
aherring@mines.edu

CLEMENS HESKE
ASSOCIATE PROFESSOR
CHEMISTRY DEPARTMENT
UNIVERSITY OF NEVADA LAS VEGAS
4505 MARYLAND PARKWAY
LOS VEGAS, NV 89154 -4003
Tel: (702) 895-2694
Fax: (702) 895-4072
heske@unlv.nevada.edu

TAKEHITO HIRAKI
CAREM
HOKKAIDO UNIVERSITY
KITA13 NISHI8 KITA-KU
SAPPORO 060-8628
JAPAN
Tel: +81 11 706 6842
Fax: +81 11 726 0731
thiraki@eng.hokudai.ac.jp

JENS S. HUMMELSHØJ
PhD STUDENT
CAMD
TECHNICAL UNIVERSITY OF
DENMARK
FREDERIKSGÅRDS ALLÉ 14, 2
VANLØSE 2720
DENMARK
Tel: 61714745
Fac: 61714745
strabo@fysik.dtu.dk

KARL JACKSON
DEPARTMENT OF CHEMISTRY
VIRGINIA COMMONWEALTH UNIV.
1001 W. MAIN ST.
RICHMOND, VA 23284
Tel: (804) 828-1298
Fax: (804) 828-8599
jacksonjt@vcu.edu

PANCHATAPA JASH
PhD STUDENT
CHEMISTRY
UNIVERSITY OF ILLINOIS AT
CHICAGO
845 W. TAYLOR, SES 4500
CHICAGO, IL 60607
Tel: (312) 996-5424
Fax: (312) 996-0431
pjash2@uic.edu

PURU JENA
DISTINGUISHED PROFESSOR
PHYSICS DEPARTMENT
VIRGINIA COMMONWEALTH
UNIVERSITY
RICHMOND, VA 23284-2000
Tel: (804) 828-8991
Fax: (804) 828-7073
pjena@vcu.edu

BÖRJE M. JOHANSSON
PROFESSOR
DEPARTMENT OF MATERIALS SCIENCE
ROYAL INSTITUTE OF TECHNOLOGY
BRINELLVÄGEM 23
STOCKHOLM SE 10044
SWEDEN
+46704175452
borje.johannsson@fysik.uu.se

KARL JOHNSON
PROFESSOR
CHEMICAL ENGINEERING
UNIVERSITY OF PITTSBURGH
1249 BENEDUM HALL
PITTSBURGH, PA 15261
Tel: (412) 624-5644
Fax: (412) 624-9639
karlj@pitt.edu

SCOTT W. JORGENSEN
SENIOR STAFF RESEARCHER
CHEMICAL AND ENVIRONMENTAL SCIENCES LAB.
GM RESEARCH & DEVELOPMENT
480-106-160
30500 MOUND ROAD
WARREN, MI 48090
Tel: (586) 986-1915
Fax: (586) 986-2094
scott.w.jorgensen@gm.com

JAN-OLE JOSWIG
RESEARCH ASSOCIATE
PHYSICAL CHEMISTRY
TECHNICAL UNIVERSITY DRESDEN
BERGSTR. 66 B
DRESDEN 01062
GERMANY
Tel: +49 351 463 39299
Fax: +49 351 463 35953
Jan-ole.joswig@chemie.tu-dresden.de

BHARGAV KANCHIBOTLA
RESEARCH ASSISTANT
DEPARTMENT OF ELECTRICAL ENG.
VIRGINIA COMMONWEALTH UNIV.
601 W. MAIN ST, #241
RICHMOND, VA 23284
Tel: (804) 827-7040
kanchibotlbv@vcu.edu

ANIL K. KANDALAM
ASSISTANT PROFESSOR
PHYSICS DEPARTMENT
VIRGINIA COMMONWEALTH UNIVERSITY
1020 W. MAIN ST.
RICHMOND, VA 23284-2000
Tel: (804) 828-7079
Fax: (804) 828-7073
akkandalam@vcu.edu

ABHIJEET KARKAMKAR
SCIENTIST
CHEMICAL & NORTHWEST NATIONAL LAB.
PACIFIC NORTHWEST NATIONAL LAB.
PO BOX 999, K1-83
RICHLAND, WA 99352
Tel: (509) 372-6359
Fax: (509) 375-4381

RYUTA KAWAGUCHI
SENIOR PROJECT ENGINEER
EV & FCV
NISSAN TECHNICAL CENTER N.A.
37581 EMERALD FOREST DR
FARMING HILLS, MI 48331
Tel: (248) 488-8559
Fax: (248) 488-3908
ryuta.kawaguchi@nissan-usa.com

EUNJA KIM
ASSISTANT RESEARCH PROFESSOR
PHYSICS AND ASTRONOMY
UNIVERSITY OF NEVADA, LAS VEGAS
4505 S. MARYLAND PARKWAY
LAS VEGAS, NV 89154-4003
Tel: (702) 895-1706
Fax: (702) 895 0804
kimej@physics.unlv.edu

CARTER KITTRELL
RESEARCH SCIENTIST
CHEMISTRY DEPARTMENT
RICE UNIVERSITY
MS 600, POB 1892
HOUSTON, TX 77251-1892
Tel: (713) 348-5108
Fax: (713) 348-5320
kittrell@rice.edu

YOSHITSUGU KOJIMA
PROFESSOR
INSTITUTE FOR ADVANCED MATERIALS RESEARCH
HIROSHIMA UNIVERSITY
1-3-1 KAGAMIYAMA
HIGASHI-HIROSHIMA 739-8530
HIROSHIMA, JAPAN
Tel: +81-82-424-3904
Fax: +81-82-424-5744
kojimay@hiroshima-u.ac.jp

ATHANASIOS G. KONSTANDOPOULOS
DIRECTOR
AEROSOL & PARTICLE TECHNOLOGY LAB.
CPERI/CERTH & ARISTOTLE UNIVERSITY
THERMI, THESSALONIKI 57001
GREECE
Tel: +30 2310 498192
Fax: +30 2310 498190
agk@cperi.certh.gr

PAUL S. KORINKO
FELLOW SCIENTIST
MATERIALS SCIENCE & TECHNOLOGY
SAVANNAH RIVER NATIONAL LAB
BLD 773-A
AIKEN, SC 29808
Tel: (803) 725-3390
Fax: (803) 725-7369
paul.korinko@srnl.doe.gov

GREGORY J. KUBAS
LABORATORY FELLOW
CHEMISTRY DEPARTMENT
LOS ALAMOS NATIONAL LAB
LOS ALAMOS, NM
Tel: (505) 667-5767
Fax: (505) 667-0440
kubas@lanl.gov

AKIHIKO KUDO
PROFESSOR
DEPARTMENT OF APPLIED CHEMISTRY
TOKYO UNIVERSITY OF SCIENCE
1-3 KAGURAZAKA, SHINJUKU-KU
TOKYO 162-8601 JAPAN
T: +81-3-5228-8267
F: +81-3-5261-4631
a-kudo@rs.kagu.tus.ac.jp

ZEYNEP KURBAN
EngD (PhD) STUDENT
PHYSICS & ASTRONOMY
UNIVERSITY COLLEGE LONDON
GOWER STREET
LONDON WC1E 6BT
UK
z.kurban@ucl.ac.uk

SA LI
POSTDOC
PHYSICS DEPARTMENT
VIRGINIA COMMONWEALTH UNIVERSITY
RICHMOND, VA 23284-2000
Tel: (804) 828-2770
Fax: (804) 828-7073
lisakite@gmail.com

ISABEL LLAMAS-JANSA
INSTITUTE FOR METALLIC MATERIALS
LEIBNIZ INST. FOR SOLID STATE & MATERIALS RESEARCH
PF 27 01 16
DRESDEN, SACHSEN 01171
GERMANY
Tel: +49 (351) 46 59-669
Fax: +49 (351) 46 59-540
i.llamas.jansa@ifw-dresden.de

ARTHUR LOVELL
PhD STUDENT
PHYSICS AND ASTRONOMY
UCL
GOWER STREET
LONDON WC1E 6BT
UK
Tel: +44 207 679 3409
Fax: +44 207 679 7145
Arthur.lovell@ucl.ac.uk

REBECCA L. LOWTON
INORGANIC CHEMISTRY DEPT.
UNIVERSITY OF OXFORD
SOUTH PARKS ROAD
OXFORD
OX13QR UK
Tel: 01865 272 600
Fax: 01865 272 690
rebecca.lowton@chem.ox.ac.uk

GUSTAVO LOZANO
GKSS RESEARCH CENTER
NANOTECHNOLOGY INST. OF
MATERIAL RES.
MAX-PLANCK STR.1
GEESTHACHT
SCHLEWWIG-HOLSTEN
21502 GERMANY
Tel: +49-41 5287-2643
Fax: +49-41 5287-2625
gustavo.lozano@gkss.de

ANDREW M. MANCE
GM R&D CENTER
30500 MOUND ROAD
MC480-106-710
WARREN, MI 48090-9055
Tel: 586-986-0705
Fax: 586-986-2094
andrew.m.mance@gm.com

NATHAN D. MARSH
CHEMICAL ENGINEER
BUILDING AND FIRE RESEARCH LAB
NIST
100 BUREAU DRIVE
GAITHERSBURG, MD 20899
Tel: (301) 975-5441
nathan.marsh@nist.gov

TOMOYA MATSUNAGA
MATERIAL ENGINEERING DIV.3
TOYOTA MOTOR CORPORATION
1200, MISHUKU, SUSONO
SHIZUOKA 410-1193
JAPAN
Tel: +81-55-997-7086
Fax: +81-55-997-7879
tomoya@matsunaga.tec.toyota.co.jp

MANOS MAVRIKAKIS
PROFESSOR
CHEMICAL AND BIOLOGICAL
ENGINEERING
UNIVERSITY OF WISCONSIN-
MADISON
MADISON, WI 53718
T: (608) 262-9053
F: (608) 262-9053
manos@engr.wisc.edu

TASIOS MELIS
PROFESSOR
PLANT & MICROBIAL BIOLOGY
UNIVERSITY OF CALIFORNIA-
BERKELEY
111 KOSHLAND HALL
BERKELEY, CA 94720-3102
Tel: (510) 642-8166
Fax: (510) 642-4995
melis@nature.berkeley.edu

JEREMY P. MEYERS
ASSISTANT PROFESSOR
MECHANICAL ENGINEERING
MATERIAL SCIENCE & ENGINEERING
COCKRELL SCHOOL OF
ENGINEERING
THE UNIVERSITY OF TEXAS AT
AUSTIN
ETC 9.154
1 UNIVERSITY STATION, C2200
AUSTIN, TEXAS 78712-0292
Tel: (512) 232-5276
Cell: (512) 964-4288
jeremypmeyers@mail.utexas.edu

SHINJI NAGATA
ASSOCIATE PROFESSOR
INSTITUTE FOR MATERIALS
RESEARCH
TOHOKU UNIVERSITY
2-1-1, KATAHIRA, AOBA-KU
SENDAI 980-8577
JAPAN
Tel: +81 -22 215 2062
Fax: +81 -22 215 2061
nagata@imr.tohoku.ac.jp

EIJI NAKAMURA
MANAGER
KURARAY AMERICA, INC.
11500 BAY AREA BLVD.
PASADENA, TX 77507
Tel: (281) 474-1579
Fax: (281) 474-1572
eiji_nakamura@kuraray.co.jp

DIANA E. NANU
RESEARCHER
MATERIALS SCIENCE AND ENGINEERING
DELFT UNIVERSITY OF TECHNOLOGY
MEKELWEG 2
DELFT 2628CD
THE NETHERLANDS
Tel: +31 – 15 2782264
Fax: +31 – 15 2786730
d.e.nanu@tudelft.nl

SAROJ K NAYAK
DEPARTMENT OF PHYSICS, APPLIED PHYSICS AND ASTRONOMY
RENSSELAER POLYTECHNIC INSTITUTE
110 8TH STREET
SCIENCE CENTER, 1C25
TROY, NY 12180-3590
T: 518/276-2932
F: 518/276-6680
nayaks@rpi.edu

ANNE NICKELS
RESEARCH STUDENT
INORGANIC CHEMISTRY LABORATORY
UNIVERSITY OF OXFORD
SOUTH PARKS ROAD
OXFORD
OX1 3QR UK
Tel: +44 (0) 1865 272640
anne.nickels@chem.ox.ac.uk

SHIN-ICHI ORIMO
ASSOCIATE PROFESSOR
INST. FOR MATERIAL RESEARCH
TOHOKU UNIVERSITY
SENDAI 980-8577
JAPAN
Tel: +81-22-215-2093
Fax: +81-22-215-2091
orimo@imr.tohoku.ac.jp

KEVIN C. OTT
LOS ALAMOS NATIONAL LABORATORY
MS J514
LOS ALAMOS, NM 87544
Tel: (505) 667-4600
kcott@lanl.gov

CHONGCHAO PAN
PhD CANDIDATE
SCHOOL OF MATERIALS SCIENCE AND ENGINEERING
TSINGHUA UNIVERSITY
516 ROOM YIFU BUILDING
BEIJING 100084
P.R. CHINA
Tel: +86-010-62772620
Fax: +86-010-62771160
pcc05@mails.tsinghua.edu.cn

GEORGE D. PARKS
RESEARCH FELLOW
RESEARCH & DEVELOPMENT
CONOCOPHILLIPS
344A PL BTC
BARTLESVILLE, OK 74004
Tel: (918) 661-7780
Fax: (918) 662-1097
george.d.parks@conocophillips.com

MARK D. PASTER
ACTING TEAM LEADER
HYDROGEN PRODUCTION & DELIVERY
HYDROGEN PROGRAM
U.S. DEPARTMENT OF ENERGY
WASHINGTON, D.C.
Tel: (202) 586-2821
mark.paster@ee.doe.gov

SRIDHAR PATIBANDLA
RESEARCH ASSISTANT
ELECTRICAL ENGINEERING DEPT.
VIRGINIA COMMONWEALTH UNIV.
601 E. MAIN ST.
RICHMOND, VA 23284
Tel: (804) 827-7040 x 615
sridharttu@gmail.com

VITALIJ K. PECHARSKY
PROFESSOR
MATERIALS SCIENCE &
ENDINEERING & AMES LAB
IOWA STATE UNIVERSITY
253 SPEDDING
AMES, IA
Tel: (515) 294-8220
Fax: (515) 294-9579
vitkp@ameslab.gov

GUIDO PEZ
CHIEF SCIENTIST
AIR PRODUCTS AND CHEMICALS
MATERIALS RESEARCH CENTER
7201 HAMILTON BLVD.
ALLENTOWN, PA 18195
Tel: (610) 481-4271
Fax: (610) 481-7719
pezgp@airproducts.com

ADAM PHILLIPS
UNIVERSITY OF VIRGINIA
382 McCORMICK ROAD
CHARLOTTESVILLE, VA 22903
Tel: (434) 924-7683
abp6q@virginia.edu

CLAUDIO PISTIDDA
NANOTECHNOLOGY DEPARTMENT
GKSS RESEARCH CENTRE
GESSTHACHT GmbH
Max Planck Strasse 1
GEESTHACHT 21502
GERMANY
claudio.pistidda@gkss.de

ASTRID PUNDT
UNIVERSITÄT GÖTTINGEN
FRIEDRICH-HUND-PLATZ 1
37077 GÖTTINGEN
GERMANY
Tel: +49 551 39 5002
Fax: +49 551 39 5012
apundt@ump.gwdg.de

ALI RAISSI
DIRECTOR, HYDROGEN R&D
DIVISION
FLORIDA SOLAR ENERGY CENTER
UNIVERSITY OF CENTRAL FLORIDA
1679 CLEARLAKE ROAD
COCOA, FLORIDA 32922-5703
Tel: (321) 638-1446
Fax: (321) 504-3438
Cell: (321) 536-4888
ali@fsec.ucf.edu

SIVAKUMAR RAMANATHAN
GRADUATE STUDENT
ELECTRICAL & COMPUTER ENG.
VIRGINIA COMMONWEALTH UNIV.
601 W. MAIN ST.
RICHMOND, VA 23284
Tel: (804) 714-7494
ramanathans@vcu.edu

ANIBAL J. RAMIREZ-CUESTA
SCIENTIST
ISIS FACILITY
STFC RUTHERFORD APPLETON LAB
ROOM 1-43
CHILTON OXON OX11 0QX
UK
Tel: +44 1235 446510
a.j.ramirez-cuesta@rl.ac.uk

THOMAS REICH
GRADUATE STUDENT
DEPARTMENT OF CHEMISTRY
VIRGINIA COMMONWEALTH UNIV.
1001 W. MAIN ST.
RICHMOND, VA 23284
Tel: (804) 828-1298
Fax: (804) 828-8599
reichte@vcu.edu

ROBERT J. REMICK
DIRECTOR
HYDROGEN TECHNOLOGIES &
SYSTEMS CENTER
NATIONAL RENEWABLE ENERGY
LAB
1617 COLE BLVD.
GOLDEN CO 80401-3393
Tel: (303) 275-3820
Fax: (303) 275-2905
Robert_remick@nrel.gov

EWA RONNEBRO
TECHNICAL STAFF
ENERGY SYSTEMS DEPARTMENT
SANDIA NATIONAL LABORATORIES
7011 EAST AVE. MAILSTOP 9161
LIVERMORE, CA 94551
Tel: (925) 294-6493
ecronne@sandia.gov

MONIKA RUCHALA
GRADUATE STUDENT
PHYSICS DEPT.
VIRGINIA COMMONWEALTH UNIV.
RICHMOND, VA 23221
Tel: (804) 484-0735
monikaruchala@hotmail.com

CHRIS SAN MARCHI
SENIOR MEMBER, TECHNICAL STAFF
HYDROGEN SCIENCES
SANDIA NATIONAL LABORATORIES
7011 EAST AVENUE MS 9402
LIVERMORE CA 94550
Tel: (925) 294 4880
Fax: (925) 294 3410
cwsanma@sandia.gov

SUNITA SATYAPAL
HYDROGEN STORAGE TEAM LEADER
DOE HYDROGEN PROGRAM
HYDROGEN, FUEL CELLS, &
INFRASTRUCTURE TECHNOLOGIES
U.S. DEPARTMENT OF ENERGY, EE-2H
1000 INDEPENDENCE AVENUE,
WASHINGTON, DC 20585-0121
Tel: 202-586-2336
Fax: 202-586-1637
sunita.satyapal@ee.doe.go

RALPH H. SCHEICHER
PHYSICS DEPARTMENT
UPPSALA UNIVERSITY
BOX 530
SE-751 21 UPPSALA
SWEDEN
Tel: +46 18 471 5865
ralph.scheicher@fysik.uu.se

GREGORY K. SCHENTER
SCIENTIST
CHEMICAL & MATERIALS SCIENCES
DIV.
PACIFIC NORTHWEST NATIONAL LAB
PO BOX 999, K1-83
RICHLAND, WA 99352
Tel: (509) 375-4334
Fax: (509) 375-4381

EBERHARD SCHMIDT-IHN
FUEL CELL DRIVE SYSTEM
DEVELOPMENT
DAIMLER AG
NEUE STRASSE 95
KIRCHHEIM/TECK-NABERN D73230
GERMANY
Tel: +49 7021 89 4610
Fax: +49 711 3052 114244
eberhard.schmidt-ihn@daimler.com

TOM SCHNEIDER
TECH MANAGER
ITT
2560 HUNTINGTON AVENUE
ALEXANDRIA, VA 22303
Tel: (703) 682-4394
tom.schneider@itt.com

UNCHARAT SETTHANAN
POSTDOCTORAL FELLOW
CHEMISTRY DEPARTMENT
UNIVERSITY OF NEW BRUNSWICK
FREDERICTON, NB e3b 6e2
CANADA
Tel: (506) 447-3162
Fax: (505) 453-4981
uncharat@unb.ca

BELLAVE S. SHIVARAM
PROFESSOR
PHYSICS DEPARTMENT
UNIVERSITY OF VIRGINIA
MCCORMICK ROAD
RICHMOND, VA 22901
Tel: (434) 924-6818
bss2d@virginia.edu

ROSHAN SHRESTHA
MATERIALS PHYSICS & APPLICATIONS
LOS ALAMOS NATIONAL LABORATORY
MAIL STOP J514
LOS ALAMOS, NM 87545
Tel: (505) 667-3588
Fax: (505) 667-9905
rshrestha@lanl.gov

SUBHASH C. SINGHAL
BATELLE FELLOW & DIRECTOR, FUEL CELLS
PACIFIC NORTHWEST NATIONAL LABORATORY
902 BATTELLE BLVD
RICHLAND, WA 99352
Tel: (509) 375-6738
Fax: (509) 375-4300
singhal@pnl.gov

CHRIS SMITH
STUDENT
INORGANIC CHEMISTRY
UNIVERSITY OF OXFORD
SOUTH PARKS ROAD
OXFORD, OXON
OX1 EQR
UK
Tel: +44 1865 272640
Fax: +44 1865 272690
Christopher.smith@chem.ox.ac.uk

MARCO SOMMARIVA
POST DOCTORAL RESEARCH ASSISTANT
ISIS FACILITY-RUTHERFORD APPLETON LAB
STFC
RUTHERFORD APPLETON LAB, CHILTON
DIDCOT, OXON, OX11 0QX
UK
Tel: +441235445116
Fax: +441235445720
m.sommariva@rl.ac.uk

QIANG SUN
PROFESSOR
DEPARTMENT OF ADVANCED MATERIALS AND NANOTECHNOLOGY
PEKING UNIVERSITY
BEIJING 100871, CHINA
Tel: (10)6275-2043
Fax: (10)6275-2043
sunqiang@pku.edu.cn

HITOSHI TAKAMURA
ASSOCIATE PROFESSOR
DEPARTMENT OF MATERIALS SCIENCE
TOHOKU UNIVERSITY
6-6-11-301-2 ARAMAKI AZA AOBA
SENDAI MIYAGI 980-8579
JAPAN
Tel: +81-22-795-3938
Fax: +81-22-795-3938
takamura@material.tohoku.ac.jp

GEORGE J. THOMAS
SANDIA NATIONAL LABS (RET)
18124 WEDGE PKWY #433
RENO, NV 89511
george.thomas@sbcglobal.net

AASHANI D. TILLEKARATNE
STUDENT
CHEMISTRY DEPARTMENT
UNIVERSITY OF ILLINOIS AT CHICAGO
845 W. TAYLOR ST., ROOM 4500 SES
CHICAGO, IL 60607
Tel: (312) 498-8146
ktille2@uic.edu

MIKE TRENARY
PROFESSOR
CHEMISTRY DEPARTMENT
UNIVERSITY OF ILLINOIS AT CHICAGO
845 W. TAYLOR ST. SES RM. 4500
CHICAGO, IL 60607
Tel: (312) 996-0777
Fax: (312) 996-0431
mtrenary@uic.edu

BUN TSUCHIYA
ASSISTANT PROFESSOR
INSTITUTE FOR MATERIALS RESEARCH
TOHOKU UNIVERSITY
2-1-1, KATAHIRA, AOBA-KU
SENDAI 980-8577
JAPAN
Tel: +81-22-215-2063
Fax: +81-22-215-2061
tsuchiya@imr.tohoku.ac.jp

JOHN TURNER
PRINCIPAL SCIENTIST
NATIONAL RENEWABLE ENERGY LAB.
HYDROGEN TECHNOLOGIES & SYSTEMS CENTER
GOLDEN, CO 80401
Tel: (303) 275-4270
Fax: (303) 275-2905
jturner@nrel.gov

TERRENCE J. UDOVIC
SENIOR SCIENTIST
CENTER FOR NEUTRON RESEARCH
NIST
100 BUREAU DR., MS 6102
GAITHERSBURG, MD 20899-6102
Tel: (301) 975-6241
Fax: (301) 921-9847
udovic@nist.gov

JOHN J. VAJO
MEMBER TECHNICAL STAFF
HRL LABORATORIES
ENERGY TECHNOLOGIES
3011 MALIBU CANYON ROAD
MALIBU, CA 90265
T: (310) 317-5745
F: (310) 317-5483
vajo@hrl.com

RONNIE VANG
POSTDOC
INTERDISCIPLINARY NANOSCIENCE CENTER
UNIVERSITY OF AARHUS
NY MUNKEGADE BLDG. 1521
AARHUS C 8000
DENMARK
rtv@inano.dk

QIAN WANG
RESEARCH ASSOCIATE PROFESSOR
PHYSICS DEPARTMENT
VIRGINIA COMMONWEALTH UNIVERSITY
1020 W. MAIN ST.
RICHMOND, VA 23284-2000
Tel: (804) 828-2770
Fax: (804) 828-7079
qwang@vcu.edu

CARLOS WEXLER
ASSOCIATE PROFESSOR
PHYSICS AND ASTONOMY
UNIVERSITY OF MISSOURI
223 PHYSICS
COLUMBIA, MO 65211
Tel: (573) 882-8241
Fax: (573) 882-4195
wexlerc@missouri.edu

ERIN WHITNEY
NREL
1617 COLE BLVD.
GOLDEN, CO 80401
Tel: (303) 384-6619
Fax: (303) 384-6655
erin_whitney@nrel.gov

MARY WILLIS
PhD CANDIDATE
PHYSICS DEPARTMENT
VIRGINIA COMMONWEALTH UNIVERSITY
1020 W. MAIN ST.
RICHMOND, VA 23284-2000
Tel: (804) 828-2770
Fax: (804) 828-7073
mil.willis@gmail.com

CHOI DONG WOONG
GRADUATE STUDENT
DEPARTMENT OF CHEMICAL & BIOLOGICAL ENGINEERING
KOREA UNIVERSITY
ANAM-DONG 5GA, SUNGBUK-GU
SEOUL 136-713
KOREA
Tel: 082-02-3290-3725
apolon7908@hanmail.net

OMAR M. YAGHI
PROFESSOR
CHEMISTRY & BIOCHEMISTRY
UCLA
607 CHARLES E. YOUNG DRIVE, E.
LOS ANGELES, CA 90095
T: (310) 206-3182
F: (310) 206-5891
yaghi@chem.ucla.edu

SHENYUAN YANG
DEPARTMENT OF PHYSICS &
ASTRONOMY
THE UNIVERSITY OF TENNESSEE
401 NIELSEN PHYSICS BLDG.
KNOXVILLE, TN 37996
Tel: (865) 974-4553
syang13@utk.edu

TANER YILDIRIM
CENTER FOR NEUTRON RESEARCH
NIST
100 BUREAU DRIVE
GAITHERSBURG, MD 20899
Tel: (301) 975-6228
Fax: (301) 921-9847
taner@nist.gov

GLEB YUSHIN
ASSISTANT PROFESSOR
MATERIALS SCIENCE &
ENGINEERING
GEORGIA INSTITUTE OF
TECHNOLOGY
771 FERST DRIVE N.W.
ATLANTA, GA
T: (404) 385-3261
F: (404) 894=9140
yushin@gatech.edu

SHENGBAI ZHANG
SENIOR SCIENTIST
NATIONAL RENEWABLE ENERGY
LAB.
1617 COLE BLVD.
GOLDEN, CO 80401
Tel: (303) 384-6622
Fax: (303) 384-6432
shengbai_zhang@nrel.gov

RAGAIY ZIDAN
ADVISORY SCIENTIST
ENERGY SECURITY DEPARTMENT
SAVANNAH RIVER NATIONAL
LABORATORY
999-2W ROOM 121
AIKEN, SC 29803
Tel: (803) 646-8876
Fax: (803) 652-8137
ragaiy.zidan@srnl.doe.gov

AUTHOR INDEX